Progress in Systems and Control Theory
Volume 20

Series Editor
Christopher I. Byrnes, Washington University

Computation and Control IV

Proceedings of the Fourth
Bozeman Conference,
Bozeman, Montana
August 3-9, 1994

K.L. Bowers and J. Lund
Editors

1995

Birkhäuser
Boston • Basel • Berlin

Kenneth Bowers
Department of Mathematics
Montana State University
Bozeman, Montana 59717

John Lund
Department of Mathematics
Montana State University
Bozeman, Montana 59717

Printed on acid-free paper
© 1995 Birkhäuser Boston
Softcover reprint of the hardcover 1st edition 1995

Birkhäuser

ISBN-13: 978-1-4612-7586-2   e-ISBN-13: 978-1-4612-2574-4
DOI: 10.1007/978-1-4612-2574-4

Typeset by the Authors in Latex

9 8 7 6 5 4 3 2 1

CONTENTS

PREFACE

The fourth Conference on Computation and Control was held at Montana State University in Bozeman, Montana from August 3–9, 1994. These proceedings represent the continued evolution of the cross-disciplinary dialogue begun at the 1988 conference (Volume 1 of PSCT) and continued on a biennial basis in 1990 and 1992. Those proceedings are housed in Volume 11 of PSCT and Volume 15 of PSCT, respectively.

In recent years considerable effort has been devoted to the problem of developing rigorous numerical methods and computational tools for control design and analysis. Although computational problems in control are extremely complex, these problems often have special structures that can be exploited to obtain both theoretical and computational results. Moreover, problems that arise from applications are best addressed by interdisciplinary approaches where experts in various disciplines (e.g. numerical analysis, control theory, fluid dynamics) come together and create numerical algorithms specifically for use in control design. This communication not only creates new mathematical tools, it often produces new research problems in the individual disciplines. This conference continues to bring together leading experts in control theory, numerical mathematics and various application areas to discuss recent developments in interdisciplinary approaches to computational control.

The organizers acknowledge the generous support of the Montana State University Foundation, the National Science Foundation's Division of Mathematical Sciences and the EPSCoR Program (Montana Group for Nonlinear Science). Montana State University and the city of Bozeman provided intellectual and aesthetic environment for this conference and the many individuals who gave their time are gratefully acknowledged. For more patient understanding that can be briefly summarized, the organizers thank their wives Sandy and Nancy.

Kenneth L. Bowers and John Lund
Bozeman, March 17, 1995

CONTRIBUTORS

Tom Banks, Center for Research in Scientific Computation, North Carolina State University, Raleigh, North Carolina 27695-8205

Ken Bowers, Department of Mathematical Sciences, Montana State University, Bozeman, Montana 59717-0240

John Burkardt, Interdisciplinary Center for Applied Mathematics Virginia Polytechnic Institute and State University, Blacksburg, Virginia, 24061

John Burns, Center for Optimal Design and Control, Interdisciplinary Center for Applied Mathematics, Virginia Polytechnic Institute and State University, Blacksburg, Virginia 24061

Chris Byrnes, Department of Systems Science and Mathematics, Washington University, St. Louis, Missouri, 63130

Ruth Curtain, Mathematics Institute, University of Groningen, Netherlands

Alisa DeStefano, Department of Mathematics, College of the Holy Cross, Worcester, Massachusetts 01610

Per Enqvist, Department of Mathematics, Texas Tech University, Lubbock, TX 79409

Mark A. Erickson, RelMan, Inc., Mountain View, CA 94041

Rich Fabiano, Department of Mathematics, Texas A&M University, College Station, Texas 77843

Ben Fitzpatrick, Department of Mathematics and Center for Research in Scientific Computation, North Carolina State University, Raleigh, NC 27695-8205

Dave Gilliam, Department of Mathematics, Texas Tech University, Lubbock, Texas 79409

Max Gunzburger, Interdisciplinary Center for Applied Mathematics, Department of Mathematics, Virginia Polytechnic Institute and State University, Blacksburg, Virginia 24061

Robert Hoar, Department of Mathematical Sciences, Montana State University, Bozeman, MT 59717

Mary Jarratt, Department of Mathematics and Computer Science, Boise State University, Boise, Idaho 83725

Steven P. Kaliszewski, Department of Mathematics, University of Newcastle, Newcastle, NSW 2308 Australia

Belinda King, Department of Mathematics, Oregon State University, Corvallis, Oregon 97331

Julius King, Department of Mathematics and Center for Research in Scientific Computation, North Carolina State University, Raleigh, NC 27695-8205

Alan Laub, Dept. of Electrical and Computer Engineering, University of California, Santa Barbara, CA 93106-9560

Pamela Lockwood, Department of Mathematics, Texas Tech University Lubbock, TX 79409-1042

Siyuan Lu, Department of Mathematics, Texas Tech University, Lubbock TX 79409-1042

Nancy Lybeck, Center for Research in Scientific Computation, North Carolina State University, Raleigh, NC 27695-8205

Bernard Mair, Department of Mathematics, University of Florida, Gainesville, Florida 32611

Clyde Martin, Department of Mathematics, Texas Tech University, Lubbock, Texas 79409

Anne Morlet, Department of Mathematics, Ohio State University, Columbus, OH 43210-1174

Janet Peterson, Interdisciplinary Center for Applied Mathematics, Department of Mathematics, Virginia Polytechnic Institute and State University, Blacksburg, Virginia 24061

F. H. Ruymgaart, Department of Mathematics, Texas Tech University Lubbock, TX 79409

Victor Shubov, Department of Mathematics, Texas Tech University Lubbock, TX 79409

Ralph Smith, Institute for Computer Applications in Science and Engineering, NASA Langley Research Center, Hampton, Virginia 23681

Frank Stenger, Department of Computer Science, University of Utah, Salt Lake City, Utah 84112

Steve Taylor, Department of Mathematical Sciences, Montana State University, Bozeman, Montana 59717-0240

John Tomlinson, Department of Mathematics, Texas Tech University Lubbock, TX 79409

Curt Vogel, Department of Mathematical Sciences, Montana State University, Bozeman, Montana 59717-0240

Gordon Wade, Institute for Scientific Computation, Texas A&M University, College Station, Texas 77843

Dorothy Wallace, Department of Mathematics and Computer Science, Dartmouth College, Hanover, New Hampshire 03755

Yun Wang, Mathematical Products Division, Armstrong Laboratory Brooks AFB, TX 78235

Zaichao Xu, Departmnt of Mathematics, Texas Tech University, Lubbock TX 79409

Zhimin Zhang, Department of Mathematics, Texas Tech University, Lubbock TX 79409

Yishao Zhou, Mathematics Institute, University of Groningen, Netherlands

WELL-POSEDNESS FOR A ONE DIMENSIONAL NONLINEAR BEAM

H.T. Banks
Center for Research in Scientific Computation
North Carolina State University
Raleigh, NC 27695 *

David S. Gilliam, Victor I. Shubov
Department of Mathematics
Texas Tech University
Lubbock, TX 79409 †

1 Introduction

In this note we present well-posedness results for a class of nonlinear beam models with linear damping and general external time dependent forcing in the Sobolev space H^{-2}. Our efforts here are motivated by the eventual goal of developing computational methodologies for the control of smart material composites undergoing large deformations and/or deformations that fall within the regime of nonlinear stress-strain laws. It is known [2] in engineering applications (especially with curved and twisted structures) that large deflections can occur even when strain levels remain relatively low. On the other hand in composites and certain types of elastomers one encounters a nonlinear stress-strain relationship even in the case of small deformations [2], [8]. The equations considered here include the important class of nonlinear but monotone stress-strain laws encountered in the derivation of the transverse bending equation for a beam (cf. [10], [11]). The external applied forces (or controls $f = Bu$) are precisely the class that arise in current versions of "smart material" structures when beams or plates are loaded with piezoceramic actuators and sensors (e.g. see [3], [4] and the references therein).

The techniques used to establish our results on well-posedness combine the powerful constructive functional analytic arguments used for linear systems in [3], [13], [12], [6] with equally powerful monotonicity methods usually encountered in nonlinear equations [5] (see also [1, 7, 9] for nonlinear equations technologies).

The basic results given in this report can be considerably generalized and are applicable to a wide range of nonlinear system problems. Some of these extensions and generalizations are outlined in a forthcoming manuscript.

---

*Research supported in part by the Air Force Office of Scientific Research under grant AFOSR-F49620-93-0198.

†Supported in part by AFOSR and Texas ARP

A more abstract formulation for rather general nonlinear systems, which includes the results of this report as special cases, is currently in preparation by the authors.

2   Mathematical Formulation and Statement of Main Result

We consider the nonlinear equation

$$w_{tt} + \kappa_1 w_{xxxx} + \kappa_2 w_{xxxxt} + g(w_{xx})_{xx} = f \tag{2.1}$$
$$w_x(0,t) = w(0,t) = 0 \tag{2.2}$$
$$w_x(1,t) = w(1,t) = 0$$
$$w(\cdot,0) = \phi_0 \in H_0^2(0,1) \tag{2.3}$$
$$w_t(\cdot,0) = \phi_1 \in L^2(0,1) \tag{2.4}$$

Assumption 2.1 *We assume that*

$$G(\xi) = \int_0^\xi g(\tau)\,d\tau, \quad g(\xi) = G'(\xi) \tag{2.5}$$

satisfying the following

1. *There exist positive constants C_j for $j = 1,2,3$ such that*

$$-\frac{1}{2}(\kappa_1 - \epsilon)|\xi|^2 - C_1 \le G(\xi) \le C_2|\xi|^2 + C_3 \tag{2.6}$$

   for $\epsilon > 0$.

2. *There are positive constants $\widetilde{C}_j$, $j = 1,2$ such that*

$$|g(\xi)| \le \widetilde{C}_1|\xi| + \widetilde{C}_2 \tag{2.7}$$

3. *g is a nondecreasing function, i.e.,*

$$g'(\xi) \ge 0, \quad \forall\, \xi \in \mathcal{R} \tag{2.8}$$

To establish the existence of a weak solution we only will use (2.6), (2.7) and (2.8) plus the assumption that $f \in L^\infty((0,T), H^{-2})$. To obtain strong solvability in $H^{-2}(0,1)$ we should also assume that there is a positive constant $\widetilde{C}_3$ such that

$$|g'(\xi)| \le \widetilde{C}_3. \tag{2.9}$$

Our primary concern is establishing the existence of a weak solution to (2.1). To this end we define the notion of a weak solution as follows.

Definition 2.1 *We denote by $\mathcal{L}_T$ the Banach space of functions defined on the rectangle $Q_T = [0,1] \times [0,T]$ and having the following properties:*

1. *For each $t \in [0,T]$ there exist weak derivatives*

$$w_t(\cdot, t), \; w_{xx}(\cdot, t) \in L^2(0,1) \tag{2.10}$$

2. *There exists the weak derivative*

$$w_{xxt} \in L^2(Q_T) \tag{2.11}$$

The norm in this space is given by

$$\|w\|_{\mathcal{L}_T} = ess \; sup(\|w_t(\cdot,t)\| + \|w_{xx}(\cdot,t)\|) + \|w_t\|_{L^2(Q_T)}$$

Definition 2.2 *A function $w \in \mathcal{L}_T$ is a weak solution of (2.1)-(2.4) if it satisfies the following identity for every $t \in [0,T]$*

$$\int_{Q_t} (-w_t\eta_t + \kappa_1 w_{xx}\eta_{xx} + \kappa_2 w_{xxt}\eta_{xx} + g(w_{xx})\eta_{xx}) \, dx \, dt$$

$$+ \int_0^1 w_t(x,t)\eta(x,t) \, dx$$

$$= \int_0^1 \phi_1(x)\eta(x,0) \, dx + \int_{Q_t} f\eta \, dx \, dt \tag{2.12}$$

for all $\eta \in \mathcal{L}_T$.

In addition w must also satisfy

$$w(x,0) = \phi_0(x), \quad a.e. \;\; x \in [0,1] \tag{2.13}$$

Theorem 2.3 *The equations (2.1) has a unique weak solution.*

3 Preliminary Results

We define the operator

$$A = \frac{d^4}{dx^4}$$

with dense domain in $L_2(0,1)$

$$\mathcal{D}(A) = \{\varphi \in H^4(0,1) : \; \varphi'(0) = \varphi(0) = 0, \; \varphi'(1) = \varphi(1) = 0\}.$$

Then A is a strictly positive self adjoint operator. Defining $\lambda = \mu^4$ it is a simple computation to obtain the characteristic equation providing the eigenvalues (spectrum of A)

$$\cos(\mu)\cosh(\mu) = 1.$$

This equation has infinitely many zeros $\{\mu_j\}_{j=1}^{\infty}$ satisfying $0 < \mu_1 < \pi/2$, and $\mu_j - j\pi/2 \to 0$ as $j \to \infty$. Corresponding to the eigenvalues $\lambda_j = \mu_j^4$ for A we have the complete orthonormal system of eigenfunctions $\{\psi_j\}$ in $L_2(0,1)$. The eigenvalues $\lambda_j = \mu_j^4$, are simple and for all $\varphi \in L_2(0,1)$ we have

$$\varphi = \sum_{j=1}^{\infty} \varphi_j \psi_j, \quad \varphi_j = <\varphi, \psi_j>,$$

$$D(A) = \{\varphi \in L_2(0,1) : \sum_{j=1}^{\infty} \lambda_j^2 |\varphi_j|^2 < \infty\}$$

and for $\varphi \in D(A)$

$$A\varphi = \sum_{j=1}^{\infty} \lambda_j \varphi_j \psi_j.$$

In the usual way, the operator A defines an infinite scale of Hilbert spaces $\mathcal{H}^s$ $(s \in \mathcal{R})$ with norms

$$\|w\|_s = \|A^{s/4} w\|$$

where $\|\cdot\|$ denotes the norm in $L_2(0,1)$. It is easy to see that the norm in $\mathcal{H}^2$ can be represented by

$$\|\varphi\|_2^2 = \int_0^1 |\varphi_{xx}|^2 \, dx = \|\phi_{xx}\|^2.$$

Proposition 3.1 *For each $\varphi \in \mathcal{H}^2$ we have*

$$\|\varphi\|^2 \leq \lambda_1^{-1} \|\varphi\|_2^2,$$

where $0 < \lambda_1 < \pi^2/4$ is the first eigenvalue of A. Furthermore, the norm $\|\cdot\|_2$ is equivalent to the $H_0^2(0,1)$ Sobolev norm and, therefore $\mathcal{H}^2 = H_0^2(0,1)$ with

$$a\|\varphi\|_{H_0^2} \leq \|\varphi\|_2 \leq b\|\varphi\|_{H_0^2}$$

4   A Priori Estimates

We begin with the derivation of the main *a priori* estimate.

Theorem 4.1 *The following estimate holds*

$$\|w_t(t)\|^2 + \epsilon\|w_{xx}(t)\|^2 + \kappa_2 \int_0^t \|w_{xxt}(\tau)\|^2 \, d\tau \leq C$$

where

$$C \equiv C(\|\phi_1\|, \|\phi_{0xx}\|, \|f\|_{H^{-2}}, T) =$$

$$= \|\phi_1\|^2 + C_4\|\phi_{0xx}\|^2 + C_5 + \frac{1}{2\kappa_2}T\|f\|_{L_2((0,T),H^{-2})} \quad (4.1)$$

Proof: Take the L^2-inner product of (2.1) with w_t to obtain

$$(w_{tt}, w_t) + \kappa_1(w_{xxxx}, w_t) + \kappa_2(w_{txxxx}, w_t) = (g(w_{xx})_{xx}, w_t) + (f, w_t), (4.2)$$

where $(\cdot, \cdot)$ denotes the inner product in $L_2(0,1)$.

Now notice that

$$(g(w_{xx})_{xx}, w_t) = (g(w_{xx}), w_{xxt}) = \int_0^1 g(w_{xx})w_{xxt}\, dx$$

$$= \frac{d}{dt}\int_0^1 G(w_{xx})\, dx \quad (4.3)$$

Integrating by parts in the second and third terms in (4.2) and using (4.3), we obtain

$$\frac{d}{dt}\left[\frac{1}{2}\|w_t\|^2 + \frac{\kappa_1}{2}\|w_{xx}\|^2 + \int_0^1 G(w_{xx})\, dx - (f, w)\right] + \kappa_2\|w_{txx}\|^2 = 0. \quad (4.4)$$

Integrating (4.4) from 0 to t we get

$$\frac{1}{2}\|w_t(t)\|^2 + \frac{\kappa_1}{2}\|w_{xx}(t)\|^2 + \int_0^1 G(w_{xx}(t))\, dx + \kappa_2\int_0^t \|w_{xxt}(\tau)\|^2\, d\tau$$

$$= \frac{1}{2}\|\phi_1\|^2 + \frac{\kappa_1}{2}\|\phi_{0xx}\|^2 + \int_0^1 G(\phi_{0xx})\, dx + \int_0^t (f, w_t)\, d\tau \quad (4.5)$$

For the last term we have

$$\left|\int_0^t (f, w_t)\, d\tau\right| \leq \delta\int_0^t \|w_{xxt}\|^2\, d\tau + \frac{1}{4\delta}\int_0^t \|f\|_{H^{-2}}^2\, d\tau. \quad (4.6)$$

Taking into account (4.6) and (2.6) we obtain from (4.5) that

$$\frac{1}{2}\|w_t\|^2 + \frac{\epsilon}{2}\|w_{xx}\|^2 + \frac{\kappa_2}{2}\int_0^t \|w_{xxt}\|^2\, d\tau \leq \quad (4.7)$$

$$\leq \frac{1}{2}\|\phi_1\|^2 + \frac{\kappa_1}{2}\|\phi_{0xx}\|^2 + C_2\|\phi_{0xx}\|^2 + C_3 + C_1 + \frac{t}{2\kappa_2}\|f\|_{L_2((0,t),H^{-2})}^2$$

Here we have used $\delta = \kappa_2/2$ and also the inequalities

$$\int_0^1 G(w_{xx})dx \geq -\frac{1}{2}(\kappa_1 - \epsilon)\|w_{xx}\|^2 - C_1 \tag{4.8}$$

and

$$\int_0^1 G(\phi_{0xx})dx \leq C_2\|\phi_{0xx}\|^2 + C_3 \tag{4.9}$$

which follow immediately from (2.6). $\square$

5   Galerkin Approximations

We seek approximate solutions in the form

$$w^N(x,t) = \sum_{k=1}^N C_k^N(t)\psi_k(x), \quad C_k^N(t) = (w^N(\cdot,t),\psi_k). \tag{5.1}$$

According to the Galerkin procedure we seek $\{C_k^N(t)\}$ such that

$$\frac{d^2}{dt^2}C_k^N(t) + \kappa_1\lambda_k C_k^N(t) + \kappa_2\lambda_k\frac{d}{dt}C_k^N(t) + (g(w_{xx}^N),\psi_{kxx}) = (f,\psi_\kappa), \tag{5.2}$$

for $k = 1,\cdots,N$ and

$$C_k^N(0) = (\phi_0,\psi_k), \quad \frac{d}{dt}C_k^N(0) = (\phi_1,\psi_k). \tag{5.3}$$

The above equation is equivalent to

$$\frac{d^2}{dt^2}(w^N(\cdot,t),\psi_k) + \kappa_1\lambda_k(w^N(\cdot,t),\psi_k) +$$

$$+\kappa_2\lambda_k\frac{d}{dt}(w^N(\cdot,t),\psi_k) + (g(w_{xx}^N),\psi_{kxx}) = (f(t),\psi). \tag{5.4}$$

Multiplying (5.4) by $\frac{d}{dt}C_k^N(t)$ and summing over $k = 1,\cdots N$ and then integrating over $[0,t]$ we obtain an estimate similar to (4.1) for Galerkin approximates.

Namely, we have

$$\|w_t^N(t)\|^2 + \epsilon\|w_{xx}^N(t)\|^2 + \kappa_2\int_0^t \|w_{xxt}^N(\tau)\|^2 \, d\tau$$

$$\leq C \equiv C(\|\phi_1\|,\|\phi_{0xx}\|,\|f\|_{L_2((0,T),H^{-2})},T) \tag{5.5}$$

for all $N = 1,2,\cdots$ and $t \in [0,T]$.

6   Convergence of the Galerkin Approximations

Lemma 6.1 *The set* $\{w_{xx}^N\}_{N=1}^\infty$ *is uniformly bounded and equicontinuous in* $C([0,T], L^2(0,1))$ *with norm* $\sup_{t\in[0,T]} \|w(\cdot,t)\|$.

Proof: It is clear from (5.5) that

$$\max_{t\in[0,T]} \|w_{xx}^N(t)\|^2 \le \epsilon^{-1}C. \tag{6.1}$$

Further we have

$$\|w_{xx}^N(t+\Delta t) - w_{xx}^N(t)\|^2 = \int_0^1 |w_{xx}^N(x,t+\Delta t) - w_{xx}^N(x,t)|^2 \, dx \tag{6.2}$$

$$= \int_0^1 \left| \int_t^{t+\Delta t} w_{xxt}^N(x,\tau) \, d\tau \right|^2 dx \le \Delta t \int_0^1 \int_t^{t+\Delta t} |w_{xxt}^N(x,\tau)|^2 \, d\tau dx$$

$$= \Delta t \int_t^{t+\Delta t} \|w_{xxt}(\tau)\|^2 \, d\tau \le \Delta t \kappa_2^{-1} C,$$

where in the first inequality we applied the Cauchy-Schwartz inequality and in the second inequality we used (5.5).   □

Lemma 6.2 *The set* $\{w_x^N\}_{N=1}^\infty$ *is relatively compact in* $C([0,T], L^2(0,1))$.

Proof: Uniform boundedness and equicontinuity follow from Lemma 6.1. In addition, the set $\{w_x^N(t)\}_{N=1}^\infty$ is relatively compact in $L^2(0,1)$ for all $t \in [0,T]$. This follows from the estimate $\|w_{xx}^N(t)\|^2 \le C$ and the compactness of the embedding $H^1(0,1) \subset L^2(0,1)$.   □

Remark 6.1 Throughout the rest of the paper we will denote all subsequences by the same set $\{w^N\}$.

Corollary 6.3 *There exists a function* $w \in C([0,T], H_0^1(0,1))$ *and a subsequence* $\{w^N\}$ *such that*

$$\max_{t\in[0,T]} \|w_x^N(t) - w_x(t)\| \to 0, \quad N \to \infty \tag{6.3}$$

$$\max_{t\in[0,T]} \|w^N(t) - w(t)\|_{H^1(0,1)} \to 0, \quad N \to \infty. \tag{6.4}$$

The fact that in (6.3) we have precisely the weak derivative w_x *follows from the theory of weak derivatives.*

Lemma 6.4 *The function w has the weak derivative $w_{xx}(t) \in L^2(0,1)$ for every $t \in [0,T]$. For a certain subsequence*

$$w_{xx}^N(\cdot, t) \to w_{xx}(\cdot, t), \quad \text{weakly in } L^2(0,1) \text{ for all } t \in [0,T] \qquad (6.5)$$

(the convergence is uniform in t).

Proof: Consider the set of scalar functions of $t \in [0,T]$,

$$\left\{ \left(w_{xx}^N(\cdot, t), \psi \right) \right\}_{N=1}^{\infty}$$

where ψ is any function in $L^2(0,1)$. This set is uniformly bounded (due to (5.5)) and equicontinuous (due to Lemma 6.1). Therefore, it is compact in $C([0,T])$. For some subsequence we have

$$\max_{t \in [0,T]} \left| \left(w_{xx}^N(\cdot, t), \psi \right) - \Psi(t) \right| \to 0, \; N \to \infty \qquad (6.6)$$

where $\Psi \in C([0,T])$. Now, from the weak sequential completeness of $L^2(0,1)$ and the theory of weak derivatives $\Psi(t)$ must be given precisely by:

$$\Psi(t) = (w_{xx}(\cdot, t), \psi) \qquad (6.7)$$

where w is the function from Lemma 6.3.   □

Lemma 6.5 *For any fixed $k = 1, 2, \cdots$, the set of functions*

$$\left\{ \frac{d^2}{dt^2} C_k^N(t) = \frac{d^2}{dt^2} (w^N(t), \psi_k) \right\}_{N=k}^{\infty}$$

is uniformly bounded in $C([0,T])$.

Proof: From (5.4) we have

$$\left| \frac{d^2}{dt^2} (w^N(t), \psi_k) \right| \le \kappa_1 \lambda_k \left| (w^N(\cdot, t), \psi_k) \right| + \kappa_2 \lambda_k \left| (w_t^N(\cdot, t), \psi_k) \right|$$

$$+ |(g(w_{xx}^N), \psi_{kxx})| + |(f(t), \psi_k)| \le \kappa_1 \|w_{xx}^N(t)\| + \kappa_2 \lambda_k \|w_t^N(t)\| \quad (6.8)$$

$$+ \max_{x \in [0,1]} |\psi_{kxx}(x)| \int_0^1 |g(w_{xx}^N)| \, dx + \lambda_k^{1/2} \|f(t)\|_{H^{-2}}$$

where we have used

$$|(\psi, \psi_k)| \le \lambda_k^{-1/2} |(\psi_{xx}, \psi_k)|, \quad \text{for all } \psi \in H_0^2(0,1)$$

and

$$|(f(t), \psi_k)| \le \lambda_k^{1/2} \left(\sum_{k=1}^{\infty} \lambda_k^{-1} |(f(t), \psi_k)|^2 \right)^{1/2} = \lambda_k^{1/2} \|f(t)\|_{H^{-2}}.$$

Recall that due to (2.7) and (5.5):

$$\int_0^1 |g(w_{xx}^N)| \, dx \le \tilde{C}_1 \|w_{xx}^N\| + \tilde{C}_2 \le \tilde{C}_1 C^{1/2} + \tilde{C}_2 \qquad (6.9)$$

So, we have (using (5.5))

$$\left| \frac{d^2}{dt^2}(w^N(t), \psi_k) \right| \le \kappa_1 \lambda_k^{1/2} \epsilon^{-1/2} C^{1/2} + \kappa_2 \lambda_k C^{1/2} + \qquad (6.10)$$

$$\max_{x \in [0,1]} |\psi_{kxx}(x)|(\tilde{C}_1 C^{1/2} + \tilde{C}_2) + \lambda_k^{1/2} \|f\|_{L^{\infty}((0,T), H^{-2})} \equiv M_k$$

$\square$

Corollary 6.6 *For any* $k = 1, 2, \cdots$ *the set* $\left\{ C_k^N(t) \right\}_{N=k}^{\infty}$ *is relatively compact in* $C^1[0, T]$.

Proof: Since the embedding $C^2[0, T] \subset C^1[0, T]$ is compact, it is sufficient to show that the above set is bounded in $C^2[0, T]$. The boundedness of the second derivatives follows from Lemma 6.4. We have only to check that the sets $\left\{ C_k^N(t) \right\}_{N=k}^{\infty}$ and $\left\{ \frac{d}{dt} C_k^N(t) \right\}_{N=k}^{\infty}$ are bounded in $C[0, T]$. Taking into account the second initial condition (5.3), we obtain

$$\frac{d}{dt} C_k^N(t) = (\phi_1, \psi_k) + \int_0^t \frac{d^2}{dt^2} C_k^N(\tau) \, d\tau.$$

This equality together with the estimate (6.10) implies

$$\left| \frac{d}{dt} C_k^N(t) \right| \le \|\phi_1\| + T M_k, \ \forall t \in [0, T].$$

Taking into account the last estimate and the first initial condition in (5.3) we obtain

$$C_k^N(t) = (\phi_0, \psi_k) + \int_0^t \frac{d}{dt} C_k^N(\tau) \, d\tau,$$

and

$$|C_k^N(t)| \le \|\phi_0\| + (\|\phi_1\| + T M_k)T, \ \forall t \in [0, T].$$

$\square$

Lemma 6.7 *The function w has the weak derivative $w_t(\cdot, t) \in L^2(0,1)$ for all $t \in [0, T]$ and*

$$w_t^N(\cdot, t) \rightharpoonup w_t(\cdot, t), \text{ weakly in } L^2(0,1) \text{ uniformly in } t \in [0, T]. \quad (6.11)$$

Proof: It follows from Corollary 6.7 that for each $k = 1, 2, \cdots$ there exists a subsequence

$$\left\{ C_k^{N_{m,k}}(t) \right\}_{m=1}^{\infty} \subset \left\{ C_k^N(t) \right\}_{N=k}^{\infty},$$

which converges in the space $C^1[0, T]$ to a certain function $C_k(t)$, i.e.,

$$\left\| C_k^{N_{m,k}} - C_k \right\|_{C^1[0,T]} \to 0, \quad m \to \infty.$$

We show that the numbers $N_{m,k}$ can be chosen so that they do not depend on k. Notice, that the sequences $\{N_{m,k}\}$ for $k = 1, 2, \cdots$ can be chosen in such a way that

$$\{N_{m,k+1}\}_{m=1}^{\infty} \subset \{N_{m,k}\}_{m=1}^{\infty}$$

for all $k = 1, 2, \cdots$. We now take the diagonal sequence $\{N_{m,m}\}_{m=1}^{\infty}$ and denote it by $\{N_m\}_{m=1}^{\infty}$. It is clear that for each $k = 1, 2, \cdots$ the sequence $\left\{ C_k^{N_m}(t) \right\}_{m=1}^{\infty}$ converges to the function $C_k(t)$ in $C^1[0, T]$ as $m \to \infty$. Now we can recall Remark 6.1 and assume that for each $k = 1, 2, \cdots$ the whole sequence of Galerkin coefficients $\left\{ C_k^N(t) \right\}_{N=k}^{\infty}$ converge to $C_k(t)$ in $C^1[0, T]$ as $N \to \infty$.

Let us now take any function ψ of the form $\psi = \sum_{k=1}^{M} a_k \psi_k$, where $\{a_k\}$ are arbitrary real numbers and M is an arbitrary positive integer. It is easy to see that the sequence of functions

$$(w_t^N, \psi) = \left(\sum_{\ell=1}^{N} \frac{d}{dt} C_\ell^N(t) \psi_\ell, \sum_{k=1}^{M} a_k \psi_k \right) = \sum_{k=1}^{M} a_k \frac{d}{dt} C_k^N(t), \quad M \leq N$$

converges in $C[0, T]$ to the function

$$\Psi(t) = \sum_{k=1}^{M} a_k \frac{d}{dt} C_k(t)$$

as $N \to \infty$. Now notice that the functions ψ form a dense set in $L^2(0,1)$. On the other hand we know from (5.5) that

$$\|w_t^N(t)\|^2 \leq C. \quad (6.12)$$

Therefore by the Banach-Steinhaus Theorem

$$w_t^N(t) \rightharpoonup v(t) \text{ weakly in } L^2(0,1) \text{ uniformly on } [0,T]. \tag{6.13}$$

Here $v(t) \in L^2(0,1)$ for all $t \in [0,T]$ is such that $(v(t), \psi) = \Psi(t)$ due to weak completeness of $L^2(0,1)$. Finally we notice that $v(t)$ must coincide with $w_t(t)$ according to the theory of weak derivatives.   □

Remark 6.2 The proof of the above Lemma 6.5 can be completed in a different way that gives additional information about the relation between the function w and the functions $\{C_k(t)\}_{k=1}^\infty$. Namely, in addition to the statement of Lemma 6.5, the following observations can be made.

For each $t \in [0,T]$ the functions $w(\cdot,t)$ and $w_t(\cdot,t)$ can be represented in the form:

$$w(x,t) = \sum_{k=1}^\infty C_k(t)\psi_k(x), \ \ w_t(x,t) = \sum_{k=1}^\infty \frac{d}{dt} C_k(t)\psi_k(x) \tag{6.14}$$

where both series are convergent in $L^2(0,1)$. In other words, the formulas (6.14) are just the expansions of $w(\cdot,t)$ and $w_t(\cdot,t)$ with respect to the orthonormal basis $\{\psi_k\}$. Let us prove Lemma 6.5 again and also obtain (6.14).

First of all we show that for each $t \in [0,T]$

$$\sum_{k=1}^\infty |C_k(t)|^2 \le \tilde{C} < \infty, \ \ \sum_{k=1}^\infty |\frac{d}{dt} C_k(t)|^2 \le C < \infty \tag{6.15}$$

where C is the constant from (5.5) and $\tilde{C} = \left(\|\phi\| + T\sqrt{C} \right)^2$. We will show only the second inequality in (6.15). The first one can be obtained in a similar way, if one takes into account that (5.5) implies the estimate

$$\|w(t)\|^2 = \|\phi + \int_0^t w_t(\tau)\,d\tau\|^2 \le \tilde{C}.$$

Let us fix any integer $N > 0$. It follows from (5.5) that for $N_m \ge N$

$$\sum_{k=1}^N |\frac{d}{dt} C_k(t)|^2 \le \sum_{k=1}^{N_m} |\frac{d}{dt} C_k(t)|^2 = \|w_t^{N_m}(t)\|^2 \le C \tag{6.16}$$

(recall that N_m is the diagonal sequence). Passing to the limit as $m \to \infty$ we obtain

$$\sum_{k=1}^N \left| \frac{d}{dt} C_k(t) \right|^2 \le C$$

for any N and, therefore, the second inequality in (6.15) holds. We conclude that for each $t \in [0, T]$ there exists a function $v(\cdot, t) \in L^2(0, 1)$ such that

$$v(x, t) = \sum_{k=1}^{\infty} \frac{d}{dt} C_k(t) \psi_k(x) \in L^2(0, 1) \quad \text{and} \quad \|v(t)\| \leq C. \qquad (6.17)$$

We now show that for arbitrary $\Phi \in L^2(0, 1)$

$$\max_{t \in [0, T]} \left| \left(w_t^{Nm}(\cdot, t) - v(\cdot, t), \Phi \right) \right| \to 0, \quad m \to \infty. \qquad (6.18)$$

This means that $w_t^{Nm}(\cdot, t)$ converges to $v(t)$ weakly in $L^2(0, 1)$ uniformly with respect to $t \in [0, T]$, i.e., (6.18) is equivalent to (6.13). If we show (6.18) then we can conclude that $v = w_t$ and (6.14) will be justified.

We have

$$\left(w_t^{Nm}(\cdot, t) - v(\cdot, t), \Phi \right) = \left(w_t^{Nm}(\cdot, t) - v(\cdot, t), \sum_{k=1}^{\infty} (\Phi, \psi_k) \psi_k \right)$$

$$\sum_{k=1}^{s} (\Phi, \psi_k)(w_t^{Nm}(\cdot, t) - v(\cdot, \cdot, t), \psi_k)$$

$$+ (w_t^{Nm}(\cdot, t) - v(\cdot, t), \sum_{k=s+1}^{\infty} (\Phi, \psi_k) \psi_k).$$

For the second term we have the estimate

$$\left(w_t^{Nm}(\cdot, t) - v(\cdot, t), \sum_{k=s+1}^{\infty} (\Phi, \psi_k) \psi_k \right) \leq$$

$$\|w_t^{Nm} - v\| \left(\sum_{k=s+1}^{\infty} |(\Phi, \psi_k)|^2 \right)^{1/2}$$

$$\leq (\|w_t^{Nm}\| + \|v\|) \left(\sum_{k=s+1}^{\infty} |(\Phi, \psi_k)|^2 \right)^{1/2}$$

$$\leq 2C \left(\sum_{k=s+1}^{\infty} |(\Phi, \psi_k)|^2 \right)^{1/2} \equiv 2C\alpha(s)$$

where $\alpha(s) \to 0$ as $s \to \infty$. Let $\delta > 0$ be given and take s so large that $2C\alpha(s) \leq \delta/2$. Then we have

$$\left| \sum_{k=1}^{s} (\Phi, \psi_k)(w_t^{Nm}(\cdot, t) - v(\cdot, \cdot, t), \psi_k) \right|$$

$$= \left| \sum_{k=1}^{s} (\Phi, \psi_k)(\frac{d}{dt} C_k^{N_m}(t) - \frac{d}{dt} C_k(t)) \right|$$

$$\leq \sum_{k=1}^{s} |(\Phi, \psi_k)| \max_{\substack{k=1,\cdot,s \\ t\in[0,T]}} \left| \frac{d}{dt} C_k^{N_m}(t) - \frac{d}{dt} C_k(t) \right|$$

$$\leq \sqrt{s} \|\Phi\| \max_{\substack{k=1,\cdot,s \\ t\in[0,T]}} \left| \frac{d}{dt} C_k^{N_m}(t) - \frac{d}{dt} C_k(t) \right| \leq \frac{\delta}{2}$$

for m sufficiently large and the result follows.

Lemma 6.8 *For the same subsequence as in Lemma 6.5 we have*

$$w_t^N \to w_t \text{ strongly in } L^2(Q_t) = L^2([0,1] \times [0,t]). \tag{6.19}$$

In order to prove this result we first need to recall the following well known result.

Lemma 6.9 *For any $\psi \in H_0^2(0,1)$ and $\delta > 0$ there exists N_δ such that*

$$\|\psi\|^2 \leq \sum_{k=1}^{N_\delta} |(\psi, \psi_k)|^2 + \delta \|\psi_{xx}\|^2. \tag{6.20}$$

Proof of Lemma 6.6: Apply Lemma 6.7 to the function $\psi = w_t^N(\cdot, t) - w_t^M(\cdot, t)$ to obtain

$$\|w_t^N(t) - w_t^M(t)\|^2 \leq$$

$$\sum_{k=1}^{N_\delta} \left| \left(w_t^N(t) - w_t^M(t), \psi_k \right) \right|^2 + \delta \|w_{xxt}^N(t) - w_{xxt}^M(t)\|^2.$$

Integrating in t from 0 to t and taking into account the definition of w^N, w^M and the fact that $\{\psi_k\}$ are orthonormal, we have

$$\|w_t^N - w_t^M\|_{L^2(Q_t)}^2 = \int_0^t \|w_t^N(t) - w_t^M(t)\|^2 \, dt \tag{6.21}$$

$$\leq \sum_{k=1}^{N_\delta} \int_0^t \left| \frac{d}{dt} C_k^N(t) - \frac{d}{dt} C_k^M(t) \right|^2 \, dt$$

$$+ 2\delta \int_0^t \left(\|w_{xxt}^N(t)\|^2 + \|w_{xxt}^M(t)\|^2 \right) \, dt$$

$$\leq N_\delta t \max_{\substack{k=1,\cdots,N_\delta \\ \tau\in[0,t]}} \left| \frac{d}{dt} C_k^N(\tau) - \frac{d}{dt} C_k^M(\tau) \right|^2 + 4\delta C$$

where C is the constant in (5.5). Now fix any $\theta > 0$ and take $\delta > 0$ such that

$$4\delta C \leq \theta/2.$$

Then, taking M and N sufficiently large we can make the first term less than or equal to $\theta/2$ since from Corollary 6.2 the sequence $\{\frac{d}{dt}C_k^N(t)\}_{N=k}^{\infty}$ converges uniformly on $[0, T]$.   □

Lemma 6.10 *The set* $\{w_{xxt}^N\}_{N=1}^{\infty}$ *is weakly compact in* $L^2(Q_t)$ *and, therefore,*

$$w_{xxt}^N \rightharpoonup w_{xxt} \text{ weakly in } L^2(Q_t) \tag{6.22}$$

for a certain subsequence.

Proof: This is obvious from (5.5) since $\|w_{xxt}^N\|_{L^2(Q_t)}^2 \leq C$.   □

Lemma 6.11 *The set of functions*

$$\{g(w_{xx}^N(x, t))\}_{N=1}^{\infty}$$

is bounded, and therefore, relatively weakly compact in $L^2(Q_t)$ *for all* $t \in [0, T]$.

Proof: From (2.7) we have

$$\|g(w_{xx}^N)\|_{L^2(Q_t)}^2 = \int_0^t \int_0^1 |g(w_{xx}^N(x, t))|^2 \, dxdt \leq$$

$$\int_0^t \int_0^1 \left(\tilde{C}_1 |w_{xx}^N(x, t)| + \tilde{C}_2\right)^2 \, dxdt$$

$$\leq 2 \int_0^t \int_0^1 \left(\tilde{C}_1^2 |w_{xx}^N(x, t)|^2 + \tilde{C}_2^2\right) \, dxdt$$

$$= 2\tilde{C}_1^2 \int_0^t \|w_{xx}^N(t)\|^2 \, dt + 2\tilde{C}_2^2 t$$

$$\leq 2\tilde{C}_1^2 Ct + 2\tilde{C}_2^2 t \leq 2(\tilde{C}_1^2 C + \tilde{C}_2^2)T, \tag{6.23}$$

where we have used (5.5) to estimate $\|w_{xx}^N(t)\|^2$.   □

Corollary 6.12 *There exists a function* $\tilde{g} \in L^2(Q_t)$ *for all* $t \in [0, T]$ *such that*

$$g(w_{xx}^N) \rightharpoonup \tilde{g} \text{ weakly in } L^2(Q_t) \ \forall \ t \in [0, T] \tag{6.24}$$

Remark 6.3 Using (2.9) it is not difficult to show that, in fact,

$$g(w_{xx}^N) \rightharpoonup \tilde{g} \text{ weakly in } L^2(0,1) \text{ uniformly on } [0,T]. \tag{6.25}$$

We will not use (6.25) in the proof of existence.

7 Passing to the limit in the integral identity

We first collect the results established to this point.

1.
$$w^N(t) \rightharpoonup w(t) \tag{7.1}$$

weakly in $H_0^2(0,1)$ uniformly on $[0,T]$. This follows from (6.3), (6.3), (6.5).

2.
$$w_t^N(t) \rightharpoonup w_t(t) \tag{7.2}$$

weakly in $L^2(0,1)$ uniformly on $[0,T]$. This is just (6.9).

3.
$$w_t^N \rightarrow w_t \tag{7.3}$$

strongly in $L^2(Q_t)$ for all $t \in [0,T]$. This is (6.19).

4.
$$w_{xxt}^N \rightharpoonup w_{xxt} \tag{7.4}$$

weakly in $L^2(Q_t)$ for all $t \in [0,T]$. This is just (6.22).

5.
$$g(w_{xx}^N) \rightharpoonup \tilde{g} \in L^2(Q_t) \tag{7.5}$$

weakly in $L^2(Q_t)$ for all $t \in [0,T]$. This is (6.24).

Now let us denote by $\mathcal{P}_M$ ($M = 1, 2, \cdots$) the class of functions $\eta(x,t)$ which can be written in the form

$$\eta(x,t) = \sum_{k=1}^{M} a_k(t)\psi_k(x), \tag{7.6}$$

where $a_k(t)$ are arbitrary C^1 smooth functions on $[0,T]$. Let

$$\mathcal{P} = \cup_{M=1}^{\infty} \mathcal{P}_M. \tag{7.7}$$

It is obvious that $\mathcal{P}$ is dense in the class $\mathcal{L}_T$.

Now return to the equations (5.4) which define the Galerkin approximations. Multiply the kth equation by an arbitrary smooth function $a_k(t)$, take the sum from $k = 1$ to M and integrate over the rectangle Q_t. Integrating by parts with respect to t in the first term and with respect to x (twice) in the second and third terms and taking into account the initial condition (5.3) and the boundary conditions for ψ_k we obtain:

$$\int_{Q_t} \left[-w_t^N \eta_t + \kappa_1 w_{xx}^N \eta_{xx} + \kappa_2 w_{xxt}^N \eta_{xx} + g(w_{xx}^N)\eta_{xx} \right] dxdt$$

$$+ \int_0^1 w_t^N \eta \, dx \Big|_{t=0}^{t=t} = \int_{Q_t} f\eta \, dxdt, \qquad (7.8)$$

which is satisfied for any $\eta \in \mathcal{P}_M$ with $M \leq N$, i.e., η has the form (7.6).

Now let us fix $\eta \in \mathcal{P}_M$ with $M \leq N$. Based on (7.1)-(7.5) we can pass to the limit $N \to \infty$ in (7.8) and obtain (for all $t \in [0,T]$)

$$\int_{Q_t} \left[-w_t \eta_t + \kappa_1 w_{xx} \eta_{xx} + \kappa_2 w_{xxt} \eta_{xx} + \widetilde{g}\eta_{xx} \right] dxdt +$$

$$\int_0^1 w_t \eta \, dx \Big|_{t=0}^{t=t} = \int_{Q_t} f\eta \, dxdt. \qquad (7.9)$$

Here (7.9) is satisfied for all $\eta \in \mathcal{P}_M$ where M is an arbitrary positive integer, and therefore, for any $\eta \in \mathcal{L}_T$ since $\mathcal{P}$ is a dense set in $\mathcal{L}_T$.

Notice that the only difference between (7.9) and the definition (2.12) of the weak solution is that in (7.9) we have $\widetilde{g}$ (which is a certain unknown function in $L^2(Q_t)$) instead of $g(w_{xx})$ in (2.12). Therefore, the proof of the existence theorem will be complete if we prove the following result.

Lemma 7.1 *For any $\eta \in \mathcal{L}_T$ and for all $t \in [0,T]$*

$$\int_{Q_t} g(w_{xx})\eta_{xx} \, dxdt = \int_{Q_t} \widetilde{g}\eta_{xx} \, dxdt, \qquad (7.10)$$

and, therefore, $g(w_{xx}(x,t)) = \widetilde{g}(x,t)$ a.e. in Q_t.

The proof of this lemma is the most difficult part of the proof of the existence result. In this proof the assumption (2.8) will play a crucial role. We divide the proof of Lemma 7.1 into several parts.

Lemma 7.2 *Due to (2.8), the following inequality holds*

$$\int_{Q_t} [g(u) - g(v)](u - v) \, dxdt \geq 0 \qquad (7.11)$$

for all u, v in $L^2(Q_t)$ for all $t \in [0,T]$.

Proof: We have

$$\int_{Q_t} [g(u) - g(v)](u - v) \, dxdt =$$

$$\int_{Q_t} \left[\int_0^1 ds \, \frac{d}{ds} g(su(x,t) + (1-s)v(x,t)) \right] (u - v) \, dxdt$$

$$= \int_{Q_t} \left[\int_0^1 ds \, g'(su(x,t) + (1-s)v(x,t)) \right] (u - v)^2 \, dxdt \geq 0.$$

$\square$

Now let us consider (7.11) with $u = w_{xx}^N$ and v replaced by v_{xx} where v is an arbitrary function from $\mathcal{P}_M$ (see (7.6)). We have

$$\int_{Q_t} [g(w_{xx}^N) - g(v_{xx})] \, (w_{xx}^N - v_{xx}) \, dxdt \geq 0 \qquad (7.12)$$

for all $v \in \mathcal{P}_M$ with $M \leq N$. Rewrite (7.12) in the form

$$\int_{Q_t} g(w_{xx}^N)(w_{xx}^N - v_{xx}) \, dxdt -$$

$$\int_{Q_t} g(v_{xx})(w_{xx}^N - v_{xx}) \, dxdt \geq 0. \qquad (7.13)$$

We return to (7.8) and consider (7.8) with $\eta = w^N - v$ (notice that this is possible since $w^N \in \mathcal{P}_N$ and $v \in \mathcal{P}_M$ ($M \leq N$) by our assumption). In the obtained relation we will have the term $\int_{Q_t} g(w_{xx}^N)(w_{xx}^N - v_{xx}) \, dxdt$. Solving for this term we obtain

$$\int_{Q_t} g(w_{xx}^N)(w_{xx}^N - v_{xx}) \, dxdt$$

$$= \int_{Q_t} \left[w_t^N(w_t^N - v_t) - \kappa_1 w_{xx}^N(w_{xx}^N - v_{xx}) \right. \qquad (7.14)$$

$$\left. - \kappa_2 w_{xxt}^N(w_{xx}^N - v_{xx}) \right] dxdt$$

$$- \int_0^1 w_t^N(w^N - v) \Big|_{t=0}^{t=t} + \int_{Q_t} f(w^N - v) \, dxdt.$$

Substituting (7.14) into (7.13) we obtain, after straightforward simplification: for all $v \in \mathcal{P}_M$ ($M \leq N$)

$$\int_{Q_t} [|w_t^N|^2 - w_t^N v_t - \kappa_1 |w_{xx}^N|^2 +$$

$$\kappa_1 w_{xx}^N v_{xx} - \frac{1}{2}\kappa_2 \frac{d}{dt}(w_{xx}^N)^2 + \kappa_2 w_{xxt}^N v_{xx} \Big] \, dxdt \qquad (7.15)$$

$$- \int_0^1 w_t^N (w^N - v) \, dx \Big|_{t=0}^{t=t}$$

$$+ \int_{Q_t} f(w^N - v) \, dxdt$$

$$- \int_{Q_t} g(v_{xx})(w_{xx}^N - v_{xx}) \, dxdt \geq 0.$$

or

$$\|w_t^N\|_{L^2(Q_t)}^2 - \kappa_1 \|w_{xx}^N\|_{L^2(Q_t)}^2 -$$
$$\frac{\kappa_2}{2}\|w_{xx}^N(t)\|^2 + \frac{\kappa_2}{2}\|w_{xx}^N(0)\|^2 \qquad (7.16)$$

$$+ \int_{Q_t} \Big[-w_t^N v_t + \kappa_1 w_{xx}^N v_{xx} +$$
$$+ \kappa_2 w_{xxt}^N v_{xx} + f(w^N - v) - g(v_{xx})(w_{xx}^N - v_{xx}) \Big] \, dxdt$$

$$- \int_0^1 w_t^N (w^N - v) \, dx \Big|_{t=0}^{t=t} \geq 0.$$

Now the most important observation is that we can pass to the limit as $N \to \infty$ in (7.16) and obtain

$$\|w_t\|_{L^2(Q_t)}^2 - \kappa_1 \|w_{xx}\|_{L^2(Q_t)}^2 -$$
$$\frac{\kappa_2}{2}\|w_{xx}(t)\|^2 + \frac{\kappa_2}{2}\|w_{xx}(0)\|^2 \qquad (7.17)$$

$$+ \int_{Q_t} \Big[-w_t v_t + \kappa_1 w_{xx} v_{xx} + \kappa_2 w_{xxt} v_{xx}$$
$$+ f(w - v) - g(v_{xx})(w_{xx} - v_{xx}) \Big] \, dxdt$$

$$- \int_0^1 w_t (w - v) \, dx \Big|_{t=0}^{t=t} \geq 0.$$

The inequality (7.17) requires some discussion. In the first term we can pass to the limit due to the strong convergence (7.3). In the fourth term we can pass to the limit because $w_{xx}^N(0) \to w_{xx}(0) = \phi_{0xx}$ strongly in $L^2(0,1)$. In all terms under the integral over Q_t we can pass to the limit due to the weak convergence (7.1), (7.4) and the strong convergence (7.3). In the terms under the integral on $[0,1]$ we can pass to the limit due to the weak convergence $w_t^N(t) \to w_t(t)$ (see (7.2)) and the strong convergence $w^N(t) \to w(t)$. It remains to explain why we can pass to the limit in the second and third terms in (7.16), which follows from weak lower semi continuity of norms in Hilbert spaces. Here we have only the weak convergence $w_{xx}^N \rightharpoonup w_{xx}$ in $L^2(Q_t)$ (this follows from the stronger fact (7.1)) and the weak convergence $w_{xx}^N(t) \rightharpoonup w_{xx}(t)$ in $L^2(0,1)$ for all $t \in [0,T]$ (see (7.1)). These weak convergences and the weak lower semicontinuity of the norms in $L^2(Q_t)$ and $L^2(0,1)$ suffice to obtain the desired limit inequality (7.17).

Notice that (7.17) is valid for any $v \in \mathcal{P}$ and, therefore, for any $v \in \mathcal{L}_T$.

Now we return to (7.9). Notice that in this relation we can set $\eta = w$, because $w \in \mathcal{L}_T$ and (7.9) is valid for $\eta \in \mathcal{L}_T$. As a result we obtain

$$\int_{Q_t} \left[-|w_t|^2 + \kappa_1 |w_{xx}|^2 + \frac{\kappa_2}{2} \frac{d}{dt}(w_{xx})^2 + \tilde{g}w_{xx} \right] dxdt +$$
$$+ \int_0^1 w_t w \, dx \Big|_{t=0}^{t=t} - \int_{Q_t} fw \, dxdt = 0 \tag{7.18}$$

or

$$-\|w_t\|_{L^2(Q_t)}^2 + \kappa_1 \|w_{xx}\|_{L^2(Q_t)}^2 + \frac{\kappa_2}{2}\|w_{xx}(t)\|^2$$
$$-\frac{\kappa_2}{2}\|w_{xx}(0)\|^2$$
$$+ \int_{Q_t} [\tilde{g}w_{xx} - fw] \, dxdt + \int_0^1 w_t w \, dx \Big|_{t=0}^{t=t} = 0. \tag{7.19}$$

Let us consider (7.9) with $\eta = -v$, where v is is exactly the same function from (7.17). We have

$$\int_{Q_t} [w_t v_t - \kappa_1 w_{xx} v_{xx} - \kappa_2 w_{xxt} v_{xx} - \tilde{g}v_{xx}] \, dxdt -$$
$$\int_0^1 w_t v \, dx \Big|_{t=0}^{t=t} + \int_{Q_t} fv \, dxdt = 0. \tag{7.20}$$

We now add the inequality (7.17) and the relations (7.19) and (7.20). After considerable cancellation we arrive at

$$\int_{Q_t} [\tilde{g} - g(v_{xx})] (w_{xx} - v_{xx}) \, dxdt \geq 0, \tag{7.21}$$

for all $v \in \mathcal{L}_T$.

Now let us take any $\theta > 0$ and let $\zeta(x,t)$ be any function in $\mathcal{L}_T$. Select

$$v(x,t) = w(x,t) - \theta\zeta(x,t). \tag{7.22}$$

From (7.21) and (7.22) we have

$$\int_{Q_t} [\tilde{g} - g(w_{xx} - \theta\zeta_{xx})] \zeta_{xx} \, dxdt \geq 0, \tag{7.23}$$

for all $\zeta \in \mathcal{L}_T$, for all $\theta > 0$.

Since (7.23) is true for all $\theta > 0$ we can pass to the limit $\theta \to 0$ (recall that g is a continuous function) to obtain

$$\int_{Q_t} [\tilde{g} - g(w_{xx})] \zeta_{xx} \, dx dt \geq 0. \tag{7.24}$$

However, inequality (7.24) holds for all $\zeta \in \mathcal{L}_T$ only if it holds for equality. Indeed, suppose that for some ζ

$$\int_{Q_t} [\tilde{g} - g(w_{xx})] \zeta_{xx} \, dx dt > 0.$$

Then, replacing ζ by $\zeta_0 = -\zeta$ we obtain

$$\int_{Q_t} [\tilde{g} - g(w_{xx})] \zeta_{0xx} \, dx dt < 0.$$

This is a contradiction to (7.24). Thus we have

$$\int_{Q_t} [\tilde{g} - g(w_{xx})] \zeta_{xx} \, dx dt = 0, \tag{7.25}$$

for all $\zeta \in \mathcal{L}_T$.

We see that (7.25) coincides with (7.10). Lemma 7.1 is thus established and the proof of the existence theorem is complete. $\square$

References

[1] F. BROWDER, "Nonlinear monotone operators and convex sets in Banach spaces," Bull. Amer. Math. Soc., 71 (1965), 780-785.

[2] O.A. BANCHAU and C.H. HONG, "Nonlinear composite beam theory," *ASME J. Applied Mechanics*, 55, (1988), 156-163.

[3] H.T. BANKS, K.ITO, Y. WANG, "Well posedness for damped second order systems with bounded input operators" *CRSC TR 93-10*, June, 1991, N.C.S.U., to appear in *Differential and Integral Equations*.

[4] H.T. BANKS and R.C. SMITH, "Models for control in smart material structures," *Identification and Control in Systems Governed by Partial Differential Equations, SIAM*, Philadelphia, 1993, 26-44.

[5] O.A. LADYZHENSKAYA, V.A. SOLONNIKOV, N. URAL'CEVA, *Linear and Quasilinear Equations of Parabolic Type*, Translations of AMS, Vol. 23, 1968.

[6] O.A. LADYZHENSKAYA, *The Boundary Value Problems of Mathematical Physics*, Springer-Verlag, New York, 1984.

[7] O.A. LADYZHENSKAYA, *Attractors for Semigroups and Evolution Equations*, Cambridge Univ. Press, 1991.

[8] S-Y LUO and T-S CHOU, "Finite deformation and nonlinear elastic behavior of flexible composites," *ASME J. Applied Mechanics*, 55, (1988), 149-155.

[9] G. MINTY, "Monotone (nonlinear) operators in Hilbert spaces," Duke Math. J., 29 (1962), 341-346.

[10] E.P. POPOV, *Introduction to Mechanics of Solids*, Prentice-Hall Englewood Cliffs, N.J., 1968.

[11] I.H. SHAMES and F.A. COZZARELLI, *Elastic and Inelastic Stress Analysis*, Prentice Hall, Englewood Cliffs, N.J. 1992.

[12] H. TANABE, *Equations of Evolution*, Pitman Publishing LTD, London, 1979.

[13] J. WLOKA, *Partial Differential Equations*, Cambridge Univ. Press, 1992.

MODELING AND PARAMETER ESTIMATION FOR AN IMPERFECTLY CLAMPED PLATE *

H.T. Banks
Center for Research in Scientific Computation
North Carolina State University
Raleigh, NC 27695

R.C. Smith	Yun Wang
Department of Mathematics	*Mathematical Products Division*
Iowa State University	*Armstrong Laboratory*
Ames, IA 50011	*Brooks AFB, TX 78235*

1 Introduction

An important consideration in the modeling of structural and structural acoustic systems involves the determination of appropriate boundary conditions for the vibrating structure. In many applications, the clamped nature of the structure leads to the use of clamped or fixed boundary conditions, in which case, it is assumed that zero displacements and slopes are maintained at the boundaries. For example, this can be an appropriate assumption when using shell equations to model a fuselage, plate equations to model panels in a transformer or beam equations to model a helicopter blade. In the first case, the experimental shell structures are often supported by heavy clamps at the ends, thus leading to the use of fixed boundary conditions. In a similar manner, the bonding of panels to an underlying substructure or the attachment of the blades to a central hub lead to models which involve fixed boundary conditions.

In both experiments and physical applications, however, it is impossible to maintain truly fixed or clamped boundary conditions. This can be due to physical properties of the system being modeled, limitations in the clamping mechanisms, or material properties which prevent perfect clamps (for example, slipping will occur if the material in the structure being clamped is significantly softer than the clamps). One manner in which the energy loss due to movement in the boundary conditions is manifested is through a lowering of the natural frequencies for the structure or structural acoustic system. This shift in frequencies can range from levels of $1 - 2\%$ in

---

*The research of H.T.B. and Y.W. was supported in part by the Air Force Office of Scientific Research under grant AFOSR-F49620-93-1-0198. This research was also supported by the National Aeronautics and Space Administration under NASA Contract Numbers NAS1-18605 and NAS1-19480 while the authors were visiting scientists at the Institute for Computer Applications in Science and Engineering (ICASE), NASA Langley Research Center, Hampton, VA 23681. Additional support was also provided in part under NASA grant NAG-1-1600.

well-clamped systems to $20 - 30\%$ in loosely clamped systems [11]. These loosely-clamped systems still support a moment, however, so that simply-supported or pinned boundary conditions (zero displacement and zero moment) are not appropriate.

To determine when the assumption of fixed boundary conditions (zero displacement and slope) yields a sufficiently accurate model, and to provide a means of accounting for the energy loss when this assumption is not appropriate, it is necessary to provide a more comprehensive model for imperfectly clamped boundary conditions.

In this paper, we discuss a model for imperfectly clamped boundary conditions that is derived under the assumption that there is some variance in both the displacement and slope. To account for this variance, boundary moment expressions involving four undetermined parameters are employed. In applications, these parameters, as well as other physical parameters are determined through fit-to-data techniques. The modeling of these boundary conditions in the context of a loosely-clamped circular plate, to which piezoceramic patches are bonded, is presented in the second section of this paper. The presentation includes both the strong and weak forms of the equations of motion as well as a discussion pertaining to when each is useful. In Section 3, the modeling equations are posed in terms of sesquilinear forms and well-posedness issues are discussed. State approximation is addressed in Section 4 and parameter estimation is discussed in Section 5. Finally, we point out that while this model for the boundary conditions is presented in the context of a clamped circular plate, analogous expressions follow for structures involving beams and shells.

2   Modeling Equations

The derivation which follows will be for a circular plate although analogous arguments can be used for plates having other geometries, as well as shells and beams. We will consider a plate of radius a and thickness h as depicted in Figure 2.1. The region occupied by the unstrained neutral surface of the plate will be denoted by Γ_0. We also consider the possibility of piezoceramic patches being bonded to the plate in either axisymmetric or general configurations. The piecewise continuous density, Young's modulus, Poisson ratio, and damping coefficient for the combined structure are denoted ρ, E, ν and c_D, respectively.

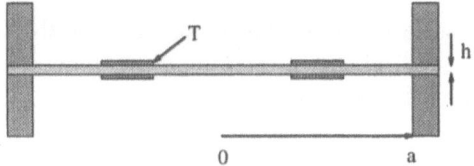

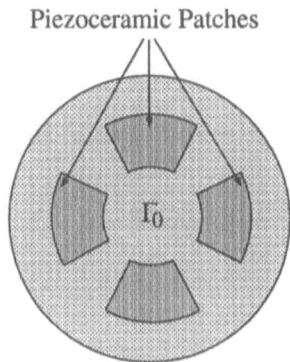

Piezoceramic Patches

Figure 2.1: A thin circular plate with piezoceramic patches bonded in pairs to its surface.

2.1 Plate Equations

To specify moments and forces acting on various regions of the plate, we let $\mathcal{M}_r$ and $\mathcal{M}_\theta$ denote general moment resultants with respect to θ and r, respectively, and take Q_r and Q_θ as corresponding shear resultants. The twisting moments are denoted by $\mathcal{M}_{r\theta}$ and $\mathcal{M}_{\theta r}$. As detailed for the axisymmetric case in [14] and the general nonaxisymmetric problem in [10] and (A1), (A3) and (A4) of [5], moment and force balancing for the circular plate then yields

$$Q_r = \frac{1}{r}\mathcal{M}_r + \frac{\partial \mathcal{M}_r}{\partial r} - \frac{1}{r}\mathcal{M}_\theta + \frac{1}{r}\frac{\partial \mathcal{M}_{\theta r}}{\partial \theta}$$

$$Q_\theta = \frac{2}{r}\mathcal{M}_{r\theta} + \frac{\partial \mathcal{M}_{r\theta}}{\partial r} + \frac{1}{r}\frac{\partial \mathcal{M}_\theta}{\partial \theta}$$

(2.1)

and

$$Q_r + \frac{\partial Q_r}{\partial r}r + \frac{\partial Q_\theta}{\partial \theta} = -rf + r\rho h\frac{\partial^2 w}{\partial t^2}$$

(2.2)

where f denotes a surface load on the plate and w is the transverse plate displacement. The sense of the internal moments and forces is illustrated in Figure 2.2. Elimination of shear terms then yields

$$\rho h \frac{\partial^2 w}{\partial t^2} - \frac{\partial^2 \mathcal{M}_r}{\partial r^2} - \frac{2}{r}\frac{\partial \mathcal{M}_r}{\partial r} + \frac{1}{r}\frac{\partial \mathcal{M}_\theta}{\partial r} - \frac{2}{r}\frac{\partial^2 \mathcal{M}_{r\theta}}{\partial r \partial \theta}$$
$$- \frac{2}{r^2}\frac{\partial \mathcal{M}_{r\theta}}{\partial \theta} - \frac{1}{r^2}\frac{\partial^2 \mathcal{M}_\theta}{\partial \theta^2} = f(t, r, \theta) .$$

(2.3)

The general moments are given by

$$\begin{aligned} \mathcal{M}_r &= M_r - (M_r)_{pe} \\ \mathcal{M}_\theta &= M_\theta - (M_\theta)_{pe} \\ \mathcal{M}_{r\theta} &= M_{r\theta} \end{aligned}$$

(2.4)

where

$$M_r = -D\left(\frac{\partial^2 w}{\partial r^2} + \frac{\nu}{r}\frac{\partial w}{\partial r} + \frac{\nu}{r^2}\frac{\partial^2 w}{\partial \theta^2}\right) - c_D\frac{\partial}{\partial t}\left(\frac{\partial^2 w}{\partial r^2} + \frac{\nu}{r}\frac{\partial w}{\partial r} + \frac{\nu}{r^2}\frac{\partial^2 w}{\partial \theta^2}\right)$$

$$M_\theta = -D\left(\frac{1}{r}\frac{\partial w}{\partial r} + \frac{1}{r^2}\frac{\partial^2 w}{\partial \theta^2} + \nu\frac{\partial^2 w}{\partial r^2}\right) - c_D\frac{\partial}{\partial t}\left(\frac{1}{r}\frac{\partial w}{\partial r} + \frac{1}{r^2}\frac{\partial^2 w}{\partial \theta^2} + \nu\frac{\partial^2 w}{\partial r^2}\right)$$

(2.5)

$$M_{r\theta} = -D(1-\nu)\left(\frac{1}{r}\frac{\partial^2 w}{\partial r \partial \theta} - \frac{1}{r^2}\frac{\partial w}{\partial \theta}\right) - c_D(1-\nu)\frac{\partial}{\partial t}\left(\frac{1}{r}\frac{\partial^2 w}{\partial r \partial \theta} - \frac{1}{r^2}\frac{\partial w}{\partial \theta}\right)$$

are the internal plate moments (see [7, 10] for derivations of these moment expressions). As detailed in [7], the density ρ in (2.3) and parameters $D = \frac{Eh^3}{12(1-\nu^2)}, \nu$ and c_D in the moment expressions are discontinuous due to the geometrical and material changes resulting from the bonding of the patches to the plate. The external moments

$$(M_r)_{pe} = (M_\theta)_{pe} = -\sum_{i=1}^{s} \mathcal{K}_i^B u_i(t)\chi_i(r, \theta)$$

(2.6)

are generated by s pairs of patches in response to applied voltages. Here $\chi_i(r, \theta)$ denotes the characteristic function which has a value of 1 in the region covered by the ith patch and is 0 elsewhere. Moreover, $u_i(t)$ is the voltage into the ith patch and $\mathcal{K}_i^B$ is a parameter which depends on the geometry, piezoceramic material properties and piezoelectric strain constants (see [7] for details). We point out that the piezoceramic material parameters $\mathcal{K}_i^B, i = 1, \cdots, s$ as well as the plate parameters ρ, D, c_D and ν should be considered as unknown and in applications must be estimated using data fitting techniques. The piecewise nature of the material parameters and input moments is one motivation for approximating the problem in a weak or variational form.

We also point out that in general, the patches are not bonded near the edge of the plate. Hence $\mathcal{M}_r = M_r$ and $\mathcal{M}_\theta = M_\theta$ at $r = a$ since there are no contributing external moments.

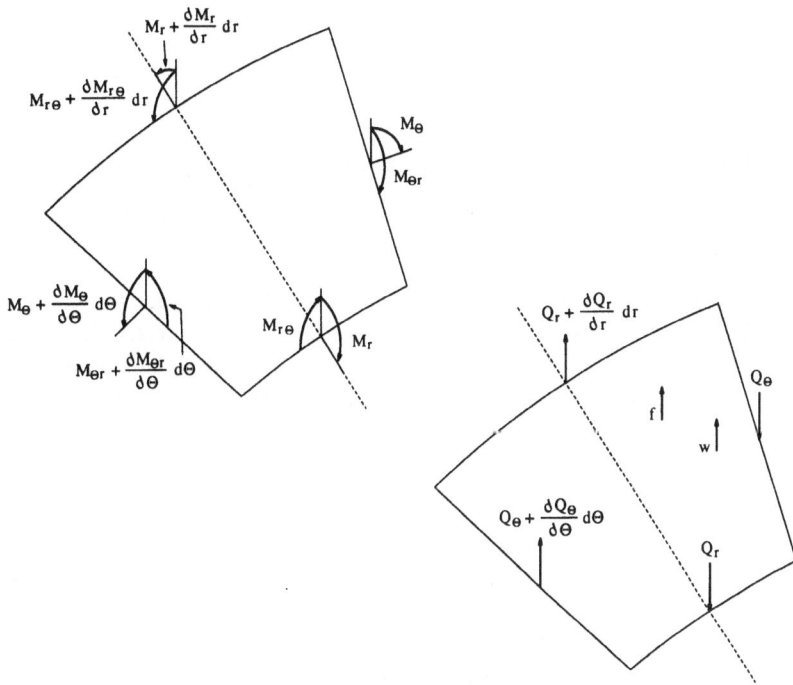

Figure 2.2: The internal moments and forces for the circular plate.

2.2 Fixed Boundary Conditions

A common set of boundary conditions that are used for clamped plates are the fixed-end conditions

$$w(a, \theta) = \frac{\partial w}{\partial r}(a, \theta) = 0 \qquad (2.7)$$

[10, 13] in which it is assumed that zero displacements and slopes are maintained at the boundary of the plate. As noted in the introduction, however, it is impossible to maintain truly fixed boundary conditions in physical applications. This motivates the investigation of imperfectly clamped conditions which allow for rotations at the boundary as well as small transverse boundary displacements.

2.3 Imperfectly Clamped Conditions

When developing the imperfectly clamped conditions for the plate, we make the following assumptions:

1. The transverse boundary displacement and rotation are independent of θ. In applications, this would be the case when clamping is not truly fixed but bolts are uniformly torqued along the boundary of the plate. Since a torque wrench is usually used to tighten the bolts, this is a reasonable assumption in many applications. In cases where uneven clamping along the boundary is suspected, analysis with θ-dependent parameters can be used in a manner analogous to that presented here.

2. The bending moment $\mathcal{M}_\theta = M_\theta$ and twisting moment $\mathcal{M}_{r\theta} = M_{r\theta}$ are 0 at $r = a$. This assumption is based on the premise that, even with weakly clamped boundary conditions, the rotation at the boundary will be primarily in the radial direction.

To model the transverse motion at the clamped end, it is assumed that this boundary motion can be modeled using a spring-dashpot system on either side of the beam centerline as depicted in Figure 2.3. Here $\tilde{k}_t$ and $\tilde{c}_t$ are used to denote the equivalent spring coefficient and viscous damping coefficient, respectively. The balancing of forces then yields

$$-\rho h \frac{\partial^2 w}{\partial t^2}(t, a, \theta) = Q_r(t, a, \theta) + 2\tilde{k}_t w(t, a, \theta) + 2\tilde{c}_t \frac{\partial w}{\partial t}(t, a, \theta) \ .$$

Under the assumption that $\mathcal{M}_\theta = \mathcal{M}_{r\theta} = 0$ at $r = a$, (2.1) can be used to relate the shear resultant Q_r to the moments through the expression

$$Q_r = \frac{1}{a} M_r + \frac{\partial M_r}{\partial r}$$

(recall that $\mathcal{M}_r = M_r$ at the plate boundary). The displacement at $r = a$ then admits the differential equation

$$\frac{1}{a} M_r(t, a, \theta) + \frac{\partial M_r}{\partial r}(t, a, \theta)$$

$$= -k_t w(t, a, \theta) - c_t \frac{\partial w}{\partial t}(t, a, \theta) - \rho h \frac{\partial^2 w}{\partial t^2}(t, a, \theta) .$$

(2.8)

The change in slope at the boundary is modeled by assuming that rotation at the clamp is constrained by a compressive moment with magnitude proportional to the tangent angle as depicted in Figure 2.4. Under the assumption of small rotations, $\tan \theta \approx \frac{\partial w}{\partial x}$ at $r = a$ which yields

$$- M_r(t, a, \theta) = -k_p \frac{\partial w}{\partial r}(t, a, \theta) - c_p \frac{\partial^2 w}{\partial r \partial t}(t, a, \theta) .$$

(2.9)

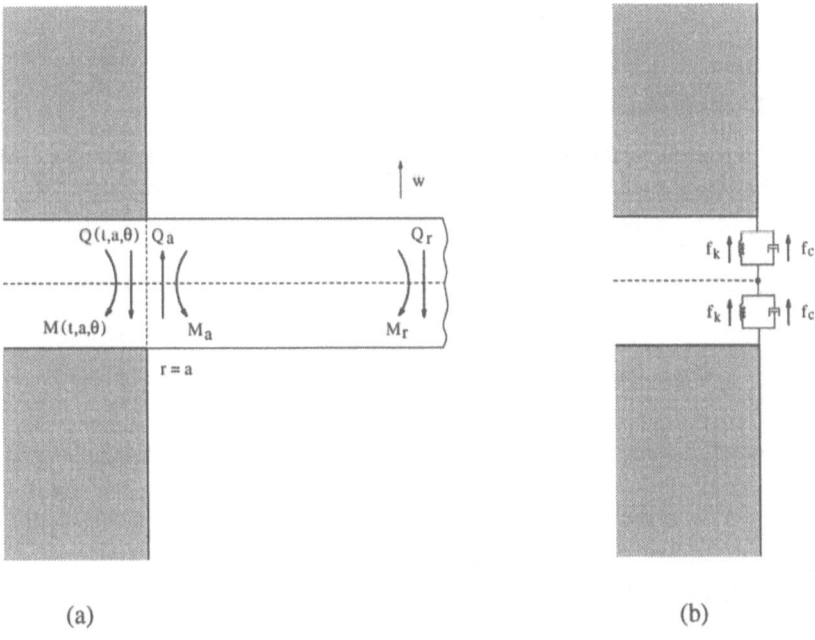

(a)   (b)

Figure 2.3: (a) Orientation of force and moment resultants. Note that $Q(t, a, \theta) = -Q_a$ and $M(t, a, \theta) = -M_a$. (b) Transverse boundary motion modeled as a spring-dashpot system. The forces are given by $f_k = \tilde{k}_t w(t, a, \theta)$ and $f_c = \tilde{c}_t \frac{\partial w}{\partial t}(t, a, \theta)$.

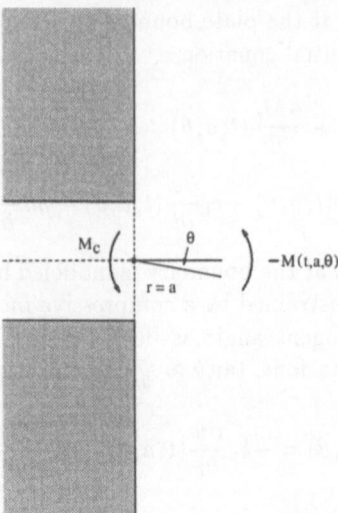

Figure 2.4: Boundary rotation and resulting moments.

2.4   Strong Form of the Plate Equations

The differential equation (2.3) and boundary conditions (2.8) and (2.9) can then be combined with initial conditions to yield the strong form of the circular plate equations

$$
\begin{cases}
\begin{aligned}
& \rho h \frac{\partial^2 w}{\partial t^2} - \frac{\partial^2 \mathcal{M}_r}{\partial r^2} - \frac{2}{r}\frac{\partial \mathcal{M}_r}{\partial r} + \frac{1}{r}\frac{\partial \mathcal{M}_\theta}{\partial r} - \frac{2}{r}\frac{\partial^2 \mathcal{M}_{r\theta}}{\partial r \partial \theta} \\
& \qquad - \frac{2}{r^2}\frac{\partial \mathcal{M}_{r\theta}}{\partial \theta} - \frac{1}{r^2}\frac{\partial^2 \mathcal{M}_\theta}{\partial \theta^2} = f(t,r,\theta)
\end{aligned}
& \begin{aligned}
& 0 < \theta \le 2\pi \\
& 0 \le r < a
\end{aligned}
\end{cases}
$$

$$
\begin{cases}
\begin{aligned}
& \frac{1}{a}M_r(t,a,\theta) + \frac{\partial M_r}{\partial r}(t,a,\theta) \\
& \qquad = -k_t w(t,a,\theta) - c_t \frac{\partial w}{\partial t}(t,a,\theta) - \rho h \frac{\partial^2 w}{\partial t^2}(t,a,\theta) \\
& M_r(t,a,\theta) = k_p \frac{\partial w}{\partial r}(t,a,\theta) + c_p \frac{\partial^2 w}{\partial r \partial t}(t,a,\theta)
\end{aligned}
\end{cases}
$$

$$
\left\{ w(0,r,\theta) = w_0(r,\theta) \quad , \quad \frac{\partial w}{\partial t}(0,r,\theta) = w_1(r,\theta) \;\; . \right.
$$

$$(2.10)$$

Explicit expressions for the moments are given in (2.4) and (2.5). It should be noted that since k_p and k_t are assumed to be constants, one can divide through by these quantities in the boundary conditions. In the limits

$k_p \to \infty$ and $k_t \to \infty$, the imperfectly clamped conditions converge to the fixed-end conditions (2.7). This should be expected since increased spring stiffnesses implies that one is approaching an ideal clamp in which displacements and slopes are truly zero.

2.5   Weak Form of the Plate Equations

To define a weak form of the equations of motion for the plate with weakly-clamped edges, we must determine an appropriate state, state space and space of test functions. The state is taken to be $z = (w(a, \cdot), w(\cdot, \cdot))$ in the space $H \equiv L^2(0, 2\pi) \times L^2(\Gamma_0)$. For $\Phi = (\zeta(\cdot), \phi(\cdot, \cdot))$ and $\Psi = (\xi(\cdot), \eta(\cdot, \cdot))$, the H-inner product is given by

$$\langle \Phi, \Psi \rangle_H = \int_0^{2\pi} \rho h a \zeta(\theta) \overline{\xi(\theta)} \, d\theta + \langle \rho h \phi, \eta \rangle$$

where $\langle F, G \rangle \equiv \int_{\Gamma_0} F \overline{G} d\gamma$ with $d\gamma = r d\theta dr$. The overbar here denotes complex conjugation. To satisfy regularity requirements, the space of test functions and associated inner product are defined to be

$$V \equiv \left\{ \Psi = (\xi(\cdot), \eta(\cdot, \cdot)) \in H : \eta \in H^2(\Gamma_0), \eta(a, \theta) = \xi(\theta) \right\}$$

$$\begin{aligned}
\langle z, \Psi \rangle_V &= -\left\langle M_r, \frac{\partial^2 \eta}{\partial r^2} \right\rangle - \left\langle \frac{1}{r} M_\theta, \frac{\partial \eta}{\partial r} \right\rangle - \left\langle \frac{1}{r^2} M_\theta, \frac{\partial^2 \eta}{\partial \theta^2} \right\rangle \\
&\quad - 2 \left\langle \frac{1}{r} M_{r\theta}, \frac{\partial^2 \eta}{\partial r \partial \theta} \right\rangle + 2 \left\langle \frac{1}{r^2} M_{r\theta}, \frac{\partial \eta}{\partial \theta} \right\rangle \\
&\quad + \int_0^{2\pi} a k_p \frac{\partial w}{\partial r}(a, \theta) \overline{\frac{\partial \eta}{\partial r}}(a, \theta) d\theta + \int_0^{2\pi} a k_t w(a, \theta) \overline{\eta}(a, \theta) d\theta
\end{aligned}$$

where the moments are defined in (2.5) with $c_D = 0$ in the respective definitions.

Integration of the internal moments in the second and third terms of (2.10) against the test functions yields

$$-\int_{\Gamma_0} \left[\frac{\partial^2 M_r}{\partial r^2} + \frac{2}{r} \frac{\partial M_r}{\partial r} \right] \overline{\eta} d\gamma$$

$$= -\int_{\Gamma_0} M_r \frac{\overline{\partial^2 \eta}}{\partial r^2} d\gamma - \int_0^{2\pi} \left[\left(\frac{\partial M_r}{\partial r} r + M_r \right) \overline{\eta} - M_r r \frac{\overline{\partial \eta}}{\partial r} \right] \Bigg|_{r=a} d\theta .$$

In reducing the boundary term, we have used the fact that $M_r(t, 0, \theta) = 0$ as a result of our choice of coordinate system. With the assumption that

$M_\theta = M_{r\theta} = 0$ at $r = a$, the remaining internal moment terms can be integrated by parts to yield

$$\int_{\Gamma_0} \left[\frac{1}{r} \frac{\partial M_\theta}{\partial r} - \frac{2}{r} \frac{\partial^2 M_{r\theta}}{\partial r \partial \theta} - \frac{2}{r^2} \frac{\partial M_{r\theta}}{\partial \theta} - \frac{1}{r^2} \frac{\partial^2 M_\theta}{\partial \theta^2} \right] \overline{\eta} d\gamma$$

$$= -\int_{\Gamma_0} \frac{1}{r^2} M_\theta \left[r \frac{\overline{\partial \eta}}{\partial r} + \frac{\overline{\partial^2 \eta}}{\partial \theta^2} \right] d\gamma - 2 \int_{\Gamma_0} \frac{1}{r^2} M_{r\theta} \left[r \frac{\overline{\partial^2 \eta}}{\partial r \partial \theta} - \frac{\overline{\partial \eta}}{\partial \theta} \right] d\gamma$$

for all $\Psi \in V$. Inclusion of external moments and forces and consolidation of terms then gives the weak form

$$\int_{\Gamma_0} \rho h \frac{\partial^2 w}{\partial t^2} \overline{\eta} d\gamma - \int_{\Gamma_0} M_r \frac{\overline{\partial^2 \eta}}{\partial r^2} d\gamma - \int_{\Gamma_0} \frac{1}{r^2} M_\theta \left[r \frac{\overline{\partial \eta}}{\partial r} + \frac{\overline{\partial^2 \eta}}{\partial \theta^2} \right] d\gamma$$

$$-2 \int_{\Gamma_0} \frac{1}{r^2} M_{r\theta} \left[r \frac{\overline{\partial^2 \eta}}{\partial r \partial \theta} - \frac{\overline{\partial \eta}}{\partial \theta} \right] d\gamma$$

$$+ \int_0^{2\pi} a \left[k_t w(t, a, \theta) + c_t \frac{\partial w}{\partial t}(t, a, \theta) + \rho h \frac{\partial^2 w}{\partial t^2}(t, a, \theta) \right] \overline{\eta}(a, \theta) d\theta \qquad (2.11)$$

$$+ \int_0^{2\pi} a \left[k_p \frac{\partial w}{\partial r}(t, a, \theta) + c_p \frac{\partial^2 w}{\partial r \partial t}(t, a, \theta) \right] \frac{\overline{\partial \eta}}{\partial r}(a, \theta) d\theta$$

$$= -\int_{\Gamma_0} \sum_{i=1}^{s} \mathcal{K}_i^B u_i(t) \chi_i(r, \theta) \overline{\nabla^2 \eta} d\gamma + \int_{\Gamma_0} f \overline{\eta} d\gamma$$

for all $\Psi = (\eta(a, \cdot), \eta(\cdot, \cdot)) \in V$. Again, explicit expressions for the internal moments are given in (2.5). It is easily noted that in this form, derivatives that were originally applied to moments have been transferred onto the test functions. This eliminates the difficulties associated with differentiating the piecewise constant parameters D, ν and c_D found in the internal moments as well as the discontinuous input parameters $\mathcal{K}_i^B$.

3   Model Well-Posedness

A first step in proving the existence of a unique solution to these modeling equations is to pose the problem in abstract first-order form. To do so, we first define the sesquilinear forms $\sigma_i : V \times V \to \mathbb{C}$, $i = 1, 2$ by

$$
\begin{aligned}
\sigma_1(W, \Psi) &= \int_{\Gamma_0} D \left(\frac{\partial^2 w}{\partial r^2} + \frac{\nu}{r} \frac{\partial w}{\partial r} + \frac{\nu}{r^2} \frac{\partial^2 w}{\partial \theta^2} \right) \overline{\frac{\partial^2 \eta}{\partial r^2}} \, d\gamma \\
&+ \int_{\Gamma_0} D \left(\frac{1}{r^2} \frac{\partial w}{\partial r} + \frac{1}{r^3} \frac{\partial^2 w}{\partial \theta^2} + \frac{\nu}{r} \frac{\partial^2 w}{\partial r^2} \right) \overline{\frac{\partial \eta}{\partial r}} \, d\gamma \\
&+ \int_{\Gamma_0} D \left(\frac{1}{r^3} \frac{\partial w}{\partial r} + \frac{1}{r^4} \frac{\partial^2 w}{\partial \theta^2} + \frac{\nu}{r^2} \frac{\partial^2 w}{\partial r^2} \right) \overline{\frac{\partial^2 \eta}{\partial \theta^2}} \, d\gamma \\
&+ \int_{\Gamma_0} 2(1 - \nu) D \left(\frac{1}{r^2} \frac{\partial^2 w}{\partial \theta \partial r} - \frac{1}{r^3} \frac{\partial w}{\partial \theta} \right) \overline{\frac{\partial^2 \eta}{\partial r \partial \theta}} \, d\gamma \\
&+ \int_{\Gamma_0} 2(1 - \nu) D \left(-\frac{1}{r^3} \frac{\partial^2 w}{\partial \theta \partial r} + \frac{1}{r^4} \frac{\partial w}{\partial \theta} \right) \overline{\frac{\partial \eta}{\partial \theta}} \, d\gamma \\
&+ \int_0^{2\pi} a k_p \frac{\partial w}{\partial r}(a, \theta) \overline{\frac{\partial \eta}{\partial r}}(a, \theta) d\theta + \int_0^{2\pi} a k_t w(a, \theta) \overline{\eta}(a, \theta) d\theta
\end{aligned}
$$

and

$$
\begin{aligned}
\sigma_2(W, \Psi) &= \int_{\Gamma_0} c_D \left(\frac{\partial^2 w}{\partial r^2} + \frac{\nu}{r} \frac{\partial w}{\partial r} + \frac{\nu}{r^2} \frac{\partial^2 w}{\partial \theta^2} \right) \overline{\frac{\partial^2 \eta}{\partial r^2}} \, d\gamma \\
&+ \int_{\Gamma_0} c_D \left(\frac{1}{r^2} \frac{\partial w}{\partial r} + \frac{1}{r^3} \frac{\partial^2 w}{\partial \theta^2} + \frac{\nu}{r} \frac{\partial^2 w}{\partial r^2} \right) \overline{\frac{\partial \eta}{\partial r}} \, d\gamma \\
&+ \int_{\Gamma_0} c_D \left(\frac{1}{r^3} \frac{\partial w}{\partial r} + \frac{1}{r^4} \frac{\partial^2 w}{\partial \theta^2} + \frac{\nu}{r^2} \frac{\partial^2 w}{\partial r^2} \right) \overline{\frac{\partial^2 \eta}{\partial \theta^2}} \, d\gamma \\
&+ \int_{\Gamma_0} 2(1 - \nu) c_D \left(\frac{1}{r^2} \frac{\partial^2 w}{\partial \theta \partial r} - \frac{1}{r^3} \frac{\partial w}{\partial \theta} \right) \overline{\frac{\partial^2 \eta}{\partial r \partial \theta}} \, d\gamma \\
&+ \int_{\Gamma_0} 2(1 - \nu) c_D \left(-\frac{1}{r^3} \frac{\partial^2 w}{\partial \theta \partial r} + \frac{1}{r^4} \frac{\partial w}{\partial \theta} \right) \overline{\frac{\partial \eta}{\partial \theta}} \, d\gamma \\
&+ \int_0^{2\pi} a c_p \frac{\partial w}{\partial r}(a, \theta) \overline{\frac{\partial \eta}{\partial r}}(a, \theta) d\theta + \int_0^{2\pi} a c_t w(a, \theta) \overline{\eta}(a, \theta) d\theta .
\end{aligned}
$$

With $\langle \cdot, \cdot \rangle_{V^*, V}$ denoting the usual duality product, $F = (0, \frac{f}{\rho h})$ and $z(t) = (w(t, a, \cdot), w(t, \cdot, \cdot))$, the system (2.11) containing the weak form of the modeling equations can then be written as

$$
\langle z_{tt}(t), \Psi \rangle_{V^*, V} + \sigma_2(z_t(t), \Psi) + \sigma_1(z(t), \Psi) = \langle F + B, \Psi \rangle_{V^*, V} \qquad (3.1)
$$

for Ψ in V. The input operator $B \in \mathcal{L}(\mathbb{R}^s, V^*)$ defined by

$$\langle Bu, \Psi \rangle_{V^*, V} = -\int_{\Gamma_0} \sum_{i=1}^s \mathcal{K}_i^B u_i \chi_i(r, \theta) \overline{\nabla^2 \eta} d\gamma$$

contains the contributions due to the excitation of the patches through applied voltages.

Before writing the system in first-order form and proving well-posedness, we must further examine the properties of the sesquilinear forms. It is first noted that σ_1 satisfies

$$|\sigma_1(\Phi, \Psi)| \leq c_1 |\Phi|_V |\Psi|_V \quad \text{, for some} \quad c_1 \in \mathbb{R} \qquad \text{(bounded)}$$

$$\text{Re } \sigma_1(\Phi, \Phi) \geq c_2 |\Phi|_V^2 \quad \text{, for some} \quad c_2 > 0 \qquad (V\text{-elliptic}) \qquad (3.2)$$

$$\sigma_1(\Phi, \Psi) = \overline{\sigma_1(\Psi, \Phi)} \qquad \text{(symmetric)}$$

for all $\Phi, \Psi \in V$. The boundedness follows from Schwarz's inequality for inner products in conjunction with equivalence results for various Sobolev norms. The V-ellipticity and symmetry of σ_1 follow directly from the fact that $\sigma_1(\Phi, \Psi) = \langle \Phi, \Psi \rangle_V$.

Under the assumption that the damping parameters satisfy $c_D(r, \theta) \geq \overline{c_D} > 0, c_p > 0$ and $c_t > 0$, it also follows that

$$|\sigma_2(\Phi, \Psi)| \leq c_3 |\Phi|_V |\Psi|_V \quad \text{, for some} \quad c_3 \in \mathbb{R} \qquad \text{(bounded)}$$

$$\text{Re } \sigma_2(\Phi, \Phi) \geq c_4 |\Phi|_V^2 \quad \text{, for some} \quad c_4 > 0 \qquad (V\text{-elliptic}) \ . \qquad (3.3)$$

To prove convergence in the parameter estimation problem, we also need to show continuity of the sesquilinear forms with respect to parameters. To this end, we let

$$q = (\rho, D, \nu, c_D, k_t, k_p, c_t, c_p, \mathcal{K}_1^B, \cdots, \mathcal{K}_s^B)$$

and let d denote the standard metric in the metric space $\tilde{Q} = [L_\infty(\Gamma_0)]^4 \times \mathbb{R}^4 \times \mathbb{R}^s$ (details regarding this choice of metric space will be given in the Section 5). Using Schwarz's inequality for inner products in conjunction with appropriate Sobolev inequalities, it can then be shown that for $q, \tilde{q} \in \tilde{Q}$ and $\Psi, \Phi \in V$,

$$|\sigma_1(q)(\Phi, \Psi) - \sigma_1(\tilde{q})(\Phi, \Psi)| \leq \gamma_1 d(q, \tilde{q}) |\Phi|_V |\Psi|_V$$

$$|\sigma_2(q)(\Phi, \Psi) - \sigma_2(\tilde{q})(\Phi, \Psi)| \leq \gamma_2 d(q, \tilde{q}) |\Phi|_V |\Psi|_V \ . \qquad (3.4)$$

Hence both sesquilinear forms are continuous with respect to parameters as well as satisfy the previously discussed continuity, coercivity and symmetry conditions.

To pose the system in first-order form, we form the product space terms $\mathcal{B}u(t) = (0, Bu(t))$ and $\mathcal{F}(t) = (0, F(t))$ in $\mathcal{V}^* = V \times V^*$ and define the operators $A_1, A_2 \in \mathcal{L}(V, V^*)$ by

$$\langle A_i \Phi, \Psi \rangle_{V^*, V} = \sigma_i(\Phi, \Psi)$$

for $i = 1, 2$ (the existence of A_1 and A_2 is guaranteed by the boundedness of σ_1 and σ_2). Then, for the state $\mathcal{Z}(t) = (z, \dot{z})$ in $\mathcal{H} = H \times V$, the weak form (3.1) is *formally* equivalent to the system

$$\dot{\mathcal{Z}}(t) = \mathcal{A}\mathcal{Z}(t) + \mathcal{B}u(t) + \mathcal{F}(t) \tag{3.5}$$

in $\mathcal{V}^*$ where

$$\text{dom}\,\mathcal{A} = \{\Theta = (\Upsilon, \Lambda) \in \mathcal{H} : \Lambda \in V, A_1 \Upsilon + A_2 \Lambda \in H\}$$

$$\mathcal{A} = \begin{bmatrix} 0 & I \\ -A_1 & -A_2 \end{bmatrix}. \tag{3.6}$$

(the representation is formal in the sense that the manner in which differentiation and the resulting solution exist has not yet been specified).

To argue the well-posedness of the model, it is first noted that the sesquilinear forms σ_1 and σ_2 are V-elliptic and continuous and σ_1 is symmetric. From the Lumer-Philips theorem (with further arguments found on pages 82-84 of [1]) this then implies that the operator $\mathcal{A}$ defined in (3.6) generates an analytic semigroup T on the state space $\mathcal{H}$. Moreover, the semigroup satisfies the exponential bound $|T(t)| \leq e^{\omega t}$ for $t \geq 0$ (where in fact, $\omega = 0$ due to the fact that $\mathcal{A}$ is dissipative as shown in [1]).

In the case of bounded input operators, one would then define a mild solution on $\mathcal{H}$ and show that this mild solution is indeed a strong solution to the problem. Here, however, $\mathcal{B}u(t)$ lies in $\mathcal{V}^*$ rather than $\mathcal{H}$ and one must extend the semigroup $T(t)$ on $\mathcal{H}$ to a semigroup $\tilde{T}(t)$ on a larger space $\mathcal{W}^* \supset \{0\} \times V^*$ so as to be compatible with the forcing term.

As detailed in [3, 6], the space of interest is defined in terms of dom $\mathcal{A}^*$ where

$$\text{dom}\,\mathcal{A}^* = \{\chi = (\Phi, \Psi) \in \mathcal{H} | \Psi \in V, A_1^* \Phi - A_2^* \Psi \in H\}$$

$$\mathcal{A}^* \chi = \begin{pmatrix} -\Psi \\ A_1^* \Phi - A_2^* \Psi \end{pmatrix}$$

Specifically, the space $\mathcal{W} = [\text{dom}\,\mathcal{A}^*]$ is taken to be dom $\mathcal{A}^*$ with the inner product

$$\langle \Phi, \Psi \rangle_{\mathcal{W}} = \langle (\lambda_0 - \mathcal{A}^*)\Phi, (\lambda_0 - \mathcal{A}^*)\Psi \rangle_{\mathcal{H}}$$

for some arbitrary but fixed λ_0 with $\lambda_0 > \omega$ (recall that the original solution semigroup satisfies the bound $|T(t)| \leq e^{\omega t}$). By employing "extrapolation space" ideas and arguments similar to those presented in [3, 9], one can then extend the operator $\mathcal{A} : \operatorname{dom} \mathcal{A} \subset \mathcal{H} \to \mathcal{H}$ to an operator $\tilde{\mathcal{A}}$ which is defined on all of $\mathcal{H}$. Moreover, the operator $\tilde{\mathcal{A}}$ is the infinitesimal generator of an analytic semigroup $\tilde{T}(t)$ which is an extension of T from $\mathcal{H}$ to $\mathcal{W}^*$.

Using this extension, the following theorem from [3] then guarantees the existence of a unique solution as well verifies the equivalence of the strong and weak solutions.

Theorem 3.1: Existence and Equivalence of Solutions) *Consider the system represented by (2.11) or (3.1) and suppose that the mappings $t \mapsto \{u_i(t)\}$ and $t \mapsto F(t)$ from $[0, T]$ to $\mathbb{R}^s$ and V^*, respectively, are Lipschitz continuous. Then for each $\mathcal{Z}_0 \in \mathcal{H} = \operatorname{dom} \tilde{\mathcal{A}}$, we have that (3.5) taken with $\mathcal{Z}(0) = \mathcal{Z}_0$ has a unique strong solution given by*

$$\mathcal{Z}(t) = \tilde{T}(t)\mathcal{Z}_0 + \int_0^t \tilde{T}(t - s) \begin{pmatrix} 0 \\ Bu(s) + F(s) \end{pmatrix} ds .$$

Moreover, this strong solution is equal to the weak solution of (2.11) or (3.1).

As a result of this theorem, the imperfectly clamped plate model is well-posed. We turn next to the issue of estimating parameters in this model.

4   State Approximation

Having established the existence of a solution to the imperfectly clamped plate model, we turn now to the issue of approximating the infinite dimensional system dynamics. As discussed in [5, 12], an appropriate choice for the basis and Fourier-Galerkin expansion of the plate displacement is $B_k^N(r, \theta) = r^{|\hat{m}|} B_n^m(r) e^{im\theta}$ and

$$
\begin{aligned}
w^N(t, r, \theta) &= \sum_{m=-M}^{M} \sum_{n=1}^{N^m} w_{mn}^N(t) r^{|\hat{m}|} B_n^m(r) e^{im\theta} \\
&= \sum_{k=1}^{N} w_k^N(t) B_k^N(r, \theta) .
\end{aligned}
$$

(4.1)

In the previously cited references, basis alterations were needed to accommodate the displacement and slope conditions mandated by the fixed boundary conditions. This is not necessary when considering imperfectly clamped boundary conditions since it is no longer required that the approximate solution satisfy $w^{\mathcal{N}}(t, a, \theta) = \frac{\partial w^{\mathcal{N}}}{\partial r}(t, a, \theta) = 0$. In this case, $B_n^m(r)$ denotes a cubic spline that is modified so as to satisfy the condition $\frac{dB_n^m(0)}{dr} = 0$ when $m = 0$. This guarantees differentiability at the origin and implies that

$$N^m = \begin{cases} N + 2 & , \quad m = 0 \\ N + 3 & , \quad m \neq 0 \end{cases}$$

where N denotes the number of modified cubic splines. Thus a total of $\mathcal{N} = (2M + 1)(N + 3) - 1$ basis functions are used when approximating the plate displacement. As discussed in the [5, 12], the inclusion of the weighting term $r^{|\hat{m}|}$ with

$$\hat{m} = \begin{cases} 0 & , \quad m = 0 \\ 1 & , \quad m \neq 0 \end{cases}$$

is motivated by the asymptotic behavior of the Bessel functions (which make up the analytic plate solution) as $r \to 0$. It also serves to ensure the uniqueness of the solution at the origin. The Fourier coefficient in the weight is truncated to control the conditioning of the mass and stiffness matrices (see the examples in [5]).

To obtain a matrix system, the $\mathcal{N}$ dimensional approximating subspace is taken to be $H^{\mathcal{N}} = span\{B_k^{\mathcal{N}}\}$ and the product space for the first-order system is $\mathcal{H}^{\mathcal{N}} \times \mathcal{H}^{\mathcal{N}}$. The restriction of the infinite-dimensional system (2.11), or equivalently (3.1), to the space $\mathcal{H}^{\mathcal{N}} \times \mathcal{H}^{\mathcal{N}}$ then yields a matrix system of the form

$$\begin{aligned} \dot{y}^{\mathcal{N}}(t) &= A^{\mathcal{N}} y^{\mathcal{N}}(t) + B^{\mathcal{N}} u(t) + F^{\mathcal{N}}(t) \\ y^{\mathcal{N}}(0) &= y_0^{\mathcal{N}} \end{aligned} \tag{4.2}$$

where $y^{\mathcal{N}}(t) = [w_1(t), \cdots, w_N(t), \dot{w}_1(t), \cdots, \dot{w}_N(t)]$ denotes the $2\mathcal{N} \times 1$ vector containing the generalized Fourier coefficients for the approximate displacement and velocity. Details concerning the construction of the component vectors and matrices in (4.2) can be found in [5, 12].

5   Parameter Estimation

We now turn to the problem of estimating the physical plate parameters $\rho, D, \nu, c_D, k_t, k_p, c_t, c_p, \mathcal{K}_1^B, \cdots, \mathcal{K}_s^B$ by fitting data z obtained from displacement, velocity or acceleration measurements at various locations on the plate. To formulate the problem in an optimization setting, we let

$q = (\rho, D, \nu, c_D, k_t, k_p, c_t, c_p, \mathcal{K}_1^B, \cdots, \mathcal{K}_s^B)$ and assume that $q \in Q$ where the admissible parameter space Q is taken to be a compact subset of the metric space $\tilde{Q} = [L_\infty(\Gamma_0)]^4 \times \mathbb{R}^4 \times \mathbb{R}^s$ with elements also satisfying the physical constraints $\rho > 0$, $D > 0$, $\nu > 0$ and $c_D > 0$ on Γ_0. Associated with the space $\tilde{Q}$ is the standard metric d.

For time domain estimation with data consisting of position, velocity, or acceleration measurements at points $(\hat{r}, \hat{\theta})$ on the plate, the infinite dimensional parameter estimation problem is to seek $\bar{q}$ which minimizes

$$J(q) = \sum_i \left| \frac{\partial^\nu w}{\partial t^\nu}(t_i, \hat{r}, \hat{\theta}; q) - z_i \right|^2 \quad , \quad \nu = 0, 1 \text{ or } 2 \qquad (5.1)$$

subject to w satisfying (2.11) or (3.1). As noted in [8], similar penalty functionals can be used if the data consists of accumulated strain measurements obtained from the patches.

To facilitate the estimation of the material parameters, we now make some assumptions regarding their spatial behavior. As discussed previously, the parameters ρ, D, ν and c_D are assumed to be piecewise constants in order to account for the presence and differing material properties of the piezoceramic patches. For the case in which s patches or patch pairs are bonded to the plate, these parameters can then be expressed as

$$\rho(r, \theta) = \sum_{i=1}^{s+1} c_{\rho i} \chi_i(r, \theta) \quad , \quad D(r, \theta) = \sum_{i=1}^{s+1} c_{Di} \chi_i(r, \theta)$$

$$\nu(r, \theta) = \sum_{i=1}^{s+1} c_{\nu i} \chi_i(r, \theta) \quad , \quad c_D(r, \theta) = \sum_{i=1}^{s+1} c_{c_D i} \chi_i(r, \theta)$$

where again, $\chi_i(r, \theta)$, $i = 1, \cdots, s$ is the characteristic functions over the ith patch or patch pair and χ_{s+1} is the characteristic function over the portion of the plate not covered with patches. The parameters k_t, k_p, c_t and c_p are constants which must also be estimated when using this plate model. Finally, we recall from the definition (2.6) that the patch parameters $\mathcal{K}_1^B, \cdots, \mathcal{K}_s^B$ are constants which depend on piezoelectric properties, the geometry and size of the patch, and bonding layer and patch properties.

To provide an optimization problem involving a finite dimensional state, the approximation techniques discussed in the previous section are employed. The finite dimensional minimization problem corresponding to (5.1) is to then seek $\bar{q} \in Q$ which minimizes

$$J^N(q) = \sum_i \left| \frac{\partial^\nu w^N}{\partial t^\nu}(t_i, \hat{r}, \hat{\theta}; q) - z_i \right|^2 \quad , \quad \nu = 0, 1, 2 \qquad (5.2)$$

subject to w^N satisfying the approximate plate equations (hence the coefficients $\{w_k^N\}$ of w^N must satisfy (4.2).

The following theorem, taken from [8], specifies conditions under which convergence and continuous dependence (on data) of the solutions to the finite dimensional parameter estimation problems involving the functional (5.2) can be expected.

Theorem 5.1 *Let Q be a compact subset of a metric space $\tilde{Q}$ with metric d and assume that $H^N \subset V$ approximates V in the sense that for each $\Phi \in V$, there exists $\Phi^N \in H^N$ such that*

$$\left|\Phi - \Phi^N\right|_V \leq \varepsilon(N) \to 0 \text{ as } N \to \infty . \tag{5.3}$$

Furthermore, assume that $\sigma_1(q)$ and $\sigma_2(q)$ are V-elliptic, continuous, and satisfy the continuity with respect to parameter condition

$$\left|\sigma_i(q)(\Phi, \Psi) - \sigma_i(\tilde{q})(\Phi, \Psi)\right| \leq \gamma_i d(q, \tilde{q})|\Phi|_V|\Psi|_V \quad , \text{ for } \Phi, \Psi \in V \tag{5.4}$$

for $i = 1, 2$ and $q, \tilde{q} \in Q$. Finally, assume that

$$q \mapsto (Bu + F)(t; q) \text{ is continuous from } Q \text{ to } L^2((0, T), V^*) . \tag{5.5}$$

For arbitrary q^N such that $q^N \to q$ in Q, one then has the convergence

$$\begin{aligned} z^N(t; q^N) &\to z(t; q) \quad \text{in } V \text{ norm} \\ z_t^N(t; q^N) &\to z_t(t; q) \quad \text{in } V \text{ norm} \end{aligned} \tag{5.6}$$

for $t > 0$. Here z and z_t are solutions to the system (2.11) or (3.1) and z^N and z_t^N solve the corresponding linear finite dimensional system in H^N.

In considering the hypotheses, we note that the sesquilinear terms satisfy the ellipticity and continuity conditions due to the assumption of Kelvin-Voigt damping in the plate model. As noted in (3.4), the sesquilinear terms also satisfy the parameter continuity condition. Moreover, the input term $(F + Bu)(q)$ satisfies the condition (5.5). Finally, the convergence condition specified in (5.3) is satisfied as a consequence of the approximating properties of the cubic splines and Fourier elements.

For $\nu = 0, 1$, it then follows directly that there exists a subsequence of solutions $\bar{q}^N$ minimizing (5.2) subject to (4.2) which converge to a solution $\bar{q}$ of the original minimization problem (5.1) subject to (2.11) or (3.1). The convergence in the case involving the minimization of (5.2) with acceleration data does not follow directly from this theorem but can be obtained using results from the theory of analytic semigroups in a manner analogous to that used in [4].

6   Conclusions

In this paper, Newtonian principles were used to derive a model for boundary conditions for an imperfectly clamped thin circular plate. Under the assumption that some boundary displacement and rotation occur during oscillations, boundary moments, involving parameters to be estimated through fit-to-data techniques, were included in the modeling equations for the plate. It was demonstrated that these boundary conditions converge to fixed-end boundary conditions (zero displacement and slope) as the parameters modeling boundary stiffness tend to infinity.

An equivalent weak form of the modeling equations was then derived. By formulating this weak form in terms of sesquilinear forms and corresponding bounded operators, a semigroup formulation of the problem was developed. Extrapolation techniques were used to extend the semigroup to a space compatible with the input terms at which point, well-posedness of the system model and equivalence between strong and weak solutions were demonstrated. Finally, techniques for approximating the plate dynamics were outlined and criteria leading to parameter convergence when estimating parameters (including parameters in the boundary conditions) were discussed.

This provides a physical model which can be used when investigating the mechanism for energy loss through boundary supports. It also provides a means of testing the suitability of fixed-end boundary conditions in applications in which the structure is securely clamped.

Acknowledgements

The authors would like to thank Rich Silcox and Jay Robinson, Acoustics Division, NASA Langley Research Center, for input and insights provided during discussions regarding the effects of weakly clamped boundary conditions on experimental plate data. We also thank Ken Bowers, John Lund and the staff in the Department of Mathematical Sciences at Montana State University for the organization and assistance provided during this conference.

References

[1] H.T. BANKS and K. ITO, "A Unified Framework for Approximation in Inverse Problems for Distributed Parameter Systems," *Control-Theory and Advanced Technology* v. 4, no. 1, 1988, pp. 73-90.

[2] H.T. BANKS and K. ITO, "Approximation in LQR Problems for Infinite Dimensional Systems with Unbounded Input Operators," submitted to *Journal of Mathematical Systems, Estimation and Control*.

[3] H.T. BANKS, K. ITO and Y. WANG, "Well-Posedness for Damped Second Order Systems with Unbounded Input Operators," Center for Research in Scientific Computation Technical Report, CRSC-TR93-10, North Carolina State University, to appear in *Differential and Integral Equations*, March 1995.

[4] H.T. BANKS and D.A. REBNORD, "Analytic Semigroups: Applications to Inverse Problems for Flexible Structures," *Differential Equations with Applications*, (ed. by J. Goldstein, et. al.), Marcel Dekker, 1991, pp. 21-35.

[5] H.T. BANKS and R.C. SMITH, "The Modeling and Approximation of a Structural Acoustics Problem in a Hard-Walled Cylindrical Domain," Center for Research in Scientific Computation Technical Report, CRSC-TR94-22.

[6] H.T. BANKS and R.C. SMITH, "Well-Posedness of a Model for Structural Acoustic Coupling in a Cavity Enclosed by a Thin Cylindrical Shell," ICASE Report 93-10, to appear in *Journal of Mathematical Analysis and Applications*.

[7] H.T. BANKS, R.C. SMITH and Y. WANG, "Modeling Aspects for Piezoceramic Patch Activation of Shells, Plates and Beams," Center for Research in Scientific Computation Technical Report, CRSC-TR92-12, N. C. State Univ., to appear in *Quarterly of Applied Mathematics*.

[8] H.T. BANKS, Y. WANG, D.J. INMAN and J.C. SLATER, "Approximation and Parameter Identification for Damped Second Order Systems with Unbounded Input Operators," Center for Research in Scientific Computation Technical Report, CRSC-TR93-9, North Carolina State University, to appear in *Control: Theory and Advanced Technology*.

[9] A. HARAUX, "Linear Semigroups in Banach Spaces," in *Semigroups, Theory and Applications, II* (H. Brezis, et al., eds.), Pitman Res. Notes in Math, Vol 152, Longman, London, 1986, pp. 93-135.

[10] E.H. MANSFIELD, *The Bending and Stretching of Plates*, Volume 6 in the International Series of Monographs on Aeronautics and Astronautics, The MacMillan Company, New York, 1964.

[11] J. ROBINSON, Acoustics Division, NASA Langley Research Center, personal communications.

[12] R.C. SMITH, "A Galerkin Method for Linear PDE Systems in Circular Geometries with Structural Acoustic Applications," ICASE Report No. 94-40, submitted to *SIAM Journal on Scientific Computing*.

[13] W. SOEDEL, *Vibrations of Shells and Plates*, Second Edition, Marcel Dekker, Inc., New York, 1993.

[14] S. TIMOSHENKO, *Theory of Plates and Shells*, McGraw-Hill, NY, 1940.

DISCRETIZATION OF COST AND SENSITIVITIES IN SHAPE OPTIMIZATION *

John Burkardt, Max Gunzburger, and Janet Peterson
Department of Mathematics
Interdisciplinary Center for Applied Mathematics
Virginia Polytechnic Institute and State University
Blacksburg, Virginia, 24061

1 The Physical Problem: A Forebody Simulator

We consider a problem in aircraft engine testing [1], [6]. Of special concern is the influence of the aircraft forebody on the flow that reaches the engine intake. Modern aircraft are too large to place in a wind tunnel; there may not even be room for just the forebody and engine. Resourceful engineers build small "forebody simulators" that roughly reproduce the flow disturbances known to be caused by the real forebody. A typical setup is shown in Figure 1.1. Designing an effective simulator this way is crude, tedious, and expensive, and computational guidance is desired. This paper investigates computational difficulties arising in a simplified version of this design problem.

2 Continuous Mathematical Flow Model

We model the wind tunnel problem by fluid flow in a two dimensional rectangular channel. The forebody is represented by a "bump" that partially obstructs the flow. The fluid obeys the Navier Stokes equations for steady, viscous, incompressible flow:

$$-\nu\Delta\vec{u} + \vec{u}\cdot\operatorname{grad}\vec{u} + \operatorname{grad}p \;=\; \vec{f} \qquad (2.1)$$

$$\operatorname{div}\vec{u} \;=\; 0 \qquad (2.2)$$

plus appropriate boundary conditions. Here $\vec{u}$ is the velocity vector, with components $\mathbf{u}$ and $\mathbf{v}$; ν is the kinematic viscosity; $\vec{f}$ is a given forcing function.

Our channel has opposite corners at $(0,0)$ and $(10,3)$. If we wish to include a bump, it will start at $(1,0)$ and extend to $(3,0)$, with its height defined by some given function $y = bump(x,\vec{\alpha})$, with $\vec{\alpha}$ a set of parameters.

The fluid enters at the left, with velocity $\mathbf{u}(0,y) = inflow(y,\lambda)$, $\mathbf{v} = 0$, where $inflow$ is some given function, and λ is a parameter. At the top and

*Supported by the Air Force Office of Scientific Research under grant AFOSR 93-1-0280, and the Office of Naval Research under grant N00014-91-J-1493.

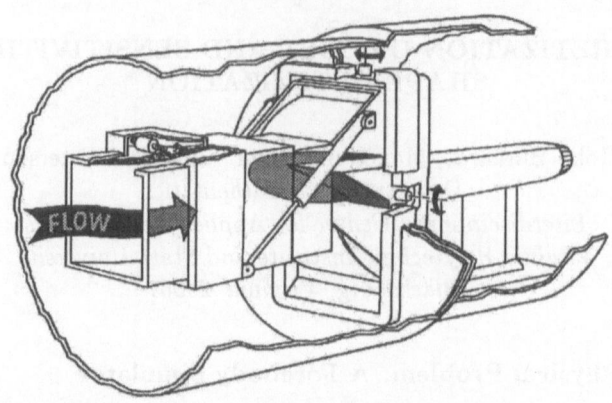

Figure 1.1: An aircraft engine and forebody simulator in a wind tunnel.

bottom of the channel we set the velocity to zero. On the right we set the usual outflow conditions $\mathbf{v} = 0$ and $\dfrac{\partial \mathbf{u}}{\partial x} = 0$.

Together, the flow equations, boundary conditions, and parameters produce a system of equations sufficient to determine the continuous quantities $(\mathbf{u}, \mathbf{v}, \mathbf{p})$ throughout the flow region, symbolized by:

$$\mathbf{G}(\mathbf{u}, \mathbf{v}, \mathbf{p}) = 0 . \tag{2.3}$$

Of course, $\mathbf{G}$ is a function of the parameters $\vec{\alpha}$ and λ, both explicitly and implicitly, through the dependence of $\mathbf{u}$, $\mathbf{v}$, and $\mathbf{p}$ on the parameters.

Once the flow is determined, we measure the state variables on a fixed vertical plane called the "sampling line", attempting to match a set of given measurements made earlier. The exact evaluation of the discrepancy will be carried out by a cost functional, to be specified.

3 Discrete Mathematical Flow Model

Equation (2.3) is not immediately amenable to computational treatment. We must formulate a discrete set of equations for data which can approximate the solution of that continuous problem. We use the weak formulation which follows [5]. We represent the region by a mesh of finite elements and approximate the continuously varying state variables $\mathbf{u}$ and $\mathbf{v}$ by coefficient vectors u and v multiplying a set of piecewise quadratic polynomials, and the pressure $\mathbf{p}$ by a set of coefficients p multiplying a set of linear polynomials. Under mild assumptions on the data, it is known that, if h is a measurement of the fineness of the finite element mesh, then the solution of the discrete equations approximates the solution of the continuous equations, as $h \to 0$.

The discretization results in a coupled system of nonlinear algebraic equations for the unknown coefficient vectors (u, v, p), which we represent by

$$G(u, v, p) = 0 .$$ (3.4)

See [7] for the formulation and convergence results for the discretized Navier Stokes equations.

4   The Optimization Problem

Suppose our flow problem is fully specified, except for the values of a set of parameters, $\vec{\beta}$. Then the specification of the parameter values determines the flow field, and hence the flow values along the sampling line, and hence the discrepancy cost functional, which we denote by J.

J will explicitly be a function of (u, v, p), but we may instead regard it as a function of the parameters $\vec{\beta}$. Our fundamental task becomes an unconstrained optimization: given a functional $J(\vec{\beta})$, its partial derivatives $\frac{\partial J}{\partial \beta_i}$, and a starting point $\vec{\beta}_0$, we seek to minimize J.

As a sample cost function, we suppose $u_s(y)$ is a given set of flow measurements along the sampling line, and consider the integral:

$$J_1 = \int_{x=x_s} \left(u(x_s, y) - u_s(y) \right)^2 \, dy .$$ (4.5)

Since the dependence of J on $\vec{\beta}$ is implicit, we cannot compute an explicit formula for the partial derivatives we will need. We might try finite differences:

$$\frac{\partial J}{\partial \beta_i} \approx \frac{\Delta J}{\Delta \beta_i} .$$ (4.6)

This method is straightforward but costly, each derivative requiring at least one additional Navier Stokes solution.

A cheaper alternative uses the *sensitivities*, derived from the implicit relationship between the state variables and parameters. We rewrite the continuous Navier Stokes equations to include the parameters:

$$\mathbf{G}(\mathbf{u}, \mathbf{v}, \mathbf{p}, \vec{\beta}) = 0 .$$ (4.7)

and if $\mathbf{G}$ is smooth, we may differentiate with respect to any β_i:

$$\frac{\partial \mathbf{G}}{\partial \mathbf{u}} \frac{\partial \mathbf{u}}{\partial \beta_i} + \frac{\partial \mathbf{G}}{\partial \mathbf{v}} \frac{\partial \mathbf{v}}{\partial \beta_i} + \frac{\partial \mathbf{G}}{\partial \mathbf{p}} \frac{\partial \mathbf{p}}{\partial \beta_i} = -\frac{\partial \mathbf{G}}{\partial \beta_i} .$$ (4.8)

Because **G** generally involves derivatives of the continuous variables **u**, **v** and **p**, we have implicitly assumed we may interchange differentiations. It is natural to consider the corresponding discrete version of Equation (4.8), which is called the *discrete sensitivity equations*:

$$\frac{\partial G}{\partial u}\widehat{\frac{\partial u}{\partial \beta_i}} + \frac{\partial G}{\partial v}\widehat{\frac{\partial v}{\partial \beta_i}} + \frac{\partial G}{\partial p}\widehat{\frac{\partial p}{\partial \beta_i}} = -\frac{\partial G}{\partial \beta_i} \tag{4.9}$$

where the quantities

$$\left(\widehat{\frac{\partial u}{\partial \beta_i}}, \widehat{\frac{\partial v}{\partial \beta_i}}, \widehat{\frac{\partial p}{\partial \beta_i}}\right) \tag{4.10}$$

are called the *(discrete) sensitivities* with respect to β_i.

It may be tempting to assume that a discrete sensitivity is equal to the derivative of the discrete state variable, but this is only true in the limit:

$$\widehat{\frac{\partial u}{\partial \beta_i}} \approx \frac{\partial \mathbf{u}}{\partial \beta_i} \approx \frac{\partial u}{\partial \beta_i} \tag{4.11}$$

where the left and right quantities approach the middle quantity (and hence, each other) as $h \to 0$.

If the mesh spacing h is suitably fine, we may use the easily computable discrete sensitivities as approximations to the unknown state derivatives, producing an approximation to the desired cost function derivative:

$$\frac{\partial J}{\partial \beta_i} \approx \frac{\partial J}{\partial u}\widehat{\frac{\partial u}{\partial \beta_i}} + \frac{\partial J}{\partial v}\widehat{\frac{\partial v}{\partial \beta_i}} + \frac{\partial J}{\partial p}\widehat{\frac{\partial p}{\partial \beta_i}} . \tag{4.12}$$

If we have just used Newton's method to solve the discrete flow equations, the Newton system has the same form as the linear system in Equation (4.9); thus sensitivities can be computed at the trivial cost of a linear solve.

5   Simple Channel Flow

The first test of our program was unobstructed channel flow. The inflow was parabolic, with a strength determined by a single parameter λ. A target solution was generated with $\lambda = 0.5$, and the flow values were measured along the sampling line $x_s = 9$.

The optimization code was then given an initial value of $\lambda = 0.0$ and was requested to minimize the functional J_1 as given in Equation (4.5). It accomplished this minimization in one step; the exact flow solution, called *Poiseuille flow*, has a linear relationship between λ and the horizontal velocity at any point (x, y). This makes the functional J_1 a quadratic function of λ, which is why we can optimize it easily.

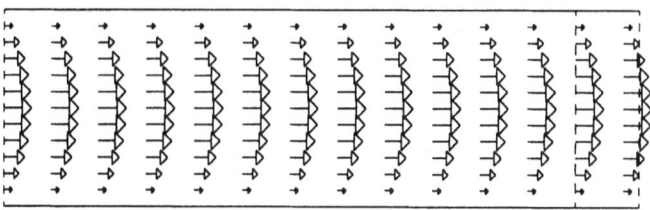

Figure 5.1: Derivative of velocity with respect to the inflow parameter.

If we plot the velocity derivatives with respect to λ, as in Figure 5.1, we can easily see that the influence of the inflow parameter extends throughout the region, dying off only near the walls. Even in problems where the channel is obstructed, the inflow parameter will continue to have this very strong global effect upon the flow. This global influence of λ could also be detected by monitoring the state variable derivatives.

6   Flow Past a Bump

We now turn to another problem, where the single parameter, α, determines the height of a parabolic bump in the channel. The inflow parameter λ will be fixed at a value of 0.5. We generate a target flow with $\alpha = 0.5$, and then begin the optimizer at $\alpha = 0$.

If we use approximate gradients based on the sensitivities, we find that the optimization does not reach the correct global minimum at $\alpha = 0.5$. Instead, the optimizer halts after 24 steps at $\alpha = 0.03$ giving the message "false convergence". Such a message generally means that the derivative data is inconsistent with the functional. We take that to mean the sensitivities aren't accurate enough, a difficulty that can be treated by refining the mesh. We will look at sensitivity failures more closely in Section 9.

However, there must be something more seriously wrong with this problem. We converted the program to use finite difference gradients, in which case the optimization came much closer to the correct answer, reaching $\alpha = 0.503$ in 17 steps. But this is hardly satisfactory for a one parameter optimization in double precision!

Why is this bump problem so different from the inflow problem? One hint comes if we look at the cost function values for the initial guess, $\alpha = 0$. The simple channel flow problem had a cost of $J_1(0) = 0.4$, but our bump problem has a cost of $J_1(0) = 10^{-8}$. Our problem is obviously very badly scaled.

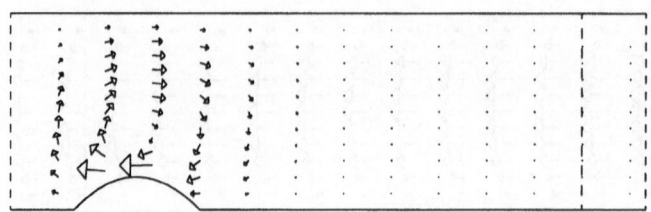

Figure 6.1: Derivatives of the velocity with respect to the bump parameter.

If we graph the field of velocity derivatives with respect to the parameter, as in Figure 6.1, we see a great deal of influence near the bump, which dies away rapidly as we move down the channel towards the sampling line. Clearly, we should try moving the sampling line closer to the bump, to make our measurements as accurately and robustly as possible.

Simply moving our sampling line to $x_s = 3$, defining a new version of our cost functional, J_2, causes the cost of the $\alpha = 0$ solution to jump to a "healthy" value of $J_2(0) = 0.009$. Our optimization converges in just 10 steps to the more accurate value $\alpha = 0.500003$. Moreover, we can return to using sensitivities in our formulation. This suggests that sensitivities on a coarse grid aren't worthless. They just aren't accurate enough to solve problems that need a great deal of resolution.

Thus, the bump problem is harder to solve than the simple channel problem, because the influence of the bump parameter on the state variable u is weak and local, a fact which we were able to deduce by looking carefully at the state derivatives.

7   Encountering a Local Minimum

We looked at problems where the bump was modeled by a cubic spline with equally spaced abscissas. We used a bump modeled by α_1, α_2, and α_3, which represented the height of the bump at each of the interior abscissas. A single parameter λ controlled the strength of the inflow. The target solution was generated with a parabolic bump described by $\alpha = (0.375, 0.5, 0.375)$ and inflow $\lambda = 0.5$.

The optimizer started from $\lambda = 0$, $\alpha = (0, 0, 0)$. Instead of reaching the target parameters, the optimizer settled down at $\lambda = 0.507$, $\alpha = (0.140, 0.539, 0.059)$, where it declared satisfactory convergence.

The cost of the zero solution was $J_2(\vec{0}) = 0.429$, so poor functional scaling was not to blame. Our next suspicion was that the cost functional

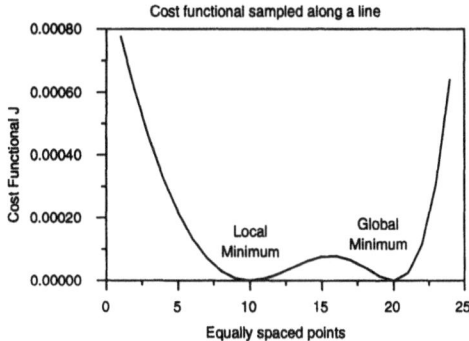

Figure 7.1: The functional between local and global minimizers.
The local minimizer has $J_2 = 0.3E - 06$, the global minimizer has $J_2 = 0$.

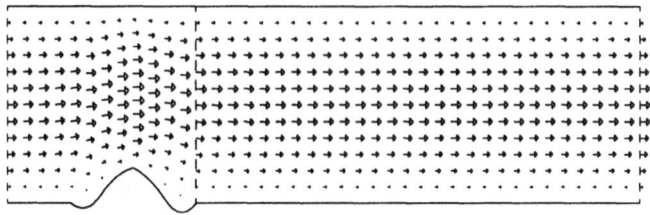

Figure 7.2: Locally minimizing flow produced by the optimizer.
The global minimizer has a parabolic bump, and no "gutters".

might be so flat between our final iterate and the target value that further
progress was not possible. But this belief was quickly dispelled when we
computed the cost functional at a series of intermediate points, and found
that it rose from $J_2 = 0.3E - 06$ to a value of $J_2 = 0.7E - 04$ before falling
to $J_2 = 0$ at the target, as shown in Figure 7.1, suggesting that we might
have reached a local minimum.

We looked at the actual flow solution, as shown in Figure 7.2, to make
sure it was acceptable and meaningful. The graph shows that the resulting
bump had roughly the same height as the target bump, but with a "gutter"
before and after it.

The question then arose as to whether this was actually a local minimum
or a *spurious* numerical solution. There are numerical reasons for doubting
the accuracy of this solution. The gutter regions are made up of elements
that have become stretched and twisted, reducing the accuracy of the finite
element discretization. This is an issue that is best addressed by a new

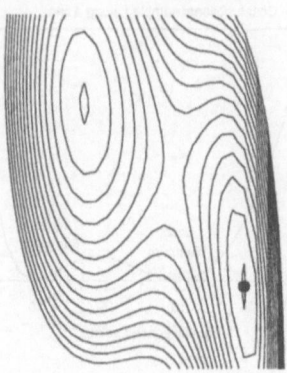

Figure 7.3: Contours for J_2 on plane including local and global minima.
The global minimum is marked.

calculation on a finer mesh.

If we used a mesh that is twice as fine, the gutters got almost twice as deep. This fact makes it unlikely that there is actually a physical solution that our data is trying to model. We therefore turned from investigating the meaning of this local minimum, and began to consider instead how we could avoid it.

8   Smoothing the Cost Functional

It's possible that a solution with very deep gutters would be unacceptable to the wind tunnel engineer: such a forebody simulator might not fit the apparatus. We leave ourselves open to such results since we haven't placed any feasibility constraints on our parameter space.

We note that it is likely that a higher Reynolds number would simplify matters. The flow should be affected in a stronger way by the details of the shape of the bump, and these effects should be passed downstream to the sampling line.

However, let us suppose that we need to solve this problem, or problems similar to it, at the given, low, Reynolds number. What changes can we make so that we are likelier to avoid the local minimum? One possibility is to add a penalty J_{bump} based on the integral of the square of the derivative of the bump. Such a penalty is zero for a flat line, low for a small parabola, and high for a curve with wiggles or severe curvature. Our formula would be:

$$J_{bump}(\vec{\alpha}) = \int_1^3 (bump_x(x, \vec{\alpha}))^2 \, dx \ . \tag{8.13}$$

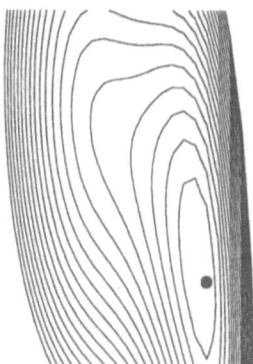

Figure 8.1: Contours of J_3 on the same plane, with $\epsilon = 0.0002$.
The local minimum evident in the previous figure has disappeared.

Then we will work with a new cost function J_3 defined by adding a "small"
multiple of the bump penalty to the original cost:

$$J_3 = J_2 + \epsilon \cdot J_{bump} . \tag{8.14}$$

To get an idea of the smoothing effect of this change in the functional,
let's consider a two dimensional plane containing the local minimum and
the global minimum. Contour lines of the functionals are displayed in
Figure 8.1, and should be compared with those drawn in Figure 7.3.

The added bump term seems to have smoothed away the local minimum.
And if we start from zero, the optimizer now finds the global minimizer.
However, the global minimizer of J_3 is *not* the minimizer of J_2, the function we actually want to minimize. What is true is that, for small ϵ, the
minimizer of J_3 is close to the minimizer of J_2. That means that, in a case
where we are bedeviled by a local minimum or other irregularities in the
functional, we can try adding such a smoothing term. Starting from a zero
initial guess, we can find the minimizer of the smoothed functional. Then
we can restart the optimization from this point, but using the unsmoothed
functional. If the minimizers of the two functions are close enough, we now
have a much better chance of converging to the desired global minimum.
For instance, an optimization of J_3 with $\epsilon = 0.0002$ converged to the point
$\lambda = 0.500$, $\alpha = (0.276, 0.495, 0.364)$. If we now reset ϵ to 0, which restores
our original optimization function J_2, and restarted the optimization, we
reached the desired target point, bypassing the local minimizer.

Of course, this doesn't settle the question. We still have to detect that
we have reached an undesired solution, and choose a smoothing function
of a "suitable" type, and a smoothing parameter ϵ of a "suitable" size.

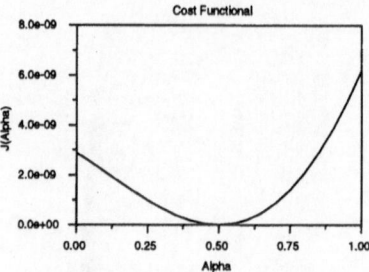

Figure 9.1: $J_1(\alpha)$, with sampling line at $x_s = 9$.

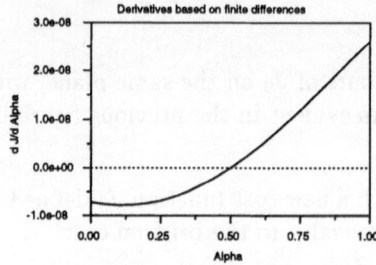

Figure 9.2: $\frac{dJ_1}{d\alpha}$ by finite differences.

For a more challenging case, we might have to smooth and restart several times, increasing the number of ad hoc choices made. In that case, a more suitable approach would be to add ϵ directly and explicitly as another parameter to the problem, and optimize once on the enlarged system. The partial derivative with respect to ϵ is trivial to compute. The only further complication is that there will be new contributions to the derivative with respect to α coming from the term $\epsilon \cdot J_{bump}$.

9   Sensitivity Failure

Throughout our discussion and computations, we have used the sensitivities to arrive at cheap approximations for the derivatives of our state variables (u, v, p) and cost functional J. Unfortunately, an optimization requires very accurate derivatives near the minimizer, precisely where the errors in the sensitivities become large, in the relative sense. We already encountered this problem in Section 6. When the optimizer returned the message "false convergence", it had reached a point where the approximate derivative of J_2 was too incorrect to use.

To get a feeling for what the optimizer was dealing with, let us look at

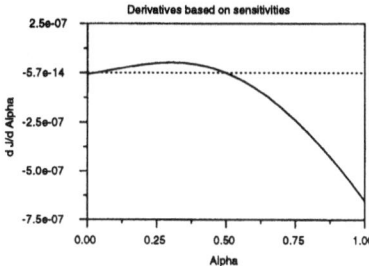

Figure 9.3: $\frac{dJ_1}{d\alpha}$ by sensitivities.

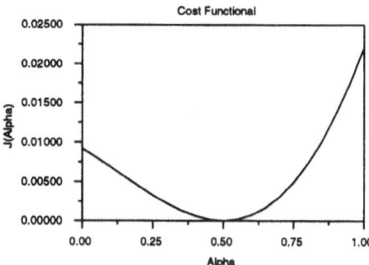

Figure 9.4: $J_2(\alpha)$ with sampling line at $x_s = 3$.

the functional J_1 and its derivative $\dfrac{dJ_1}{d\alpha}$ as approximated by finite differences and by sensitivities, in Figures 9.1 through 9.3.

The sensitivities provide an astonishingly bad "approximation". We might have been warned by the small magnitude of the quantities, though a better warning lies in the fact that the state derivatives we sample are much smaller than the same quantities elsewhere in the region, and hence are relatively poorly approximated. Figures 9.4 through 9.5 show how moving the sampling line to $x_s = 3$ corrects this problem.

We should keep in mind that the underlying data is *identical* for the two sets of plots we are comparing here. This includes the state variables and state derivatives. The difference is in the definition of the functional and its derivative, that is, in which state variables we sample.

Near the minimizer, errors in the sensitivities can become so serious that the partial derivatives are worthless. This occurred in the multiparameter bump problem, with one inflow parameter and three bump parameters, at a Reynolds number of 100. We were using a cost functional, J_4, which included the discrepancies in both horizontal and vertical velocities along

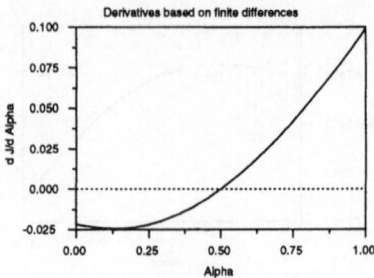

Figure 9.5: $\frac{dJ_2}{d\alpha}$ by finite differences and sensitivities.
The two calculations are now essentially identical.

the line $x_s = 3$:

$$J_4 = \int_{x=x_s} \left(u(x_s, y) - u_s(y)\right)^2 + \left(v(x_s, y) - v_s(y)\right)^2 \, dy \; . \qquad (9.15)$$

The target parameters were $\lambda = 0.500$, $\alpha = (0.375, 0.500, 0.375)$. The optimizer reached $\lambda = 0.493$, $\alpha = (-0.078, 0.476, 0.101)$ and reported "false convergence". Note that at this point, $J_4 = 0.0006$, hardly the sort of extremely small value we took as a warning earlier.

To see what was going on, consider a plane including the point where the optimizer stopped, and the global minimizer. We used sensitivities to compute the cost function gradients, projected onto the plane.

Figure 9.6 shows the functional contour lines, overlaid with the projected computed gradient field (normalized and multiplied by -1, so that it should point towards the minimizer). We can immediately see that, at least in this "slice" of parameter space, and near the minimizer, the errors of approximation are so serious that the direction field is utterly lost. It is no wonder that the optimizer is unable to find the minimizer. There can be little doubt that this problem is caused by discretization errors. If we halve the mesh size and recompute the same quantities, all the approximate gradients point inwards, towards the minimizer. Even then, small errors in the direction of the gradients plainly persist.

10 Conclusions

If a cost functional and the independent variables are only implicitly related, singularities, poor scalings, and local minima may easily occur. Diagnosis of such problems is usually possible through such means as making a plot of the state solution, a table of functional values along a line or plotting contours of the functional along two directions. Problems may be treated

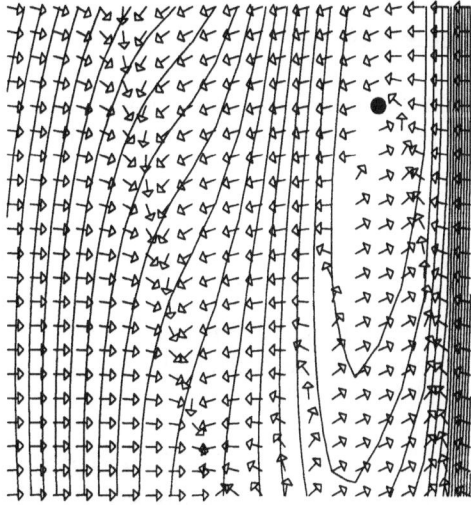

Figure 9.6: Gradients approximated by sensitivities.
The approximate gradients clearly do not match the contours.

in a variety of ways. The functional may be modified so that it depends on more state variables, or on state variables that are more sensitive to variations in the parameters. The functional may include the cost of control, or other terms that tend to "regularize" or smooth out the contour levels.

Sensitivities are a valuable method of approximating derivatives. However, their use entails an extra source of error, which must be monitored. In particular, a grid which is fine enough to get good approximations to the state variables may not be fine enough to provide approximations of the state variable derivatives.

Small errors caused by using sensitivities can also interfere with an optimization during the final steps, near a minimizer, when the gradient has dropped to a very low magnitude, essentially magnifying the significance of errors in the derivatives.

Simple checks can be applied after the failure of an optimization that uses sensitivities. These include trying a refined grid, comparing the results of approximating the derivatives with finite differences, plotting the state variable field, plotting the functional gradient field, and considering whether the relationship between the parameters and the functional is weakly or strongly mediated by the state variables.

11   Acknowledgements

The authors thank Ken Bowers and John Lund, the organizers of the fourth conference on Computation and Control at Montana State University.

References

[1] J. BORGGAARD, J. BURNS, E. CLIFF and M. GUNZBURGER, *Sensitivity Calculations for a 2D, Inviscid Supersonic Forebody Problem*, in **Identification and Control of Systems Governed by Partial Differential Equations**, H T Banks, R Fabiano, K Ito, editors, SIAM Publications, 1993.

[2] J. BURKARDT and J. PETERSON, *Control of Steady Incompressible 2D Channel Flow*, **Flow Control**, Proceedings of the IMA, Volume 68, Springer Verlag, New York, to appear 1995.

[3] C. DEBOOR, **A Practical Guide to Splines,** Springer Verlag, New York, 1978.

[4] D. GAY, *Algorithm 611, Subroutines for Unconstrained Minimization Using a Model/Trust Region Approach*, **ACM Transactions on Mathematical Software**, Volume 9, Number 4, December 1983, pages 503-524.

[5] M. GUNZBURGER and J. PETERSON, *On Conforming Finite Element Methods for the Inhomogeneous Stationary Navier-Stokes Equations*, **Numerische Mathematik**, Volume 42, pages 173-194.

[6] HUDDLESTON, *Development of a Free-Jet Forebody Simulator Design Optimization Method*, AEDC-TR-90-22, Arnold Engineering Development Center, Arnold AFB, TN, December 1990.

[7] O. KARAKASHIAN, *On a Galerkin-Lagrange Multiplier Method for the Stationary Navier-Stokes Equations*, **SIAM Journal of Numerical Analysis**, Volume 19, Number 5, October 1982, pages 909-923.

REPRESENTATION OF FEEDBACK OPERATORS FOR HYPERBOLIC SYSTEMS

John A. Burns *
Center for Optimal Design and Control
Interdisciplinary Center for Applied Mathematics
Virginia Tech
Blacksburg, Virginia 24061

Belinda B. King, †
Department of Mathematics
Oregon State University
Corvallis, Oregon 97331

1 Introduction and Motivation

The purpose of this article is to extend the representation theorem in [1] and [7] to certain classes of damped hyperbolic systems. The original motivation for our study of hyperbolic systems comes from the work by Lupi, Chun, and Turner [8]. The approach in [8] is interesting because they make no prior assumptions regarding the form of the controls and actuators so that the gains operators produced by an optimal design could be used to make decisions about where actuators and sensors are best placed. In particular, in [8] it was assumed that the input operator was the identity and the elastic system was not damped. By solving the LQR problem with the input operator equal to the identity, one can gain insight into the type and location of practical distributed controllers for structural control. This insight comes from explicit knowledge of the kernels (so called functional gains) that describe the integral representations of feedback gain operators. Even with no damping the LQR problem has a solution since the input operator is the identity (the system is exactly controllable). However, as we see below the problem with no damping is extremely complex. Basic questions concerning the existence and smoothness of functional gains remain open and yet these issues are important in the applications proposed in

---

*This research was supported in part by the Air Force Office of Scientific Research under grant F49620-93-1-0280 and by the National Aeronautics and Space Administration under contract No. NASA-19480 while the author was a visiting scientist at the Institute for Computer Applications in Science and Engineering, NASA Langley Research Center, Hampton, VA 23681-0001.

†This research was supported in part by the Air Force Office of Scientific Research under grant F49620-93-1-0280 while the author was a visiting scientist at the Air Force Center for Optimal Design and Control, Virginia Polytechnic Institute and State University, Blacksburg, VA 24061–0531, and by the National Science Foundation under grant DMS-9409506.

[8]. Consequently, as a first step we take the middle ground and consider damped systems with distributed control.

Except for the obvious cases with bounded input operator and a finite number of controllers, the problem of obtaining explicit representations of feedback laws is more complex than one might first imagine. In the most general case, this problem is equivalent to the problem that led Grothendieck to develop the theory of topological tensor products and nuclear spaces. This theory led to the famous Schwartz Kernel Theorem. However, the fact that the operators of interest often arise as solutions to Riccati equations can be exploited to yield reasonable results. The representation problem for a parabolic problem with unbounded control operator was first considered in [1] and [7]. It was shown that as long as the input operator is bounded relative to the open loop dynamic operator, the solution to the algebraic Riccati equation is Hilbert-Schmidt. This fact was then exploited to show that the resulting feedback operator had an integral representation. For the hyperbolic case considered here, these types of results are more intricate and highly dependent on the type of damping.

In the Section 2 we present a numerical example involving the control of a hybrid cable-mass system. This example is used to motivate the model problem and to demonstrate that distributed controllers can enhance disturbance attenuation. We then concentrate on the 1D wave equation with Kelvin-Voigt damping. Although similar results can be obtained for other damping models, we present the basic theorem for this model in order to keep this paper moderately short. However, we illustrate the ideas and difficulties for other damping models with numerical examples. Finally we close with a few comments about future work and other open problems.

2   Control for a Cable-Mass System

The following system was proposed by Nayfeh, Nayfeh and Mook [11] as a simple example of a nonlinear distributed parameter system with the property that many standard discretized lumped models failed to capture the essential nonlinear behavior of the dynamic system governed by the partial differential equation. This system was also considered in [2] where it was used as a test model for MINMAX control for disturbance attenuation. However, in [2] there was only one controller and, although disturbance atteunation was achieved, we shall see below adding a distributed controller can improve performance. The MINMAX approach provides a "robust state feedback control law" which is less sensitive to disturbances and certain unmodeled dynamics than is the LQR design. The idea is to obtain a representation of the control law and then use approximation theory to compute finite dimensional suboptimal controllers. These suboptimal controllers were used to attack the problem of designing reduced order state estimators. We shall limit our discussion here to the full state feedback problem.

Consider the hybrid nonlinear distributed parameter system described by a vibrating cable held fixed at one end and with a mass attached at the other end. The mass is suspended by a spring which has nonlinear stiffening terms and is forced by a disturbance (see Figure 2.1). The equations for the hybrid system are [2]

$$\rho \frac{\partial^2}{\partial t^2} w(t,s) = \frac{\partial}{\partial s} \left[\tau \frac{\partial}{\partial s} w(t,s) + \gamma \frac{\partial^2}{\partial t \partial s} w(t,s) \right] + \rho u_1(t,s),$$

$$0 < s < \ell, \ t > 0,$$

(2.1)

$$m \frac{\partial^2}{\partial t^2} w(t,\ell) = - \left[\tau \frac{\partial}{\partial s} w(t,\ell) + \gamma \frac{\partial^2}{\partial t \partial s} w(t,\ell) \right]$$

$$- \alpha_1 w(t,\ell) - \alpha_3 [w(t,\ell)]^3 + \eta(t) + m u_2(t),$$

(2.2)

with boundary condition

$$w(t,0) = 0.$$

(2.3)

The initial conditions are given by

$$w(0,s) = w_0, \quad \frac{\partial}{\partial t} w(0,s) = w_1.$$

(2.4)

Here, $w(t,s)$ represents the displacement of the cable at time t, and position s, $w(t,\ell)$ represents the position of the mass at time t, ρ and m are the densities of the cable and mass respectively, τ is the tension in the cable, and γ is a damping coefficient. The alphas are coefficients describing the nonlinear effects of the spring. The term $\eta(t)$ is viewed as a disturbance and $u_1(t,s)$ and $u_2(t)$ are control inputs. For the moment we assume Kelvin-Voigt (internal) damping in the cable.

The problem is hybrid in that the system is described by a linear partial differential equation (the wave equation) coupled through the boundary condition to a low order nonlinear ordinary differential equation (Duffing's equation). In [2] it was assumed that the control acted exclusively on the mass (i.e. that $u_1(t,x) = 0$).

This model is often first written as a second order system in a Hilbert space H of the form

$$\ddot{y}(t) + D_0 \dot{y}(t) + A_0 y(t) + F_0(y(t)) = B_0 u(t).$$

(2.5)

For the cable-mass problem considered here, $H = L_2(0,1) \times \mathbb{R}^1$ and $y(t) = [w(t,\cdot), w(t,\ell)]^T$.

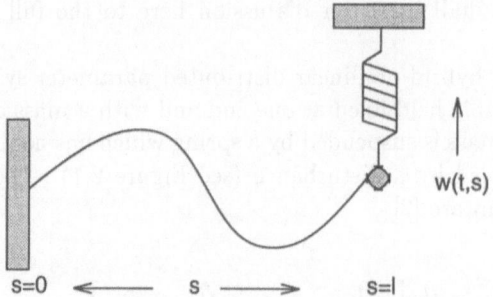

Figure 2.1: Cable-Mass System

This formal system has the advantage that it has the same appearence as
the finite dimensional case and in order to address viscous and "structural"
damping one merely replaces $D_0 = A_0$ with $D_0 = I$ and $D_0 = [A_0]^{1/2}$,
respectively. We note however, that it is more consistent with physics to
write the system in the form

$$\ddot{y}(t) + S^*(Sy(t) + \gamma T\dot{y}(t)) + F_0(y(t)) = B_0 u(t) \tag{2.6}$$

where $S = T = [A_0]^{1/2}$. Observe that $A_0 = A_0^* > 0$ and so $S^* = S$ and
$S^*S = S^*T = A_0$. Hence, (2.6) is formally obtained by factoring $[A_0]^{1/2}$
out of the expression $D_0\dot{y}(t) + A_0 y(t)$ in (2.5). Note also that (2.6) is of a
form that allows for structural damping where $S = [A_0]^{1/2}$ and $T = I$, as
well as for viscous damping where $S = [A_0]^{1/2}$ and $T = S^{-1} = [A_0]^{-1/2}$. In
addition, by writing the system in the second order form (2.6), one captures
a form that comes from balance laws and at the same time sets the stage
for a simple formulation of the problem in first order state space form.

The system governed by equations (2.1) - (2.4) can be written as a
dynamical system in an appropriate (infinite dimensional) state space. Although there are several equivalent formulations for this problem, we shall
write the governing equations as the first order system

$$\dot{x}(t) = Ax(t) + F(x(t)) + Bu(t) + D\eta(t), \qquad x(0) = x_0 \tag{2.7}$$

where at time t the state $x(t) = [y(t), \dot{y}(t)]^T$ lies in the Hilbert space $X = H_L^1 \times \mathbb{R} \times L_2 \times \mathbb{R}$. Here, H_L^1 is the subspace of the Sobolev space $H^1 = H^1[0, \ell]$ defined by $H_L^1 = \{w \in H^1 : w(0) = 0\}$, and L_2 is the standard
Lebesgue space of square integrable functions. The control $u(t)$ lies in the

control space $U = L_2 \times \mathbb{R}$. Here the inner product in X is

$$\langle [w(\cdot), \xi, v(\cdot), \mu]^T, [\hat{w}(\cdot), \hat{\xi}, \hat{v}(\cdot), \hat{\mu}]^T \rangle = \tau \int_0^\ell w'(x) \hat{w}'(x) dx + \alpha_1 \xi \hat{\xi}$$

$$+ \rho \int_0^\ell v'(x) \hat{v}'(x) dx + m \mu \hat{\mu}. \tag{2.8}$$

It is important to precisely define the system operators and their domains in order to obtain correct representations of the feedback operators that will be used to control the system. Let δ_ℓ denote the "evaluation operator" defined on $H^1[0, \ell]$ by $\delta_\ell(\phi(\cdot)) = \phi(\ell)$ and define the linear operator A on the domain $\mathcal{D}(A) \subseteq X$ by

$$\mathcal{D}(A) = \left\{ x = [w, \xi, v, \mu]^T \in X : w, v \in H_L^1, \left\{ \frac{\tau}{\rho} \frac{d}{ds} w + \frac{\gamma}{\rho} \frac{d}{ds} v \right\} \in H^1, \right.$$

$$w(\ell) = \xi, v(\ell) = \mu \}, \tag{2.9}$$

and

$$Ax = \left[v, \mu, \frac{d}{ds} \left\{ \frac{\tau}{\rho} \frac{d}{ds} w + \frac{\gamma}{\rho} \frac{d}{ds} v \right\}, -\delta_\ell \left\{ \frac{\tau}{m} \frac{d}{ds} w + \frac{\gamma}{m} \frac{d}{ds} v \right\} - \frac{\alpha_1}{m} \xi \right]^T \tag{2.10}$$

The control input operator B and the disturbance operator D are defined by

$$B = [0, I_H]^T \quad \text{and} \quad D\eta = \left[0, 0, 0, \frac{1}{m} \eta \right]^T, \tag{2.11}$$

respectively. The nonlinear operator F is defined on X by

$$F(x) = \left[0, 0, 0, -\frac{\alpha_3}{m} [\xi]^3 \right]^T = [0, F_0(y)]^T. \tag{2.12}$$

Observe that the input operator is the same as that used in [8] in their analysis.

As noted above, this problem with $u_1(t, s) = 0$ was considered in [2] where MINMAX control was used to design a low order dynamics control law. This law was based on two outputs (position and velocity of the mass) and resulted in a practical low order design. For this note, we shall consider only the full state feedback problem. However, we allow for distributed control through $u_1(t, s)$. The simplest approach is to linearize the system, use MINMAX design to obtain a feedback operator and apply this law to the full plant.

The linearized system has the form

$$\frac{d}{dt}x(t) = Ax(t) + Bu(t) + D\eta(t), \quad x(0) = x_0. \qquad (2.13)$$

For this problem, with Kelvin-Voigt damping, one can apply the MINMAX theory in [10] to obtain a feedback law of the form

$$u(t) = \left[\begin{array}{c} u_1(t,s) \\ u_2(t) \end{array} \right] = -K_\theta x(t) \qquad (2.14)$$

where for $\theta \geq 0$ the gain operator $K_\theta : X \to H$ is given by

$$K_\theta = B^* P_\theta \qquad (2.15)$$

and P_θ satisfies the algebraic Riccati equation

$$A^* P + PA - P[BB^* - \theta^2 M]P + Q = 0. \qquad (2.16)$$

Here $M = M^* \geq 0$, $Q = Q^* \geq 0$ and the system (2.16) holds in the weak sense (see [10]). When $\theta = 0$ one has the LQR design. However, when $\theta > 0$ the corresponding MINMAX controller provides additional disturbance attenuation (see [2]).

Figure 2.2 clearly illustrates the difference between the performance achieved in [2] and what one can achieve with additional distributed control. For this example, we use the same parameters and finite element scheme found in [2].

ρ	τ	γ	m	ℓ	α_1	α_2	α_3
1	1	.005	3	2	.01	0	3

When there is control only on the mass, the maximum value of θ yielding a feedback law of the form (2.15) - (2.16) was $\theta = 1.7$; this behavior is shown by the dotted line. If in addition, one allows distributed control (at $\theta = 1.7$) there is a considerable increase in attenuation as shown by the dashed line. Moreover, if both distributed control and control on the mass are allowed, then θ can be increased to $\theta = 2.5$. Thus, one increases attenuation of the disturbance and, as shown by the solid line in Figure 2.2, performance is also improved.

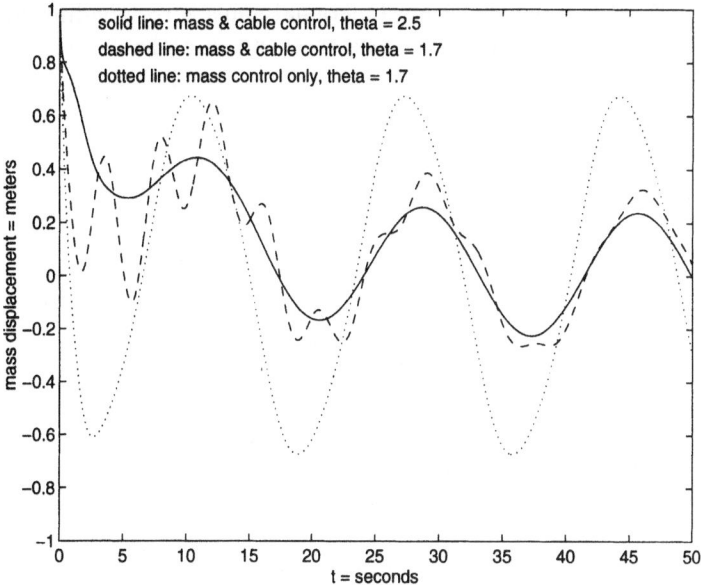

Figure 2.2: Displacement of the Mass under Three Types of Control.

This example clearly indicates the benefit of distributed control and MINMAX design. Moreover, as shown in [2] if one has explicit representation of the feedback operator K_θ, then this can be used to design practical low order dynamic controllers (nonlinear compensators). In particular, it follows that since

$$K_\theta = \begin{bmatrix} K_\theta^{11} & K_\theta^{12} & K_\theta^{13} & K_\theta^{14} \\ K_\theta^{21} & K_\theta^{22} & K_\theta^{23} & K_\theta^{24} \end{bmatrix} \qquad (2.17)$$

is bounded from X to H, the control on the mass has the form

$$\begin{aligned} u_2(t) = &-\int_0^\ell \tau k_s(z)\frac{\partial}{\partial z}w(t,z)dz - \alpha_1 k_\alpha(t,\ell) \\ &-\int_0^\ell \rho k_v(z)\frac{\partial}{\partial t}w(t,z)dz - mk_m\frac{\partial}{\partial t}w(t,\ell). \end{aligned} \qquad (2.18)$$

It is tempting to assume that one also has a representation of the form

$$\begin{aligned} u_1(t,\zeta) = &-\int_0^\ell \tau h_s(\zeta,z)\frac{\partial}{\partial z}w(t,z)dz - \alpha_1 h_\alpha(\zeta)(t,\ell) \\ &-\int_0^\ell \rho h_v(\zeta,z)\frac{\partial}{\partial t}w(t,z)dz - mh_m(\zeta)\frac{\partial}{\partial t}w(t,\ell). \end{aligned} \qquad (2.19)$$

In (2.18) $k_s(z)$ and $k_v(z)$ are functional gains corresponding to strain and velocity, respectively. Their existence and smoothness properties are assured by the Riesz Representation Theorem. For example, there is a $k_s(z) \in L_2(0, \ell)$ so that $K_\theta^{21} : H_L^1 \to \mathbb{R}$ has the form

$$\left[K_\theta^{21} \right] \phi(\cdot) = \int_0^1 \tau k_s(z) \frac{d}{dz} \phi(z) dz. \tag{2.20}$$

However, on the surface all we know about $K_\theta^{13} : L_2(0,1) \to L_2(0,1)$ is that it is bounded. As noted above, the desire to find representations of such operators as integrals led to Grothendieck's work on nuclear spaces and the Schwartz Kernel Theorem. Recall that not all bounded linear operators on $L_2(0,1)$ (even if self-adjoint) have integral representations, as illustrated by the identity operator.

In [8] this issue was avoided by assuming a representation similar to (2.19) and then allowing generalized functions as kernels. This approach proved to be satisfactory for the one dimensional case considered therein, but does not apply to more general two and three dimensional hyperbolic problems. We present some results that lead to a representation of the form (2.19) when there is suitable damping in the system. Although this approach does not apply to the undamped case, the ideas can be extended to certain higher dimensional damped elastic systems. Moreover, we conjecture that the undamped problem in 3D systems will not yield a representation even of the type considered in [8].

We turn now to the simple 1D wave equation in order to state precise results and to keep this article at a reasonable length. Moreover, we restrict our presentation to LQR design ($\theta = 0$) since we can rely on existing literature to outline the results. Extensions to higher dimensional problems with $\theta > 0$ will appear in a future paper.

3   The Wave Equation

Consider the wave equation with Kelvin-Voigt damping. Damping plays a key role in the design of controllers for hyberbolic systems. To illustrate this role, consider the LQR control problem for the wave equation defined by

$$\frac{\partial^2}{\partial t^2} w(t, s) = \frac{\partial^2}{\partial s^2} w(t, s) + \gamma \frac{\partial^3}{\partial s^2 \partial t} w(t, s) + u(t, s),$$
$$0 < s < 1, \ \ 0 < t, \tag{3.21}$$

with boundary conditions

$$w(t, 0) = 0, \ \ w(t, 1) = 0, \ \ 0 < t \tag{3.22}$$

and cost function

$$J(u) = \frac{1}{2} \int_0^\infty \int_0^1 \left[\left| \frac{\partial}{\partial s} w(t,s) \right|^2 + \left| \frac{\partial}{\partial t} w(t,s) \right|^2 + |u(t,s)|^2 \right] ds dt. \quad (3.23)$$

This problem is defined on the state space $X = H_0^1(0,1) \times L_2(0,1)$. As above, care must be used to define the system operators. In particular, let

$$\mathcal{D}(A) = \left\{ \begin{bmatrix} w \\ v \end{bmatrix} \in X : w, v \in H_0^1(0,1), w + \gamma v \in H^1(0,1) \right\} \quad (3.24)$$

and define A by

$$A \begin{bmatrix} w \\ v \end{bmatrix} = \begin{bmatrix} v(\cdot) \\ \frac{d}{ds} \left[\frac{d}{ds} w + \gamma \frac{d}{ds} v \right] \end{bmatrix}. \quad (3.25)$$

The control space is $U = L_2(0,1)$ and the control operator is defined from U into $X = H_0^1 \times L_2$ by

$$B = \begin{bmatrix} 0 \\ I_{L_2} \end{bmatrix}. \quad (3.26)$$

Here, $\theta = 0$ and $Q = R = I_X$. The LQR problem has a solution (even if $\gamma = 0$) given by

$$u(t,\cdot) = -K \begin{bmatrix} w(t,\cdot) \\ \frac{\partial}{\partial t} w(t,\cdot) \end{bmatrix}, \quad (3.27)$$

where $K : H_0^1(0,1) \times L_2(0,1)$ has the form

$$K = B^* P \quad (3.28)$$

and P satisfies the weak form of the Riccati equation (ARE) given by

$$\langle Px, Az \rangle_X + \langle Ax, Pz \rangle_X - \langle B^* Px, B^* Pz \rangle_U + \langle Cx, Cz \rangle_Y = 0, \quad (3.29)$$

for all x, y in $\mathcal{D}(A)$. For this second order system, K takes the form

$$K = [0, \ I_{L_2}]P, \quad (3.30)$$

or equivalently,

$$K = [P_{21}, \ P_{22}] \quad (3.31)$$

where

$$P = \begin{bmatrix} P_{11} & P_{12} \\ P_{21} & P_{22} \end{bmatrix}. \quad (3.32)$$

The operators $P_{21} : H_0^1(0,1) \to L_2(0,1)$ and $P_{22} : L_2(0,1) \to L_2(0,1)$ are bounded linear operators with $P_{22}^* = P_{22}$. The goal is to determine if there exist "nice" integral representations of these operators.

The following result is well known and may be found in [3].

Lemma 3.1 *The operator A generates an analytic semigroup $S(t)$ on X and there exist $M > 0, \omega > 0$ with $\|S(t)\| \leq Me^{-\omega t}$.*

Since A generates a stable analytic semigoup, we can apply Theorem 2.1 (page 36) in [9] to obtain the following regularity result for P.

Lemma 3.2 *There exists a self-adjoint, non-negative definite bounded linear transformation $P = P^*$ satisfying (3.29). Moreover, for each $\epsilon > 0$, the operators $[A^*]^{1-\epsilon}P$ belong to $\mathcal{L}(X, X)$.*

We note that Theorem 2.1 in [9] also states that ϵ can be set equal to zero if A is self-adjoint, normal, or has a Riesz basis of eigenvectors. However, it is interesting to note that A is netiher normal nor self-adjoint. For this 1D problem, A does have a Riesz basis. However, this property is not needed to establish the following representation.

Theorem 3.3 *There exist functions $k_{21}(\cdot, \cdot), k_{22}(\cdot, \cdot)$ such that*

(1) $k_{21}(\cdot, \cdot) \in L_2([0, 1] \times [0, 1]), k_{22}(\cdot, \cdot) \in L_2([0, 1] \times [0, 1])$,

(2) $k_{22}(\zeta, s) = k_{22}(s, \zeta)$,

(3) the mapping $t \to k_{21}(\xi, t)$ belongs to H_0^1 for almost all $\xi \in [0, 1]$

and one has the representations

$$[P_{21}\phi](\zeta) = \int_0^1 k_{21}(\zeta, t)\phi(t)dt, \quad \phi \in H_0^1, \tag{3.33}$$

$$[P_{22}\phi](\zeta) = \int_0^1 k_{22}(\zeta, t)\phi(t)dt, \quad \phi \in L_2. \tag{3.34}$$

The proof of Theorem 3.3 is rather tedious and will not be included here. However, we note that the proof is similar to the proof of Theorem 3.2 given in [7]. It relies on a classical theorem by Fullerton (see Theorem 6 in [5]). However, the specific structure of the A operator combined with the special B operator, $B = [0 \ I_X]^T$, is needed to carry out the proof for this hyperbolic case. Theorem 3.3 leads easily to the following representation.

Theorem 3.4 *There exist functional gains $k_v(\cdot, \cdot)$ and $k_s(\cdot, \cdot)$*

(1) $k_s(\cdot, \cdot) \in C([0, 1] \times [0, 1]), k_v(\cdot, \cdot) \in L_2([0, 1] \times [0, 1])$,

(2) $k_v(\zeta, z) = k_v(z, \zeta)$,

(3) the mapping $z \to k_s(\zeta, z)$ belongs to H^2 for almost all $\zeta \in [0, 1]$

and the LQR feedback control law has the representation

$$u(t,\zeta) = -\int_0^1 k_s(\zeta,z)\frac{\partial}{\partial z}w(t,z)dz - \int_0^1 k_v(\zeta,z)\frac{\partial}{\partial t}w(t,z)dz. \qquad (3.35)$$

Proof: Since $K : H_0^1 \times L_2 \to L_2$ is given by

$$K\left[\begin{array}{c} w(\cdot) \\ v(\cdot) \end{array}\right] = -\int_0^1 k_{21}(\zeta,z)w(z)dz - \int_0^1 k_{22}(\zeta,z)v(z)ds,$$

let $k_v(\cdot,\cdot) = k_{22}(\cdot,\cdot)$. Integration by parts on the first integral yields (for $w(\cdot) \in H_0^1$)

$$\int_0^1 k_{21}(\zeta,z)w(z)dz = -\int_0^1 \left[\int_0^z k_{21}(\zeta,t)dt\right]w'(z)dz.$$

The representation (3.35) follows, where

$$k_s(\zeta,z) = -\int_0^z k_{21}(\zeta,t)dt, \qquad (3.36)$$

and $z \to k_s(\zeta,z)$ belongs to H^2.

We turn now to some numerical experiments to illustrate the representation (3.35). In addition, we consider other damping models and present numerical results that clearly show the role that damping plays in the smoothness and existence of functional gains.

4 Numerical Experiments

We consider the wave equation with various forms of damping. We have the representation (3.35) for Kelvin-Voigt damping and similar results can be obtained for structural damping. However, as we shall see below, viscous damping is not sufficient to ensure the existence of L_2 functional gains even though the feedback operator exists and is bounded. We set $\gamma = .25$ and use standard piecewise linear finite elements to compute $k_s(\cdot,\cdot)$ and $k_v(\cdot,\cdot)$. In particular, we use finite elements to approximate the Riccati equation (3.29) and construct K as defined by (3.30-3.32).

In order to show convergence of the scheme we compute $k_s^N(\cdot,\cdot)$ and $k_v^N(\cdot,\cdot)$ for Kelvin-Voigt damping. Here, N represents the number of elements in the model (i.e., [0,1] is partitioned into (N+1) subintervals). Figure 4.1 shows the convergence of $k_s^N(\cdot,\cdot)$ and $k_v^N(\cdot,\cdot)$ to $k_s(\cdot,\cdot)$ and $k_v(\cdot,\cdot)$, respectively. Observe that the convergence of $k_s^N(\cdot,\cdot)$ is very rapid.

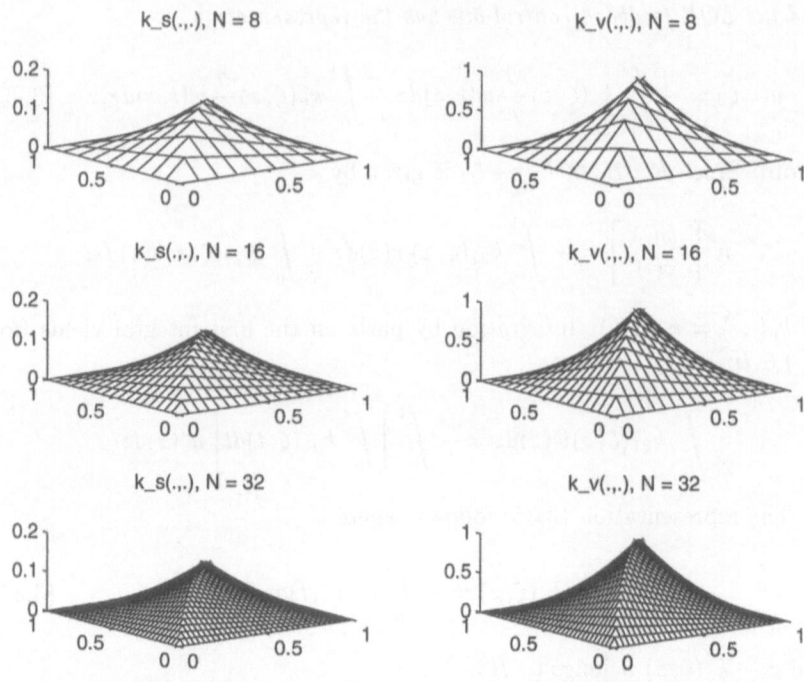

Figure 4.1: Convergence of the Functional Gains, Kelvin-Voigt Damping.

Although we presented theoretical results for the case of Kelvin-Voigt damping only, Figure 4.2 shows that similar results hold for structural damping. In particular, an integal representation exists and the finite element approximations converge. It is important to note that the functional gain for strain, $k_s(\cdot, \cdot)$, remains smooth and is the same as the gain obtained with Kelvin-Voigt damping. However, there is a marked difference in the smoothness of $k_v(\cdot, \cdot)$. This functional gain has a sharper "peak" at $\zeta = z$ than the corresponding velocity gain for Kelvin-Voigt damping. Although we expect that for structural damping, $k_v(\zeta, \cdot)$ belongs to H^1, we conjecture that $t \to k_v(\zeta, t)$ is not H^2.

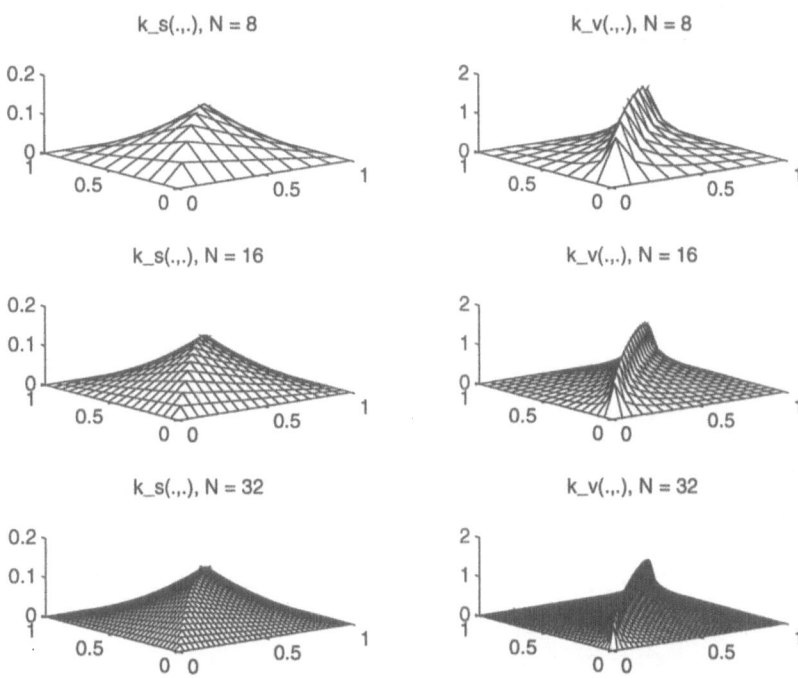

Figure 4.2: Convergence of the Functional Gains, Structural Damping.

Finally, for viscous damping, the numerical results shown in Figure 4.3 show that again, $k_s^N(\cdot, \cdot)$ is well behaved and yet $k_v^N(\cdot, \cdot)$ seems to be as singular measure concentrated at $\zeta = z$. Thus, we conjecture that the representation (3.35) does not hold for the wave equation with viscous damping for any L_2 function $k_v(\cdot, \cdot)$. Similar results were noted in [8] for undamped beam equations and in [7] for parabolic equations with highly unbounded input operators.

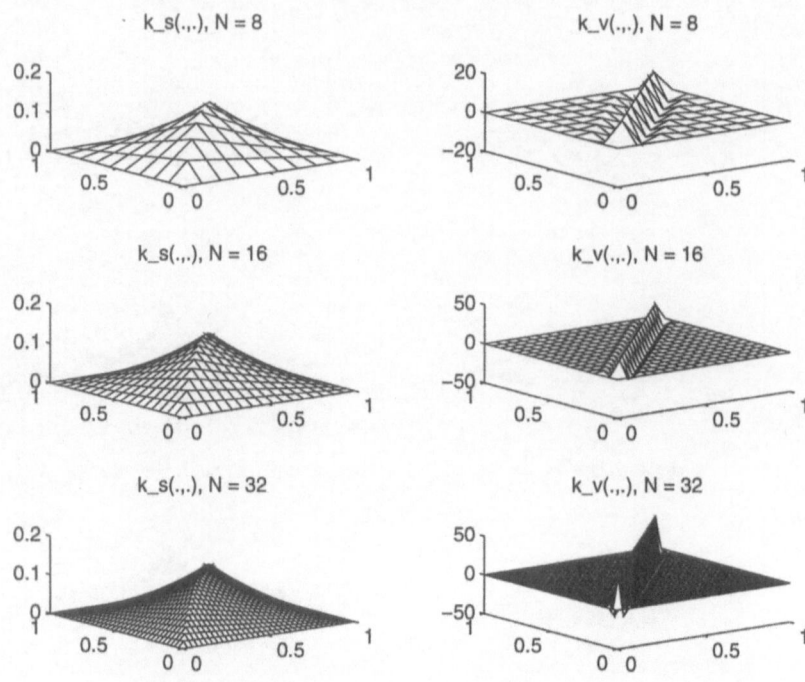

Figure 4.3: Convergence of the Functional Gains, Viscous Damping.

The numerical results also indicate that the strain functional gains are independent of the damping model. As shown in Figure 4.4, $k_s^N(\cdot, \cdot)$ does not change as the damping model changes. Thus, it seems from the numerical experiments that damping has the most impact on the existence and smoothness of $k_v(\cdot, \cdot)$.

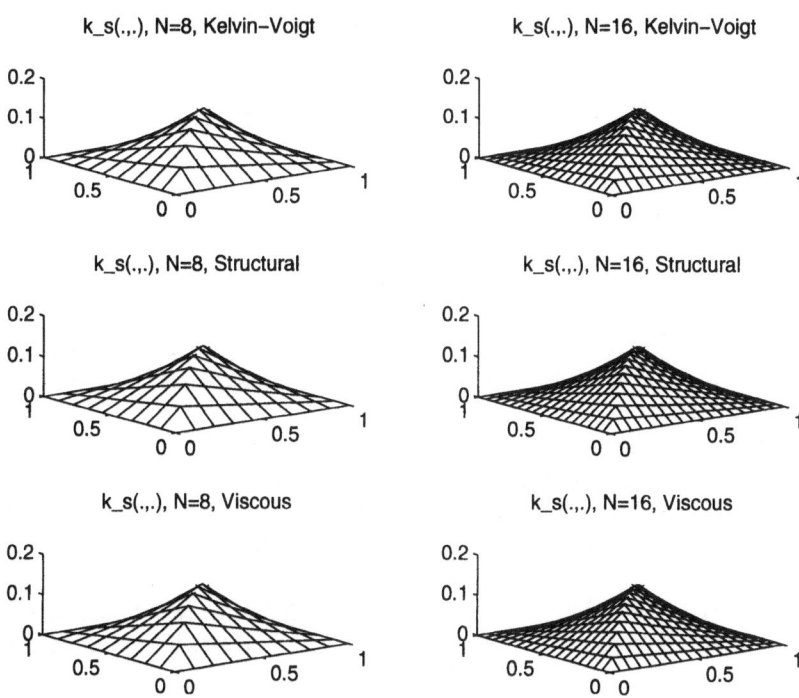

Figure 4.4: Structural Gains for Various Damping Types, N = 8,16.

5   Closing Remarks

In this paper we have provided an integral representation theorem for the LQR feedback operator for the 1D wave equation with Kelvin-Voigt damping and fully distributed control. The motivation for this effort comes from the problem of designing suboptimal low order dynamic compensators and for optimal sensor location. Although the theoretical results presented here are restricted to problems with Kelvin-Voigt damping, the numerical results suggest that more general results are available. Also, these same numerical results indicate that once the damping is insufficient to ensure the analyticity of the control system, the existence of L_2 functional gains is in doubt. In particular, the solution of the Riccati equation may not be Hilbert-Schmidt.

The LQR problem for distributed parameter systems has received considerable attention during the past ten years. However, problems in which B, Q and R are all non-compact have not been fully explored. Temam [13] considers the differential Riccati equation with $B = R = I$, but assumes that Q is Hilbert-Schmidt. Similar results are found in [4, 6, 12]. Finally, as the numerical evidence suggests, many theoretical issues are not yet settled.

References

[1] J. A. BURNS and B. B. KING, "A Note on the Regularity of Solutions of Infinite Dimensional Riccati Equations," *Appl. Math. Lett.*, v. 7, 1994, pp. 13–17.

[2] J. A. BURNS and B. B. KING, "Optimal Sensor Location for Robust Control of Distributed Parameter Systems," *Proc. of the 33rd IEEE Control and Decision Conference*, Dec. 1994, pp. 3967–3972.

[3] G. CHEN and J. ZHOU, *Vibration and Damping in Distributed Systems, Vol I. & II*, Studies in Advanced Mathematics, CRC Press, Boca Raton, 1993.

[4] A. DE SANTIS, A. GERMANI and L. JETTO, "Approximation of the Algebraic Riccati Equation in the Hilbert Space of Hilbert-Schmidt Operators," *SIAM J. Control Optim.*, v. 31, 1993, pp. 847–874.

[5] R. E. FULLERTON, "Linear Operators with Range in a Space of Differentiable Functions," *Duke Math. Journal*, v. 13, 1946, pp. 269–280.

[6] A. GERMANI, L. JETTO and M. PICCIONI, "Galerkin Approximation for Optimal Filtering of Infinite Dimensional Linear Systems," *SIAM J. Control Optim.*, v. 26, 1988, pp. 1287–1305.

[7] B. B. KING, "Existence of Functional Gains for Parabolic Control Systems" *Proc. Computation and Control IV*, this volume, 1995.

[8] V.D. LUPI, H.M. CHUN, J.D. TURNER, "Distributed Control without Mode Shapes or Frequencies," *Adv. in the Astro. Sci.*, v. 76, 1991, pp. 447–470.

[9] I. LASIECKA and R. TRIGGIANI, *Differential and Algebraic Riccati Equations with Application to Boundary/Point Control Problems: Continuous Theory and Approximation Theory*, Lecture Notes in Control and Information Sciences, v. 164, Springer-Verlag, Berlin, Heidelberg, 1991.

[10] C. MCMILLAN and R. TRIGGIANI, "Min-Max Game Theory and Algebraic Riccati Equations for Boundary Control Problems with Continuous Input-Solution Map, Part I: The Stable Case", preprint.

[11] A. H. NAYFEH, J. F. NAYFEH, and D. T. MOOK, "On Methods for Continuous Systems with Quadratic and Cubic Nonlinearities", *Nonlinear Dynamics*, v. 3, 1992, pp. 145–162.

[12] I. G. ROSEN, "On Hilbert-Schmidt Norm Convergence of Galerkin Approximation for Operator Riccati Equations," *Int. Series of Numer. Math. 91*, Birkhäuser, Basel, 1989, pp. 335–349.

[13] R. TEMAM, "Sur l'equation de Riccati associée à des opérateurs non bornés, en dimension infinie", *J. Func. Anal.*, v. 7, 1971, pp. 85–115.

STEADY STATE RESPONSE TO BURGERS' EQUATION WITH VARYING VISCOSITY

C.I. Byrnes

Systems Science and Math
Washington University
St. Louis, MO 63130 *

David S. Gilliam, Victor I. Shubov, Zaichao Xu,

Department of Mathematics
Texas Tech University
Lubbock, TX 79409 [†]

1 Introduction

Burgers' equation

$$w_t = \epsilon w_{xx} - w w_x \qquad (1.1)$$

was first introduced by J.M. Burgers about fifty years ago when he studied fluid flow, particularly, modeling turbulence. However, it was later found that the equation is a very useful mathematical model for such diverse physical problems as shock flows, traffic flow, acoustic transmission in fogs, etc., [8]. The dissipation may arise from viscosity, heat conduction, chemical reaction, mass diffusion, thermal radiation, etc. In a typical application, i.e., fluid dynamics, w is the velocity of the wave and ϵ is the viscosity of the compressible flow. However, one should understand that the parameters may have different physical meanings in different applications.

As it stands, equation (1.1) is a quasi-linear parabolic partial differential equation that describes the evolution of some dependent variable, $w(x, t)$, with time t. The equation is linear in the second x and t derivatives of w but nonlinear in w and w_x. The nonlinear term $w w_x$ is called a convective term, and ϵw_{xx} is called a diffusive term. An important feature of the equation is that it is a prototype equation for the balance between convection and diffusion.

Another very important feature of Burgers' equation is that it belongs to that rather small class of physically significant nonlinear partial differential equations for which, under very special conditions (either on the whole line or with Dirichlet or periodic boundary conditions), exact and complete solutions can be found, in terms of the initial values. More precisely, the situation is the following. The Cole-Hopf transformation [11, 15]

*Research supported in part by the AFOSR and NSF.

[†]Supported in part by AFOSR and Texas ARP

carries (1.1) into the linear heat equation. Furthermore, initial values pre-
scribed for (1.1) are transformed in a very simple manner to initial values for
the heat equation. This feature makes Burgers' equation an ideal medium
for testing and comparing various computational schemes. In the present
work we are concerned with the initial boundary value problem for Burgers'
equation with Neumann boundary conditions. In contrast with the above
mentioned cases of Dirichlet or periodic boundary conditions, the Hopf-
Cole transformation carries Neumann conditions for Burgers' equation into
complicated nonlinear boundary conditions for the heat equation. The latter circumstance makes it impossible to write the solution of the problem in
an explicit form. For this reason we apply methods of functional analysis,
center manifold and numerical methods to study this problem.

The initial boundary value problem for Burgers' equation with Neumann
boundary conditions is given by

$$w_t = \epsilon w_{xx} - w w_x \quad x \in (0, 1), \; t > 0 \tag{1.2}$$

$$w_x(0, t) = w_x(1, t) = 0$$

$$w(x, 0) = \varphi(x).$$

The linearization about zero of (1.2) is the one-dimensional heat equation with Neumann boundary conditions

$$w_t = \epsilon w_{xx} \quad x \in (0, 1), \; t > 0 \tag{1.3}$$

$$w_x(0, t) = w_x(1, t) = 0$$

$$w(x, 0) = \varphi(x).$$

Here $w(t, x)$ is the temperature distribution along a thin insulated metal bar
between 0 and 1; $\epsilon > 0$ can be interpreted as the heat diffusivity coefficient;
and $\varphi(x)$ is the initial temperature profile.

For Burgers' equation the term ϵ is related to the reciprocal of the
Reynolds number and in this paper is referred to as viscosity.

Defining the heat diffusion operator

$$A = -\frac{d^2}{dx^2} \tag{1.4}$$

with dense domain in the Hilbert space $\mathcal{H} = L^2(0, 1)$:

$$D(A) = \{f \in H^2(0, 1) : \; f'(0) = f'(1) = 0\}$$

the solution to (1.3) can be written in terms of a semigroup of transformations as

$$w(x, t) = S(t)\varphi(x) = \sum_{n=0}^{\infty} e^{-\epsilon \lambda_n t} \varphi_n \psi_n, \tag{1.5}$$

where

$$\varphi_n = \int_0^1 \varphi(x) \psi_n(x) \, dx$$

are the Fourier coefficients of the initial data φ and the spectrum of A consists of simple eigenvalues $\{\lambda_n = n^2 \pi^2\}_{n=0}^\infty$ with corresponding orthonormal eigenfunctions $\{\psi_n\}$, given by $\psi_0 = 1, \psi_n = \sqrt{2} \cos(n\pi x), \ n = 1, 2, \cdots$.

It is well known that for any $\varphi \in L^2(0, 1)$ and $\epsilon > 0$, the steady state response to (1.3) is a constant given by the first Fourier coefficient of the initial data,

$$w(x, t) \to \varphi_0 = \int_0^1 \varphi(x) \, dx, \quad t \to \infty.$$

Note that, in particular, the constant φ_0 (the mean of the initial data) is independent of ϵ. Thus the steady state is completely determined by the initial data. This is in marked contrast with Burgers' equation with Neumann boundary conditions, which forms the main concern of this work.

Obviously, for all initial data orthogonal to the first eigenvector of the operator A, the steady state of the dynamical system generated by (1.3) is 0, i.e., if $\int_0^1 \varphi(x) \, dx = 0$, $w(x, t) \to 0$ as $t \to \infty$. Conversely, if $w(x, t) \to 0$ as $t \to \infty$, the mean value of the initial data must be 0.

2 Global in Time Existence and Uniqueness for Burgers' Equation

In this section we turn to the question of long time existence of solutions to the problem (1.2) for sufficiently smooth, small initial data. First, we note that the operator A is sectorial [10] in $\mathcal{H} = L^2(0, 1)$ and there is the natural decomposition $\mathcal{H} = \mathcal{H}_0 \oplus \mathcal{H}_1$. $\mathcal{H}_0 = P\mathcal{H}$ is the subspace spanned by the first eigenvector of A, where P is the projection operator

$$P : \mathcal{H} \to \mathcal{H}_0$$

(i.e., $P\phi = < 1, \ \phi > 1$). $\mathcal{H}_1 = \mathcal{H}_0^\perp = P_1 \mathcal{H} = (I - P)\mathcal{H}$, i.e., for any function $f \in \mathcal{H}_1$, $< 1, \ f > = \int_0^1 f \, dx = 0$. B_0 is the restriction of A to $\mathcal{H}_0$, i.e.,

$$B_0 = A \mid_{\mathcal{H}_0} = 0$$

with domain $D(B_0) = \{f \in D(A) : f \text{ is constant }\}$. B_1 is the restriction of A to $\mathcal{H}_1$, i.e.,

$$B_1 = A \mid_{\mathcal{H}_1}$$

with domain $D(B_1) = \{f \in D(A) : \int_0^1 f \, dx = 0\}$.

Note that B_1 generates the C_0–semigroup [10] on the space $\mathcal{H}_1$

$$S_1(t) = e^{-\epsilon B_1 t} = \sum_{j=1}^{\infty} e^{-\lambda_j t} < \cdot, \, \psi_j > \psi_j.$$

The first eigenvector of B_1 is $\pi^2 > 0$, so B_1 is strictly positive.

B_1 also defines an infinite scale of Hilbert spaces $\mathcal{H}_1^{\alpha}$ ($\alpha \in \mathbb{R}$). If $\alpha \geq 0$ then $\mathcal{H}_1^{\alpha}$ consists of vectors $\phi \in \mathcal{H}_1$ such that

$$\|\phi\|_{\alpha} = \left(\sum_{j=1}^{\infty} \lambda_j^{\alpha}(\phi, \psi_j)^2 \right)^{1/2} < \infty. \tag{2.1}$$

Notice that $\mathcal{H}_1^0 = \mathcal{H}_1$. We will use the notation $\|\cdot\|$ instead of $\|\cdot\|_0$ for the norm in $\mathcal{H}_1^0$, which coincides with the $L^2(0,1)$-norm. These spaces can be described in a different way. Namely, the space $\mathcal{H}_1^{\alpha}$ is the domain of the operator $B_1^{\alpha/2}$ and the inner product in this space is given by

$$(\phi, \psi)_{\alpha} = \left(B_1^{\alpha/2} \phi, B_1^{\alpha/2} \psi \right) \tag{2.2}$$

which is the same as (2.1) for $\psi = \phi$. The operator $B_1^{\alpha/2}$ is defined on $\mathcal{H}_1^{\alpha}$ by the formula

$$B_1^{\alpha/2} \phi = \sum_{j=1}^{\infty} \lambda_j^{\alpha/2}(\phi, \psi_j)\psi_j. \tag{2.3}$$

The spaces $\mathcal{H}_1^{\alpha}$ have the following properties:

1. If $\beta > \alpha$ then $\mathcal{H}_1^{\beta} \subset \mathcal{H}_1^{\alpha}$ and

$$\|\phi\|_{\alpha} \leq \lambda_1^{(\alpha-\beta)/2} \|\phi\|_{\beta} \tag{2.4}$$

for all $\phi \in \mathcal{H}_1^{\beta}$; $\mathcal{H}_1^{\beta}$ is dense in $\mathcal{H}_1^{\alpha}$.

2. The imbedding $\mathcal{H}_1^{\beta} \subset \mathcal{H}_1^{\alpha}$ is compact.

We will need the following lemmas:

Lemma 2.1 *1. If $f \in \mathcal{H}_1^1$, then for every $t > 0$,*

$$\|S_1(t)f\|_1 \leq e^{-\epsilon \lambda_1 t} \|f\|_1. \tag{2.5}$$

2. For $f \in \mathcal{H}_1$ and $t > 0$, the following inequality holds

$$\|S_1(t)f\|_1 \leq \frac{C_1}{t^{1/2}} \|f\| e^{-\epsilon \lambda_1 t/2}, \tag{2.6}$$

where $C_1 = 1/\sqrt{\epsilon e}$.

PROOF: see [4]. □

Lemma 2.2 For $f \in \mathcal{H}_1^1$, $x \in [0, 1]$, the following inequality holds

$$| f(x) | \leq \|f_x\|. \tag{2.7}$$

PROOF: let $0 \leq x_1 \leq x_2 \leq 1$, then we have

$$f(x_2) - f(x_1) = \int_{x_1}^{x_2} f'(x) \, dx.$$

By integration with respect to x_1, it turns into

$$f(x_2) - \int_0^1 f(x_1) \, dx_1 = \int_0^1 \left(\int_{x_1}^{x_2} f'(x) \, dx \right) dx_1.$$

So, since f has mean zero,

$$f(x_2) = \int_0^1 \left(\int_{x_1}^{x_2} f'(x) \, dx \right) dx_1.$$

This yields

$$| f(x_2) | \leq \int_0^1 \left(\int_{x_1}^{x_2} | f'(x) | \, dx \right) dx_1$$

$$\leq \int_0^1 | f'(x) | \, dx$$

$$\leq \|f_x\|.$$

Since x_2 can be any number in $[0, 1]$, this gives

$$\|f(x)\| \leq \|f(x)\|_\infty \leq \|f_x\| \tag{2.8}$$

and the inequality (2.7). Here and below we denote by $\| \cdot \|_\infty$ the norm in $L^\infty(0,1)$. □

Note that for $\phi \in \mathcal{H}_1^1$,

$$\|\phi_x\| = \|\phi\|_1. \tag{2.9}$$

From the above facts we find that the solution of Burgers' equation can be decomposed as

$$w = c + v, \tag{2.10}$$

where $c \in \mathcal{H}_0$, $v \in \mathcal{H}_1$.

Defining the function $f(u(x,t)) = -u(x,t)u_x(x,t)$, we are now in a position to formulate the definition of solution to (1.2), i.e.,

$$\dot{w} + Aw = f(w), \tag{2.11}$$

taken in this work.

Following M. Miklavčič [13], (see also [9], page 73), we say that $w(t) = w(\cdot, t)$ is a solution to (1.2) on $[0, \tau)$ if $w : [0, \tau) \to \mathbb{C} \times \mathcal{H}_1^1$, (here $\mathbb{C}$ is the complex numbers and we recall that $\mathcal{L}_0$ is isomorphic to $\mathbb{C}$) is a continuous function of t such that $w(0) = \varphi$, $f(w(\cdot)) : [0, \tau) \to \mathcal{H}$ is continuous, $w(t) \in \mathcal{D}(A)$ and w satisfies (2.11) on $(0, \tau)$.

As in [9], we note that solutions of (2.11) coincide with with those solutions of

$$w(t) = S(t)\varphi + \int_0^t S(t - s)f(w(s))\, ds, \ 0 \le t \le \tau \tag{2.12}$$

for which $w : [0, \tau) \to \mathbb{C} \times \mathcal{H}_1^1$ is continuous and $f(w(\cdot)) : [0, \tau) \to \mathcal{H}$ is continuous.

Note that the decomposition (2.10) allows us to write the equation (2.11) in the form of a system

$$
\begin{aligned}
c_t &= Pf(c + v) = -P(cv_x) - P(vv_x) \\
&= -c < v_x, 1 > - < vv_x, 1 > \\
&= -c \int_0^1 v_x\, dx - \int_0^1 vv_x\, dx,
\end{aligned} \tag{2.13}
$$

$$
\begin{aligned}
v_t &= \epsilon B_1 v + P_1 f(c + v) \\
&= \epsilon B_1 v - P_1(cv_x) - P_1(vv_x).
\end{aligned} \tag{2.14}
$$

We will need the following estimate for (2.14):

Lemma 2.3 $\| - P_1(cv_x) - P_1(vv_x) \| \le \sqrt{2} \left(|c| + \|v\|_1 \right) \|v\|_1$

PROOF: Since $c \in \mathcal{H}_0$ is independent of x, we have

$$\| - P_1(cv_x) - P_1(vv_x) \|^2 \le 2(|c|^2 \|v_x\|^2 + \|vv_x\|^2).$$

Note that here we have used the fact that $\|P\| \le 1$ and $(a+b)^2 \le 2(a^2 + b^2)$. By using (2.7) and (2.9), we obtain

$$\|vv_x\|^2 \le \|v\|_\infty^2 \|v_x\|^2 \le \|v\|_1^4.$$

This gives the estimate.   □

Now we turn to the proof of the main result of this section.

Theorem 2.4 *If the initial data for the system (2.13), (2.14) satisfy the inequality*

$$|c(0)| + \|v(0)\|_1 < \rho/4 \tag{2.15}$$

for sufficiently small $\rho > 0$, then this system has a unique classical solution defined for all $t \in [0, \infty)$. The solution satisfies

$$|c(t)| + \|v(t)\|_1 < \rho \tag{2.16}$$

for all $t \in [0, \infty)$.

PROOF: Assume (2.15) is satisfied for some number $\rho > 0$. For this initial condition $w(\cdot, 0) = c(0) + v(\cdot, 0)$. By Theorem 4.2.2 in [9] we are guaranteed that a local unique classical solution exists on a finite time interval. Let t_1 be the largest possible time such that $|c(t)| + \|v(t)\|_1 \leq \rho$ for all $t \leq t_1$. Then, by showing that (2.16) takes place at $t = t_1$ we can draw the conclusion that $t_1 = \infty$. In the following we will use the above lemmas and estimates to show that both $|c(t)|$ and $\|v(t)\|_1$ are strictly less than $\rho/2$ at $t = t_1$ if ρ is small enough. This will give the inequality (2.16).

We first consider the term $\|v(t)\|_1$. By the variation of parameters formula, we have

$$v(t) = S_1(t)v(0) + \int_0^t S_1(t - \tau) \left[-P_1(cv_x)(\tau) - P_1(vv_x)(\tau) \right] d\tau.$$

Taking $0 < \beta < \epsilon\lambda_1/2$, we further get

$$e^{\beta t}\|B_1^{1/2}v(t)\| \leq e^{\beta t} \|B_1^{1/2}S_1(t)v(0)\|$$

$$+ \int_0^t \|B_1^{1/2}S_1(t - \tau)\| e^{\beta t} \sqrt{2} \left(|c(\tau)| + \|v(\tau)\|_1 \right) \|v(\tau)\|_1 \, d\tau$$

$$\leq e^{-(\epsilon\lambda_1 - \beta)t}\|v(0)\|_1 + \sqrt{2}\rho \int_0^t \|B_1^{1/2}S_1(t - \tau)\| e^{\beta(t-\tau)}e^{\beta\tau} \|v(\tau)\|_1 \, d\tau$$

$$\leq \|v(0)\|_1 + \sqrt{2}\rho \left(\int_0^t \|B_1^{1/2}S_1(t - \tau)\| e^{\beta(t-\tau)} \, d\tau \right) \mathcal{W}(t_1),$$

where $\mathcal{W}(t_1) = \sup_{t \in [0,t]} (e^{\beta t} \|v(t)\|_1)$. In the second inequality, we have used Lemma 2.3 and part 1 of Lemma 2.1. Now by part 2 of Lemma 2.1, we have

$$\|B_1^{1/2}S_1(t - \tau)\| \leq C_1 \frac{e^{-\epsilon\lambda_1(t-\tau)/2}}{(t - \tau)^{1/2}}.$$

So

$$e^{\beta t} \|v(t)\|_{1/2} = e^{\beta t} \|B_1^{1/2}v(t)\|$$

$$\leq \|v(0)\|_1 + \sqrt{2}\, C_1 \rho \left(\int_0^t \frac{e^{-(\frac{\epsilon\lambda_1}{2}-\beta)(t-\tau)}}{(t-\tau)^{1/2}} d\tau \right) \mathcal{W}(t_1)$$

$$= \|v(0)\|_1 + \sqrt{2}\, C_1 \rho \left(\int_0^t \frac{e^{-(\frac{\epsilon\lambda_1}{2}-\beta)s}}{s^{1/2}} ds \right) \mathcal{W}(t_1)$$

$$\leq \|v(0)\|_1 + \sqrt{2}\, C_1 \rho \left(\int_0^\infty \frac{e^{-(\frac{\epsilon\lambda_1}{2}-\beta)s}}{s^{1/2}} ds \right) \mathcal{W}(t_1)$$

$$= \|v(0)\|_1 + \sqrt{2}\, C_1 \rho \frac{\Gamma(1/2)}{(\frac{\epsilon\lambda_1}{2} - \beta)^{1/2}} \mathcal{W}(t_1),$$

where, in the last equality we have used well known results for the Laplace transform.

Note that $\sqrt{2}\, C_1 \Gamma(1/2) / (\epsilon\lambda_1 - \beta)^{1/2}$ is fixed, so we always can choose a $\rho > 0$, such that

$$\frac{\rho \sqrt{2}\, C_1 \Gamma(1/2)}{(\epsilon\lambda_1 - \beta)^{1/2}} < 1/2. \qquad (2.17)$$

Thus

$$e^{\beta t} \|v(t)\|_1 \leq \|v(0)\|_1 + \mathcal{W}(t_1)/2 < \rho/4 + \mathcal{W}(t_1)/2.$$

Since the right side is independent of t, we have

$$\mathcal{W}(t_1) = \sup_{t \in [0,t_1]} e^{\beta t} \|v(t)\|_1 < \rho/4 + \mathcal{W}(t_1)/2.$$

This yields

$$\sup_{t \in [0,t_1]} e^{\beta t} \|v(t)\|_1 < \rho/2,$$

and therefore for every $t \in [0, t_1]$,

$$\|v(t)\|_1 < \frac{\rho}{2} e^{-\beta t} < \rho/2.$$

Now we turn to $|c(t)|$. We see that (2.13) is a first-order inhomogeneous ordinary differential equation. Its solution is given by

$$c(t) = \exp \left[-\int_0^t <v_x, 1>(\tau)\, d\tau \right]$$

$$\times \left(c(0) - \int_0^t \exp \left[\int_0^\tau <v_x, 1>(\eta)\, d\eta \right] <vv_x, 1>(\tau)\, d\tau \right).$$

So we have

$$|c(t)| \leq \exp[\int_0^t |<v_x, 1>(\tau)|\, d\tau]$$

$$\times \left(\mid c(0) \mid + \int_0^t \exp\left[\int_0^\tau \mid < v_x, 1 > (\eta) \mid d\eta \right] \right.$$

$$\left. \times \mid < vv_x, 1 > (\tau) \mid d\tau \right).$$

Note that

$$\mid < v_x, 1 > \mid = \mid \int_0^1 v_x(x)\, dx \mid \le \|v_x\| = \|v(t)\|_1 \le e^{-\beta t}\rho/2$$

and

$$\mid < vv_x, 1 > \mid = \mid \int_0^1 vv_x\, dx \mid \le \|v\|\, \|v_x\| \le \|v_x\|^2 = \|v\|_1^2 \le \frac{\rho^2}{4}e^{-2\beta t},$$

where (2.8) and (2.9) are used. Therefore

$$\mid c(t) \mid \le \exp\left(\int_0^t \frac{\rho}{2}e^{-\beta t}\, d\tau \right)\left(\mid c(0) \mid + \right.$$

$$\left. \frac{\rho^2}{4}\int_0^t \exp\left(\int_0^\tau \frac{\rho}{2}e^{-\beta\eta}\, d\eta \right) \exp(-2\beta\tau)\, d\tau \right)$$

$$\le \exp\left(\frac{\rho(1 - e^{-\beta t})}{2\beta} \right)\left(\mid c(0) \mid + \right.$$

$$\left. \frac{\rho^2}{4}\int_0^t \exp\left(\frac{\rho(1 - e^{-\beta\tau})}{2\beta} \right) \exp(-2\beta\tau)\, d\tau \right)$$

$$\le \exp\left(\frac{\rho}{2\beta} \right)\left(\mid c(0) \mid + \frac{\rho^2}{4}\exp\left(\frac{\rho}{2\beta} \right)\int_0^t \exp(-2\beta\tau)\, d\tau \right)$$

$$\le \exp\left(\frac{\rho}{2\beta} \right)\left(\mid c(0) \mid + \frac{\rho^2}{4}\exp\left(\frac{\rho}{2\beta} \right) [1 - \exp(-2\beta t)]/(2\beta) \right)$$

$$\le \exp\left(\frac{\rho}{2\beta} \right)\left(\mid c(0) \mid + \frac{\rho^2}{8\beta}\exp\left(\frac{\rho}{2\beta} \right) \right)$$

$$\le \exp\left(\frac{\rho}{2\beta} \right)\left(\frac{\rho}{4} + \frac{\rho^2}{8\beta}\exp\left(\frac{\rho}{2\beta} \right) \right)$$

$$= \rho\left\{ \exp\left(\frac{\rho}{2\beta} \right)\left(1/4 + \frac{\rho}{8\beta}\exp\left(\frac{\rho}{2\beta} \right) \right) \right\}.$$

Now again we choose ρ so small that (2.17) is satisfied and

$$e^{\rho/(2\beta)}\left(1/4 + \frac{\rho}{8\beta}e^{\rho/(2\beta)} \right) < 1/2.$$

This means for every $t \in [0, t_1]$ the following inequality holds

$$| c(t) | < \rho/2.$$

Thus (2.16) holds for $t = t_1$, and t_1 must be ∞. This shows that the unique classical solution exists for all $t \geq 0$ if the initial condition is small enough. $\square$

3 Center Manifold Theorem

In this section we show that for Burgers' equation with Neumann boundary conditions, and for small initial data in $H^1(0,1)$ the solution $w(x,t)$ tends to a constant as $t \to \infty$. This result is obtained by appealing to an infinite dimensional version of the center manifold theorem [5, 10] together with the local existence and uniqueness result of the last section.

Recall the following facts concerning the abstract evolution equation (2.11). The operator A has a one dimensional null space which gives rise to the decomposition $\mathcal{H} = \mathcal{H}_0 \oplus \mathcal{H}_1$, $\mathcal{H}_0$ is the subspace spanned by the first eigenvector of A, $\mathcal{H}_0 = \mathrm{span}\{1\}$, $\mathcal{H}_1 = \mathcal{H}_0^\perp$.

The solution w to Burgers' equation can be decomposed as $w = c + v$, where $c \in \mathcal{H}_0$, $v \in \mathcal{H}_1$. The nonlinear function $f : H^1(0, 1) \to \mathcal{H}$ given by $f(u) = -uu_x$ is C^1 with $f(0) = 0$, $f'(0) = 0$, where f' denotes the Fréchet derivative of f.

Recall that P is the orthogonal projection onto $\mathcal{H}_0$ and

$$f(c, v) = PF(c + v) = <1, \ -(c+v)_x (c+v) > \tag{3.1}$$

$$= - <1, \ cv_x + vv_x > = -c \int_0^1 v_x \, dx - \int_0^1 vv_x \, dx,$$

and

$$g(c, v) = (I - P)F(c + v)$$

$$= -cv_x + c < 1, v_x > - vv_x + < 1, vv_x > . \tag{3.2}$$

So, we can write (2.11) as

$$c_t = f(c, v) \tag{3.3}$$

$$v_t = \epsilon B_1 v + g(c, v).$$

For equation (3.3), Theorem 8, Chapter 6 in [5] asserts that there exists a center manifold $v = h(c)$. Since $v \in \mathcal{H}_1$, $\int_0^1 v \, dx = 0$. This means

$\int_0^1 h(c)\, dx = 0$. Noticing that c is independent of x, we get $h(c) = 0$. Therefore the center manifold is $v = 0$.

Hence the equation on the center manifold is given by

$$u_t = f(u,\, h(u)) = f(u, 0) = 0.$$

This yields

$$u = u_0,$$

where u_0 is a constant.

By Theorem 9, Chapter 6 of [5], we can draw the following conclusions:

1. There exists a $\gamma > 0$, such that

$$c(t) = u(t) + O(e^{-\gamma t}) = u_0 + O(e^{-\gamma t}) \longrightarrow u_0$$

as $t \to \infty$.

2. There exists a constant $M > 0$, such that $\| v(x, t) \|_{H^1} < Me^{-\gamma t}$.

3. From 1 and 2, we can obtain

$$w(x, t) = c + v \longrightarrow u_0$$

as $t \to \infty$. This means that the steady state response of Burgers' equation is a constant.

The constant u_0, in general, is a function of both the initial data φ and the parameter ϵ, so from now on we will use the notation $C_{\varphi, \epsilon}$ instead of u_0 to indicate this dependence.

We note that the second statement above implies that there is a constant constant $M > 0$, such that $| v(x, t) | < Me^{-\gamma t}$, since $\sup_{x \in [0,1]} | v(x, t) | \le C\|v(t)\|_{H^1}$, where C is a positive constant.

4 Behavior of $C_{\varphi, \epsilon}$ as $\epsilon \to \infty$

Generally there is not much interest in the case in which viscosity ϵ is large. On the other hand our goal here is to specify, when possible, how the constant steady state of solutions to Burgers' equation with Neumann boundary conditions depends on viscosity. Therefore, from a mathematical point of view, it is reasonable to consider also the case in which ϵ might be large.

Theorem 4.1 *For Burgers' equation with Neumann boundary conditions (1.2), and small initial data φ, the constant steady state $C_{\varphi, \epsilon}$ satisfies*

$$C_{\varphi, \epsilon} \longrightarrow \int_0^1 \varphi(x)\, dx, \quad \epsilon \to \infty. \tag{4.1}$$

PROOF: If we introduce the transformation $\tau = \epsilon t$, then Burgers' equation becomes

$$w_\tau = w_{xx} - \frac{1}{\epsilon} w w_x.$$

Now let $\delta = 1/\epsilon$ so that Burgers' equation with Neumann boundary conditions and initial data φ can be written as

$$w_\tau = Aw - \delta w w_x, \tag{4.2}$$
$$w_x(0,\tau) = w_x(1,\tau) = 0,$$
$$w(x,0) = \varphi(x).$$

Using the notation introduced in the last section, we have

$$c_\tau(\tau, \delta) = \delta f(c, v)$$

$$= -\delta c \int_0^1 v_x \, dx - \delta \int_0^1 v v_x \, dx \tag{4.3}$$

$$v_\tau(\tau, x, \delta) = A_1 v + \delta g(c, v), \tag{4.4}$$

where, $c(\tau, \delta) \in \mathcal{H}_0$, $v(\tau, x, \delta) \in \mathcal{H}_1$, B_1 is the restriction of the operator A to $\mathcal{H}_1$, $f(c, v)$ is defined in (3.1), and $g(c, v)$ is defined in (3.2). Here we emphasized the dependence on δ in the arguments of c and v. Equation (4.3) is an inhomogeneous first-order ordinary differential equation whose solution can be written as

$$c(\tau, \delta) = e^{-\delta \int_0^\tau <v_x, 1> d\eta} \times$$
$$\left(c(0) - \delta \int_0^\tau e^{\delta \int_0^\eta <v_x, 1> d\zeta} <v, v_x> (\eta) \, d\eta \right). \tag{4.5}$$

We now show that as $\tau \to \infty$, all integrals in (4.5) are convergent and bounded by finite numbers independent of δ. Then it will be obvious that as $\delta \to 0$, $c(\infty, \delta) = C_{\varphi, \epsilon} \to c(0)$, where $c(0)$ is the projection of the initial condition $\varphi(x)$ onto $\mathcal{H}_0$, i.e.,

$$c(0) = < 1, \varphi(x) > = \int_0^1 \varphi(x) \, dx.$$

As before, denoting by $S_1(t)$ the restriction of $S(t)$ to $\mathcal{H}_1$, we know that the variation of parameter formula is satisfied by the solution of equation (4.4) and hence

$$v(x, \tau, \delta) = S_1(\tau) v(0) + \int_0^\tau S_1(\tau - \eta) \, \delta \, g(c(\eta), v(\eta)) \, d\eta. \tag{4.6}$$

From the results in the previous section, we have

$$| v(x, \tau, \delta) | < Me^{-\gamma\tau}, \tag{4.7}$$

where $M > 0$, $\gamma > 0$. Note that M is dependent on δ. Let us now show that there exists a $\gamma > 0$ independent of δ such that (4.7) holds.

Let $\delta = 1$. Then there exist $M_1 > 0$, $\gamma_1 > 0$, such that

$$| v(x, \tau, 1) | < M_1 e^{-\gamma_1\tau}.$$

Namely,

$$| S_1(\tau)v(0) + \int_0^\tau S_1(\tau - \eta)\, g(c(\eta), v(\eta))\, d\eta\, | < M_1 e^{-\gamma_1\tau}.$$

Since S_1 generates an exponentially stable semigroup, there exist $M_2 > 0$ and $\gamma_2 > 0$, such that

$$| S_1(\tau)v(0) | < M_2 e^{-\gamma_2\tau},$$

and hence

$$| \int_0^\tau S_1(\tau - \eta)\, g(c(\eta), v(\eta))\, d\eta\, | < M_1 e^{-\gamma_1\tau} + M_2 e^{-\gamma_2\tau}.$$

Note that M_1 and M_2 are all independent of δ.

Thus,

$$| v(x, \tau, \delta) | < | S_1(\tau)\, v(0) + \int_0^\tau S_1(\tau - \eta)\, \delta\, g(c(\eta), v(\eta))\, d\eta\, |$$

$$< M_2 e^{-\gamma_2\tau} + \delta\, (M_1 e^{-\gamma_1\tau} + M_2 e^{-\gamma_2\tau})$$

$$< Me^{-\gamma\tau}, \tag{4.8}$$

where $M > 0$ and $\gamma = \min(\gamma_1, \gamma_2)$. Note that the last inequality is satisfied with M and γ independent of δ when $\delta < 1$.

With the estimate (4.8), it is very clear that the following estimates hold independent of δ

$$\left| \int_0^\tau < v_x,\, 1 > d\eta\, \right| \le \int_0^\tau \left| v(1, \eta, \delta) - v(0, \eta, \delta) \right| d\eta$$

$$\le \int_0^\tau | v(1, \eta, \delta) |\, d\eta + \int_0^\tau | v(0, \eta, \delta) |\, d\eta$$

$$\le \int_0^\infty | v(1, \eta, \delta) |\, d\eta + \int_0^\infty | v(0, \eta, \delta) |\, d\eta$$

$$\le M' \tag{4.9}$$

and

$$\left| \int_0^\tau <v, \, v_x> d\eta \right| \leq \int_0^\tau \frac{1}{2} \mid v^2(1, \eta, \delta) - v^2(0, \eta, \delta) \mid d\eta$$

$$\leq \int_0^\tau \frac{1}{2} \mid v^2(1, \eta, \delta) \mid d\eta + \int_0^\tau \frac{1}{2} \mid v^2(0, \eta, \delta) \mid d\eta$$

$$\leq \int_0^\infty \frac{1}{2} \mid v^2(1, \eta, \delta) \mid d\eta + \int_0^\infty \frac{1}{2} \mid v^2(0, \eta, \delta) \mid d\eta$$

$$\leq M''. \tag{4.10}$$

Here the constants $M' > 0$, $M'' > 0$ are independent of δ.

Now return to (4.5). We first let $t \to \infty$. Then $\tau = t/\delta \to \infty$, and the solution (4.5) therefore becomes

$$C_{\varphi, \epsilon} = c(\infty, \delta) = e^{-\delta \int_0^\infty <v_x, 1> d\eta} \times$$
$$\left(c(0) - \delta \int_0^\infty e^{\delta \int_0^\eta <v_x, 1> d\zeta} <v, \, v_x> (\eta) \, d\eta \right).$$

With the estimate (4.9), we have

$$\left| \delta \int_0^\infty <v_x, \, 1> dx \right| \leq \delta M' \tag{4.11}$$

and with the estimate (4.10), we have

$$\left| \int_0^\infty \delta e^{\delta \int_0^\eta <v_x, 1> d\zeta} <v, \, v_x> (\eta) \, d\eta \right| \leq$$

$$\leq \delta \int_0^\infty e^{\delta \int_0^\eta <v_x, 1> d\zeta} \mid <v, \, v_x> (\eta) \mid d\eta$$

$$\leq \delta e^{\delta M'} \int_0^\infty \mid <v, \, v_x> \mid d\eta$$

$$\leq \delta e^{\delta M'} \int_0^\infty \mid <v, \, v_x> \mid d\eta$$

$$\leq \delta e^{\delta M'} M''. \tag{4.12}$$

We next let $\epsilon \to \infty$, i.e., $\delta \to 0$. With (4.11) and (4.12), we obtain

$$\lim_{\delta \to 0} C_{\varphi, \epsilon} = \lim_{\delta \to 0} c(\infty, \delta) = \int_0^1 \varphi(x) \, dx.$$

$$\square$$

5   A Special Class of Initial Data

As noted in the introduction, if the initial data φ has mean zero, i.e., $\int_0^1 \varphi(x)\,dx = 0$, then the solutions of linearization about zero of Burgers' equation approach zero for all ϵ. In the next section we show, by way of numerical example, that this result is no longer valid for Burgers' equation. Nevertheless, there is a stronger condition on the initial data for which we always do obtain zero steady state, independent of ϵ. In particular this result holds for initial data which are odd functions about the point $x = 1/2$ in $[0,1]$.

Theorem 5.1 *For Burgers' equation with Neumann boundary conditions, if the initial data φ is sufficiently small, and satisfies $\varphi(x) = -\varphi(1-x)$, then the steady state $C_{\varphi,\epsilon} = 0$, independent of ϵ.*

PROOF: This result will follow from the uniqueness of the solutions to Burgers' equation with Neumann boundary conditions (1.2). Using the uniqueness we will show that, if $w(x,t)$, is the solution of (1.2) for $t > 0$, then

$$w(x,t) = -w(1-x,\ t). \tag{5.1}$$

Thus, $w(1/2,\ t) = 0$, for all $t \geq 0$. On the other hand, the solution $w(x,t)$ must converge to a constant $C_{\varphi,\epsilon}$ as $t \to \infty$. Therefore, the constant $C_{\varphi,\epsilon}$ has to be zero.

Let us prove (5.1). If $w(x,t)$ satisfies Burgers' equation (1.2), then

$$z(x,t) = -w(1-x,\ t)$$

satisfies

$$
\begin{aligned}
z_t - \epsilon z_{xx} + z z_x &= \\
&= -w_t(1-x,\ t) + \epsilon w_{xx}(1-x,\ t) - w(1-x,\ t)\,w_x(1-x,\ t) \\
&= -(w_t(1-x,\ t) - \epsilon w_{xx}(1-x,\ t) + w(1-x,\ t)\,w_x(1-x,\ t)) \\
&= 0,
\end{aligned}
$$

$$
\begin{aligned}
z_x(0,t) &= -w_x(1,t) = 0, \\
z_x(1,t) &= -w_x(0,t) = 0,
\end{aligned}
$$

and

$$z(x,0) = -w(1-x,\ 0) = -\varphi(1-x) = \varphi(x).$$

So $z(x,t)$ is also a solution of Burgers' equation (1.2). Since the initial data is sufficiently small, the solution must be unique by the local existence and uniqueness theorem. Therefore,

$$z(x,t) = w(x,t),$$

which is equivalent to (5.1).   □

6   Results Based on Numerical Simulation

It is likely that there is no simple relationship in general between the constant steady state and the initial condition and viscosity. It is clear that if the initial condition is a constant, $w(x,0) = c$ then the solution for all time is $w(x,t) = c = C_{c,\epsilon}$ and we have seen that if the initial condition satisfies $\varphi(x) = -\varphi(1-x)$ then $C_{\varphi,\epsilon} = 0$. We also know that for all initial data, as ϵ tends to infinity, $C_{\varphi,\epsilon}$ tends to the mean value of the initial condition. In this section we present several unproved conjectures concerning the behavior of the constant steady state depending on various properties of the initial condition and as ϵ varies between zero and infinity. These conjectures are based on considerable numerical testing but are not supported by analytic proofs of any kind. They do reflect the behavior expected from an analysis of hyperbolic conservation laws on the whole line. For example, as discussed below, if the initial data is decreasing then as ϵ tends to zero the constant steady state appears to blow up reflecting the development of a shock.

The first two figures depict the fact that the trajectories do indeed approach a constant.

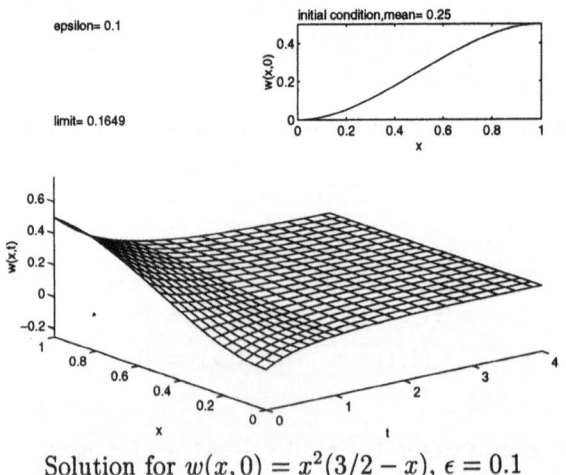

Solution for $w(x,0) = x^2(3/2 - x)$, $\epsilon = 0.1$

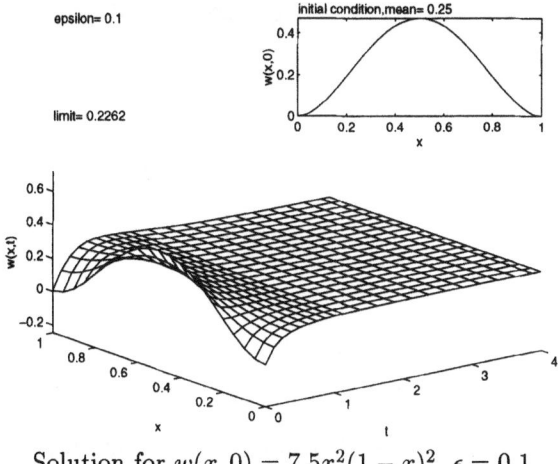

Solution for $w(x,0) = 7.5x^2(1-x)^2$, $\epsilon = 0.1$

Next we showed, by example, that the difference in the initial conditions is not preserved in their corresponding steady states. In particular, we considered the initial condition $\varphi_1(x) = \cos \pi x$. Then, with $\epsilon = .1$ fixed, we changed the initial data by adding a constant, i.e., $\varphi_2 = \varphi_1 + 0.25$, and $\varphi_3 = \varphi_1 + 1$. We found that the limit constant $C_{\varphi,\epsilon}$ changes accordingly, i.e., $C_{\varphi_1,\epsilon} \neq C_{\varphi_2,\epsilon}$, $C_{\varphi_1,\epsilon} \neq C_{\varphi_3,\epsilon}$, and $C_{\varphi_2,\epsilon} \neq C_{\varphi_3,\epsilon}$. Notice that $C_{\varphi_2,\epsilon} - C_{\varphi_1,\epsilon} \neq 0.25$, $C_{\varphi_3,\epsilon} - C_{\varphi_1,\epsilon} \neq 1$, and $C_{\varphi_3,\epsilon} - C_{\varphi_2,\epsilon} \neq 0.75$.

In the next two figures we demonstrate *the dependence of the limit constant $C_{\varphi,\epsilon}$ on ϵ.*

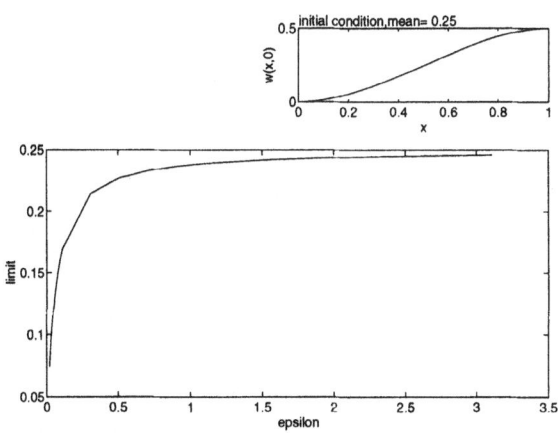

Limit $C_{\varphi,\epsilon}$ versus ϵ with $w(x,0) = x^2(3/2 - x)$

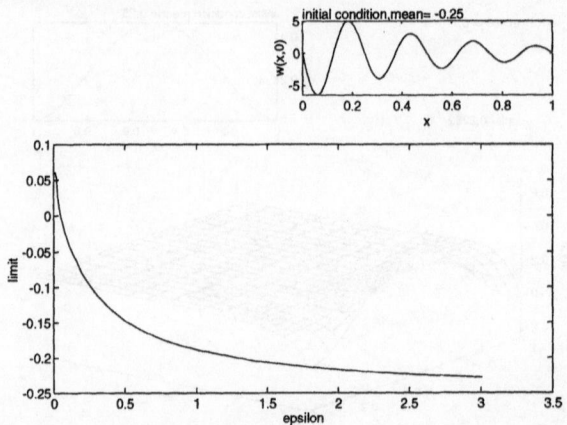

Limit $C_{\varphi,\epsilon}$ versus ϵ with $w(x,0) = -e^{-2x}\sin 8\pi x/0.1367$

In these figures, the graphs on the upper half of each figure are initial conditions. In each figure we have a fixed initial condition and let ϵ vary. For each ϵ we obtain the constant limit and the graphs on the lower half of each figure are those constants plotted versus ϵ. We see that the constant varies as ϵ changes. We also note that *for each initial condition, as ϵ increases the constant converges to the mean value of the initial condition.*

The next three figures support our result in Theorem 5.1 which states that *for any initial condition $\varphi(x)$ satisfying $\varphi(x) = -\varphi(1-x)$, $C_{\varphi,\epsilon} \equiv 0$, independent of ϵ.*

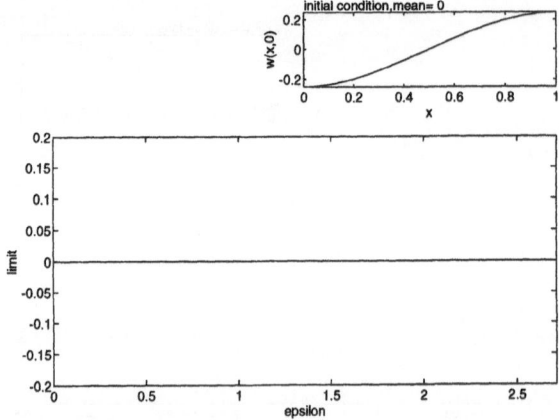

$C_{\varphi,\epsilon} = 0$ independent of ϵ with $w(x,0) = x^2(3/2 - x) - 0.25$

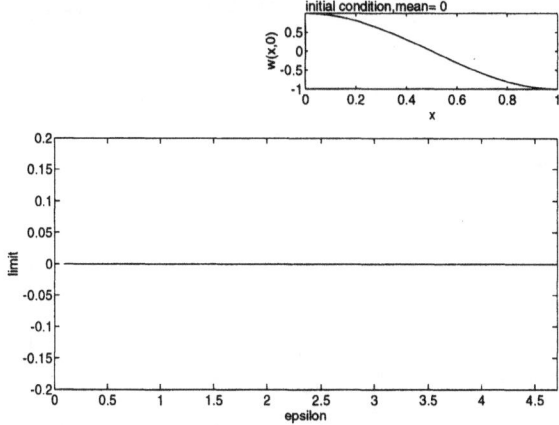

$C_{\varphi,\epsilon} = 0$ independent of ϵ with $w(x,0) = \cos \pi x$

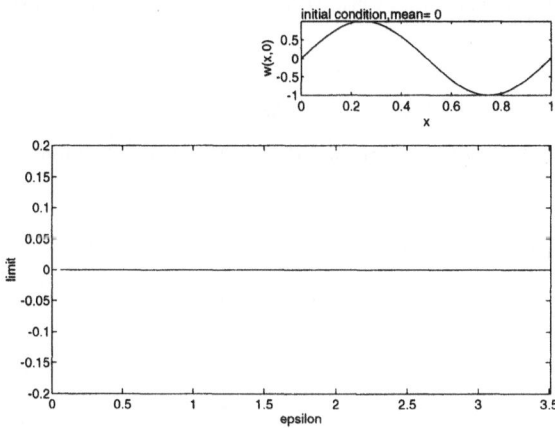

$C_{\varphi,\epsilon} = 0$ independent of ϵ with $w(x,0) = \sin 2\pi x$

The next figure shows that for an initial condition $\varphi(x)$ with mean value equal zero, i.e., $\int_0^1 \varphi(x)\,dx = 0$, if $\varphi(x) \neq -\varphi(1-x)$, the above conclusion is no longer true.

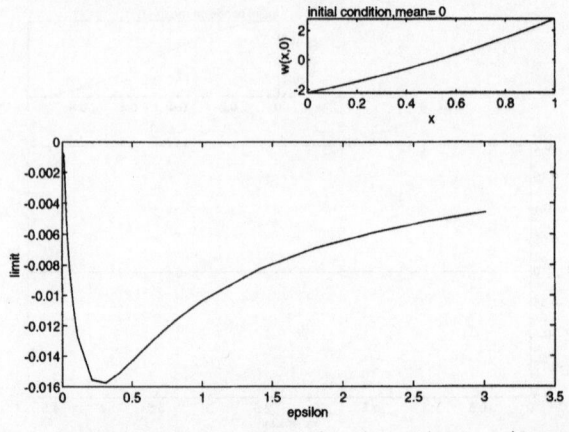

$C_{\varphi,\epsilon}$ is dependent on ϵ for $w(x,0) = 5(2^x - 1/\log 2)$

[**Conjecture 1:**] If the initial condition is monotonically increasing, and

1. its mean value is positive, then

   $C_{\varphi,\epsilon}$ decreases to 0 when ϵ decreases to 0,

   $C_{\varphi,\epsilon}$ increases to the mean of the initial condition when ϵ tends to ∞;

2. its mean value is negative, then

   $C_{\varphi,\epsilon}$ increases to 0 when ϵ decreases to 0,

   $C_{\varphi,\epsilon}$ decreases to the mean of the initial condition when ϵ increases to ∞.

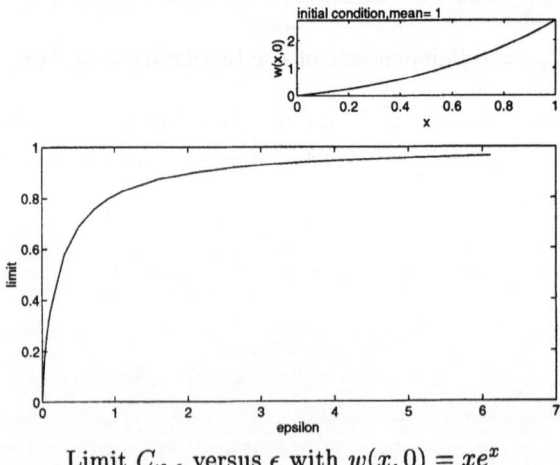

Limit $C_{\varphi,\epsilon}$ versus ϵ with $w(x,0) = xe^x$

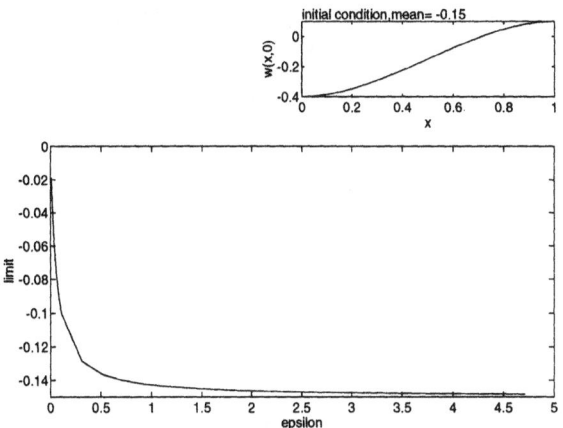

Limit $C_{\varphi,\epsilon}$ versus ϵ with $w(x,0) = x^2(3/2 - x) - 0.4$

[Conjecture 2:] If the initial condition is monotonically decreasing, and

1. its mean value is positive, then

   $C_{\varphi,\epsilon}$ increases when ϵ goes to 0,

   $C_{\varphi,\epsilon}$ decreases to the mean of the initial condition when ϵ increases to ∞;

2. its mean value is negative, then

   $C_{\varphi,\epsilon}$ decreases when ϵ goes to 0,

   $C_{\varphi,\epsilon}$ increases to the mean of the initial condition when ϵ increases to ∞.

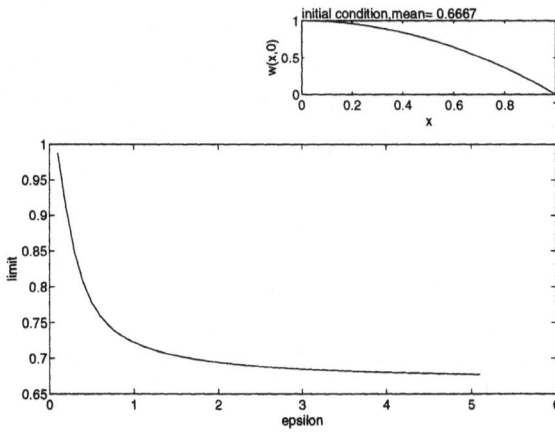

Limit $C_{\varphi,\epsilon}$ versus ϵ with $w(x,0) = 1 - x^2$

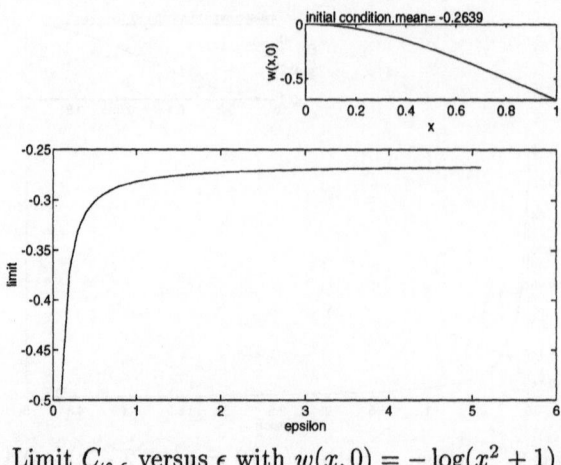

Limit $C_{\varphi,\epsilon}$ versus ϵ with $w(x,0) = -\log(x^2 + 1)$

Our simulations show that for monotonically increasing initial conditions, no matter how small the ϵ is, Burgers' equation with Neumann boundary conditions is always numerically solvable, and $C_{\varphi,\epsilon} \to 0$ as $\epsilon \to 0$. This should not be surprising, considering the well known results on the hyperbolic conservation laws. In this case, the wave convection will not cause a discontinuity. It is also observed that for non-monotonically increasing initial conditions, when ϵ is less than a certain number, the program always blows up after a finite time interval. This, no doubt, occurs due to the formation of a sharp front, which we already know should occur for $\epsilon = 0$. For ϵ very small, it is expected that a near-discontinuity (steepening) will appear because of the wave convection. This near-discontinuity may not necessarily develop into a discontinuity because, as is well known [8], the strength of the diffusive term ϵw_{xx} will prevent the near-discontinuity from developing into a discontinuity. But the near-continuity might be already too steep for numerical calculations of a derivative. This seems to be the reason why the program blows up for small ϵ. So the blow-up does not necessarily mean a global classical solution ceases to exist.

References

[1] N. U. AHMED, *Semigroup Theory with Applications to Systems and Control*, Longman Group Limited, UK, 1991 .

[2] A. BELLENI-MORANTE, *Applied Semigroups and Evolution Equations*, Oxford University Press, Cary, NC, 1979 .

[3] E. R. BENTON and G. W. PLATZMAN, "A table of solutions of the one-dimensional Burgers' equation," *Quart. Appl. Math.*, 30, 195–212

[4] C. BYRNES and D. GILLIAM and V. SHUBOV, "High Gain Limits of Trajectories and Attractors for a Boundary Controlled Viscous Burgers' Equation," *to appear in J. Math. Sys., Estimation and Control* .

[5] J. CARR, *Applications of Centre Manifold Theory*, Springer–Verlag, New York, 1980.

[6] R. CURTAIN and H. ZWART, *Distributed Parameter Systems, An Introduction*, Springer–Verlag, New York, 1994 .

[7] P. DUCHATEAU and D. ZACHMANN, *Applied Partial Differential Equations*, Harper & Row, Publishers, New York, 1989 .

[8] C. A. J. FLETCHER, "Burgers' Equation: A Model for all Reasons." *Numerical Solutions of Partial Differential Equations*, J. Noye (Editor), North–Holland Publ. Co., 1982 .

[9] J.K. HALE, "Asymptotic Behavior of Dissipative Systems," *AMS Mathematical Surveys and Monographs*, 25, 1988.

[10] D. HENRY, *Geometric Theory of Semilinear Parabolic Equations*, Springer–Verlag, New York, Lec. Notes in Math., Vol 840, 1981 .

[11] E. HOPF, "The partial differential equation $u_t + uu_x = \mu u_{xx}$," *Comm. Pure Appl. Math.*, 3 (1950), 201–230 .

[12] P. LAX, *Hyperbolic Systems of Conservation Laws and the Mathematical Theory of Shock Waves*, AMS Lec. Notes, 1972 .

[13] M. MIKLAVČIČ, "Stability for semilinear equations with noninvertible linear operator," *Pacific J. Math.* 118, 199-214, 1985.

[14] A. PAZY, *Semigroups of Linear Operators and Applications to Partial Differential Equations*, Springer–Verlag, New York, 1983 .

[15] E. Y. RODIN, "On Some Approximate and Exact Solutions of Boundary Value Problems for Burgers' Equation," *Journal Math. Anal. Appl.*, 30, 401-414, 1979 .

A Weighted Mixed-Sensitivity H_∞-Control Design for Irrational Transfer Matrices

Ruth .F. Curtain   Yishao Zhou *

Mathematics Institute
University of Groningen
P.O.Box 800
9700 AV Groningen
the Netherlands

1 Introduction

The weighted mixed-sensitivity H_∞-control problem we consider is

$$\gamma^W = \inf_{K \in \mathcal{S}} \left\| \begin{matrix} W_1(I+GK)^{-1}V \\ W_2K(I+GK)^{-1}V \end{matrix} \right\|_\infty, \tag{1.1}$$

where G is the transfer matrix of the plant, K is the transfer matrix of the controller, W_1, W_2 and V are certain rational weighting transfer functions and $\mathcal{S}$ is the class of stabilizing controllers. The weighted mixed-sensitivity H_∞-control problem is well understood for rational transfer matrices (see Kwakernaak [15] for a tutorial) and there are standard routines for its numerical solution in MATLAB [1]. It is the purpose of this paper to develop numerical solutions for a class of *irrational* transfer matrices. We suppose that G is irrational, but it has the decomposition

$$G = G_f + G_s, \tag{1.2}$$

where G_f is rational (possibly unstable) and G_s has components in H_∞. In Kwakernaak [15], it is argued convincingly that (1.1) is the most useful mixed-sensitivity design for applications, and we refer to this paper for the engineering background and for a discussion of the weights W_1, W_2 and V.

   Using a state-space approach the suboptimal version of finding a stabilizing controller

$$\bar{\gamma}^W = \inf_{K \in \mathcal{S}} \left\| \begin{matrix} W_1(I+GK)^{-1}V \\ W_2K(I+GK)^{-1}V \end{matrix} \right\|_\infty < \gamma, \quad \gamma > \gamma^W \tag{1.3}$$

was solved for the Pritchard–Salamon class of infinite-dimensional system in van Keulen and Curtain [24]. It was a special case of the H_∞-theory for infinite-dimensional systems developed in van Keulen [23] and the solution was in terms of two coupled operator H_∞-Riccati equations. While there

*Supported by Swedish Natural Science Research Council under contract NRF F-PD 06646.

are techniques for solving certain operator Riccati equations of this class
(see Itô and Tran [13]), at the time of writing this paper the theory for
H_∞-Riccati equations has not been fully developed. In Itô and Morris
[12] they prove convergence of numerical approximation scheme for a single
H_∞-Riccati equation with bounded input and output operators, but for our
problem we need convergence results for two coupled H_∞-Riccati equations
and unbounded input and output operators. At present these results are
unavailable. Moreover, the approach presented in this paper is simpler;
only finite-dimensional Riccati equations need to be solved.

Many frequency domain approaches for this and similar weighted mixed
sensitivity problems have appeared in the literature, see for example Foias *et
al* [9, 10], Flamm *et al* [8, 7], Enns, Özbay and Tannenbaum [6] and Özbay
and Tannenbaum [20]. They all have in common that you first need to
perform certain special factorizations of the original indefinite-dimensional
problem. In general, this is a difficult numerical step to perform (see
Flamm and Crow [7]), so most examples considered in the literature are
for single input single output systems for which this step is simple, e.g.
$g_\sigma(s) = e^{-sT}/(s + \sigma)$. In particular, the approach of Tannenbaum, Özbay
et al is to find the infinite-dimensional optimal controller which then needs
to be approximated. The results in Özbay [18] on approximating this controller assume very strong conditions on the plant which are not satisfied
by the ubiquitous delay example $g_\sigma(s)$ given above. By contrast, our approach does not involve any factorization step and is applicable, not only to
$g_\sigma(s)$, but to a wide class of unstable multivariable systems. The numerical
computations can be performed with standard MATLAB routines.

In this paper we propose a model reduction approach in frequency domain that is reminiscent of the theme in Curtain [2]. It is based on the
model reduction approach exploited in Curtain and Glover [3] to design
finite-dimensional robustly stabilizing controllers for infinite-dimensional
systems with *a priori* estimates. The basic assumption is that G have a
decomposition (1.2) and that $\|G_s\|_\infty$ can be made small, i.e. G is ap-
proximable by an L_∞ transfer matrix. This subject has been the subject of several papers and is well-understood; for a discussion of some approaches and references we refer to [2]. The class includes transfer matrices
which are the sum of an antistable rational part and a stable part which
is the Laplace transform of an $L_1(0, \infty)$-matrix function. By comparison,
stable Pritchard–Salamon systems (introduced in Pritchard and Salamon
[22, 21] and the topic of [23]) have an impulse response with components
in $L_1(0, \infty) \cap L_2(0, \infty)$. In addition, we assume that W_1, W_2 and V are
rational weighting transfer matrices satisfying either of the following sets
of assumptions:

A1. W_2 and $W_2^{-1} \in \mathcal{M}H_\infty$, W_1 is proper.

A2. W_2 and W_2^{-1} are stable, W_2^{-1} is strictly proper, W_1 and V are proper and $V^{-1} \in \mathcal{M}L_\infty$.

The main results of the paper relate the problems (1.1) and (1.3) to the analogous problems for the rational approximation G_f. In particular,

1. Upper and lower bounds for γ^W are given in terms of γ_f^W, the infimum of (1.1) with G_f replacing G and $\|W_1 G_s W_2^{-1}\|_\infty$.

2. Algorithms for designing rational stabilizing solutions for G with a sensitivity arbitrarily close to the optimum one γ^W are proposed. This is done by solving the suboptimal mixed weighted-sensitivity problem (1.3) for G_f.

We illustrate the results by two design examples for the ubiquitous delay system $g_\sigma(s)$ considered in [6] and [18]. Based on our experience with these and other examples, it appears to be a promising robust control design for infinite-dimensional systems. Indeed, it also offers a useful model reduction approach for finite-dimensional systems of high order.

2 Computable bounds for the weighted mixed-sensitivity H_∞-control problem

As usual, H_∞ denotes the Hardy space of functions which are bounded and holomorphic on the right half plane $\operatorname{Re} s > 0$. L_∞ denotes the functions which are measurable and bounded almost everywhere on the imaginary axis. $\mathcal{M}H_\infty$ and $\mathcal{M}L_\infty$ will be used to denote matrix functions with components in H_∞, respectively, L_∞.

The basic assumption we make is that the system has a well-posed transfer matrix satisfying the decomposition

$$G = G_f + G_s, \tag{2.1}$$

where G_f is a $p \times m$ rational proper transfer matrix (usually unstable) and $G_s \in H_\infty^{p \times m}$ is irrational, strictly proper, but stable.

We recall that a proper transfer function G has the property that $\|G(s)\|$ is bounded on $\Omega_\rho := \{\operatorname{Re} s \geq 0\} \cap \{|s| \geq \rho\}$ for some positive ρ and a strictly proper G has the additional property that $\|G(s)\| \to 0$ as $\rho \to \infty$.

The idea of this section is motivated by Curtain and Glover [3], where it was established that a rational controller K_f stabilizes the antistable rational part G_f if and only if $K = K_f(I - G_s G_f)^{-1}$ stabilizes G. By stability we mean the following.

Definition 2.1 Denote by $\mathcal{F}$ the class of transfer matrices with components in the quotient field $H_\infty \setminus H_\infty$. We say that $K \in \mathcal{F}$ stabilizes $G \in \mathcal{F}$ if

$(I+GK)$ is invertible over $\mathcal{F}$ and $(I+GK)^{-1}$, $K(I+GK)^{-1}$, $K(I+GK)^{-1}G$ and $(I+GK)^{-1}G \in \mathcal{M}\mathcal{H}_\infty$.

Notice that the above definition does not require K to be proper.

In general, one is interested in a weighted version of the mixed sensitivity problem, namely,

$$\gamma^W = \inf_{\text{stab}K} \left\| \begin{array}{c} W_1(I+GK)^{-1}V \\ W_2K(I+GK)^{-1}V \end{array} \right\|_\infty, \tag{2.2}$$

where W_1, W_2 and V are certain rational transfer functions which have been chosen as part of the design problem.

In this section we shall relate this to the analogous problem for the rational transfer matrix G_f

$$\gamma_f^W = \inf_{\text{stab}K_f} \left\| \begin{array}{c} W_1(I+G_fK_f)^{-1}V \\ W_2K_f(I+G_fK_f)^{-1}V \end{array} \right\|_\infty. \tag{2.3}$$

In Kwakernaak [15] it is argued how certain choices of W_1, W_2 and V lead to desirable performance characteristics. In particular, W_2 is often chosen so that W_2^{-1} is strictly proper. In the following lemma we show that this has the effect of forcing the stabilizing controllers to be strictly proper. This lemma utilizes the technique introduced in Krause [14].

Lemma 2.2 *Suppose that G has a decomposition (2.1), where G_f is rational, proper and stabilizable by a rational controller and $G_s \in \mathcal{M}\mathcal{H}_\infty$ is strictly proper. Suppose that W_1, W_2 and V are rational matrices satisfying either*

A1. W_2 and W_2^{-1} are stable and proper, W_1 is proper.

A2. W_2 and W_2^{-1} are stable, W_2^{-1} is strictly proper, W_1 and V are proper and $V^{-1} \in \mathcal{M}\mathcal{L}_\infty$.

Defining $\tilde{G} = GW_2^{-1}$ there holds

$$\inf_{\text{stab } K} \left\| \begin{array}{c} W_1(I+GK)^{-1}V \\ W_2K(I+GK)^{-1}V \end{array} \right\|_\infty = \inf_{\text{stab } \tilde{K}} \left\| \begin{array}{c} W_1(I+\tilde{G}\tilde{K})^{-1}V \\ \tilde{K}(I+\tilde{G}\tilde{K})^{-1}V \end{array} \right\|_\infty,$$

where we may identify $\tilde{K} = W_2K$.

Moreover, the above equality holds whether we consider only proper stabilizing controllers or allow for improper stabilizing controllers. Under assumption A1 K is (strictly) proper if and only if $\tilde{K}$ is (strictly) proper. Under assumption A2 K will always be strictly proper and $\tilde{K}$ will be proper.

Proof. See [5] for details. ∎

Lemma 2.3 *Suppose that G satisfies (2.1), where G_f is rational, proper and stabilizable by a rational controller, and $G_s \in \mathcal{M}H_\infty$ is strictly proper, and W_1, W_2 and V are rational matrices satisfying assumptions A1 or A2. Then the two weighted mixed sensitivity problems (2.2) and (2.3) have solutions, and*

$$(1 + \mu_W + \mu_W^2)^{-1/2}\gamma_f^W \leq \gamma^W \leq (1 + \mu_W + \mu_W^2)^{1/2}\gamma_f^W, \qquad (2.4)$$

where $\mu_W = \|W_1 G_s W_2^{-1}\|_\infty$.

Proof. See [5] for details. ∎

Notice that the crucial term in the bounds (2.4) is $\mu_W = \|W_1 G_s W_2^{-1}\|_\infty$ and so instead of approximating G, it makes sense to approximate $W_1 G W_2^{-1}$ in estimating γ^W. In the case that W_2^{-1} is strictly proper $W_1 G W_2^{-1}$ will be easier to approximate than G.

As before, while the stabilizing K_f in (2.3) is rational, $K = K_f(I - G_s K_f)^{-1}$ is irrational and we prefer rational controllers for G. The following lemma gives bounds for the sensitivity of a class of rational controllers stabilizing G.

Lemma 2.4 *Suppose that W_1, W_2 and V are rational transfer matrices satisfying either assumption A1 or A2. Suppose also that $G \in \mathcal{F}$ has the decomposition $G = G_f + G_2 + G_1$. Then if K_f stabilizes G_f and achieves the sensitivity $\bar{\gamma}_f^W = \left\| \begin{matrix} W_1(I+G_f K_f)^{-1}V \\ W_2 K_f(I+G_f K_f)^{-1}V \end{matrix} \right\|_\infty$ and $\|G_1 W_2^{-1}\|_\infty \bar{\gamma}_f^W < 1$, then $K_0 = K_f(I - G_2 K_f)^{-1}$ is a rational controller stabilizing G with the sensitivity*

$$\left\| \begin{matrix} W_1(I+GK_0)^{-1}V \\ W_2 K_0(I+GK_0)^{-1}V \end{matrix} \right\|_\infty = \left\| \begin{bmatrix} I & W_1 G_2 W_2^{-1} \end{bmatrix} \begin{bmatrix} W_1(I+G_f K_f)^{-1} \\ W_2 K_f(I+G_f K_f)^{-1} \end{bmatrix} FV \right\|_\infty, \qquad (2.5)$$

where $F = \left(I + G_1 K_f(I + G_f K_f)^{-1}\right)^{-1}$. In the case that both V and $V^{-1} \in \mathcal{M}L_\infty$ we obtain the upper bound

$$\left\| \begin{matrix} W_1(I+GK_0)^{-1}V \\ W_2 K_0(I+GK_0)^{-1}V \end{matrix} \right\|_\infty \leq \bar{\gamma}_f^W \frac{(1 + \|W_1 G_2 W_2^{-1}\|_\infty)\|V^{-1}\|_\infty\|V\|_\infty}{1 - \|G_1 W_2^{-1}\|_\infty \bar{\gamma}_f^W \|V^{-1}\|_\infty} \qquad (2.6)$$

Proof. See [5] for details. ∎

We remark that it is possible to calculate (2.5) numerically, once we have solved the problem for G_f. (2.6) is useful in giving insight into the important factors in achieving a performance close to γ_f^W, namely keeping $\|W_1 G_2 W_2^{-1}\|_\infty$ and $\|G_1 W_2^{-1}\|_\infty$ small.

Again, if we are interested in a low order controller, then it is wise to choose $G_2 = 0$ to obtain the following design.

Corollary 2.5 *Suppose that $G \in \mathcal{F}$ has a decomposition of the form (2.1), and that W_1, W_2 and V satisfy the conditions in Lemma 2.3. If K_f stabilizes G_f and achieves the sensitivity $\bar{\gamma}_f^W$ and $\|G_s W_2^{-1}\|_\infty \bar{\gamma}_f^W < 1$, then K_f stabilizes G and achieves the sensitivity*

$$\left\| \begin{matrix} W_1(I+GK_f)^{-1}V \\ W_2 K_f(I+GK_f)^{-1}V \end{matrix} \right\|_\infty = \left\| \begin{matrix} W_1(I+G_fK_f)^{-1}FV \\ W_2 K_f(I+G_fK_f)^{-1}FV \end{matrix} \right\|_\infty .$$

If V and $V^{-1} \in \mathcal{M}L_\infty$, then

$$\left\| \begin{matrix} W_1(I+GK_f)^{-1}V \\ W_2 K_f(I+GK_f)^{-1}V \end{matrix} \right\|_\infty \leq \frac{\bar{\gamma}_f^W \|V^{-1}\|_\infty \|V\|_\infty}{1 - \|G_s W_2^{-1}\|_\infty \bar{\gamma}_f^W \|V^{-1}\|_\infty} ,$$

where $F = (I + G_s K_f(I+G_f K_f)^{-1})^{-1}$.

3   Finite-dimensional controller design

The preceeding results suggest that under the assumptions of Lemma 2.4 we can approximately solve the weighted mixed-sensitivity problem (2.2) with a finite-dimensional controller in 5 steps.

1. Obtain a decomposition of the transfer matrix

$$G = G_f + G_s = G_f + G_2 + G_1$$

   for $\mu_W = \|W_1 G_s W_2^{-1}\|_\infty$ sufficiently small.

2. Design the weights W_1, W_2 and V for the reduced order model G_f following the procedure in Kwakernaak [15].

3. Solve the finite-dimensional minimum sensitivity problem

$$\gamma_f^W = \inf_{\text{stab } K_f} \left\| \begin{matrix} W_1(I+G_f K_f)^{-1}V \\ W_2 K_f(I+G_f K_f)^{-1}V \end{matrix} \right\|_\infty$$

   using the standard routine in MATLAB [1].

4. Obtain a good estimate of γ^W from

$$(1 + \mu_W + \mu_W^2)^{-1/2} \gamma_f^W \leq \gamma^W \leq (1 + \mu_W + \mu_W^2)^{1/2} \gamma_f^W$$

5. Using MATLAB, find a suboptimal, rational controller K_f such that

$$\bar{\gamma}_f^W = \left\| \begin{matrix} W_1(I+G_f K_f)^{-1}V \\ W_2 K_f(I+G_f K_f)^{-1}V \end{matrix} \right\|_\infty < \gamma_f^W + \varepsilon$$

   for an acceptable ε, for example, $\gamma_f^W + \varepsilon = (1 + \mu_W + \mu_W^2)^{1/2} \gamma_f^W$.

6. Calculate the performance of K_f or K_0 for G from (2.5) and decide if this is acceptable.

7. If not, obtain a better approximation and repeat the above steps.

We remark that the solution of the finite-dimensional problem at step 5 may be done using any technique, for example the polynomial method of Kwakernaak [15] or using the tricks proposed by Meisma [17].

This all sounds very plausible, but it remains to be seen how useful that is in practice. To illustrate the approach we give two design examples for the delay system examined in van Keulen and Curtain [24]. This example does have well-posed H_∞-Riccati equations for the weighted mixed-sensitivity problem. However, we do not know of any numerical scheme for solving them. Another example of the heat equation with Dirichlet boundary control and point observation is considered in [5]. This is not in the Pritchard–Salamon class and so the theory in [24] is not applicable. It is also interesting to note that the Riccati equation for this control problem falls outside the known theory in McMillan and Triggiani [16]. Fortunately, it is known that it has a well-posed transfer function (see Curtain and Weiss [4]) and that it is approximable by a rational transfer function.

4 Examples

Example 4.1 First we consider the delay example $g_\sigma(s) = e^{-sT}/(\sigma s - 1)$ which has been considered earlier in the literature, e.g. [18] and [6]. We compare our model reduction approach outlined in Section 4 with the exact infinite-dimensional solution obtained in [18] and [6]. They considered problem (1.1) for $g_\sigma(s)$ with $\sigma = 1$ and the weights $V = I$,

$$W_1 = 2\left(\frac{1 + s/\sqrt{1.01}}{1 + 10s}\right), \quad W_2 = 0.2\left(\frac{1 + s/\sqrt{1.01}}{1 + s}\right).$$

So assumption A1 is satisfied and the theory of Section 3 holds. The following table illustrates the accuracy one can obtain using Padé approximation of e^{-sT} and the approach outlined in Section 4. Although it is possible to obtain slightly better L_∞-approximations using other methods (see Glad et al [11]), Padé approximations of the type $[n, n]$ are easy to obtain and proved satisfactory.

So for small delays, we can obtain a good approximation to the optimal sensitivity using a low order controller. As would be expected, for larger delays we need to use a higher order controller. To obtain a better comparison, for the case $T = 0.37$, we compare the Nyquist plot $1 + G(j\omega)K_7(j\omega)$, with that of infinite-dimensional controller C_{opt} obtained in [6] in Figures 4.1,

Delay	0.001	0.01	0.06	0.15	0.20	0.25	0.37	0.55	0.85	1.0	1.5	2.0
Order of Padé approximation	1	1	2	2	2	3	4	4	5	5	6	8
Order of reduced model G_f	2	2	3	3	3	4	5	5	6	6	7	9
error μ_W	3.95666e-04	0.0040	0.0160	0.0401	0.0534	0.0510	0.0614	0.0912	0.1191	0.1401	0.1812	0.2133
Order of controller	4	4	5	5	5	6	7	7	8	8	8	9
sensitivity achieved	0.5095	0.5163	0.5577	0.6375	0.6852	0.7301	0.8544	1.0810	1.5640	1.8770	3.3299	5.3983
Optimal sensitivity achieved in [5]	0.5092	0.5157	0.5529	0.6244	0.6669	0.7116	0.8283	1.0326	1.4685	1.74	3.02	5.135

where the solid line corresponds to the values of $1 + G(j\omega)C_{opt}(j\omega)$ for positive values of ω and the dashed-dotted line, for negative values of ω; the dashed line corresponds to the values of $1 + G(j\omega)K_7(j\omega)$ for positive values of ω and the dotted line, for negative values of ω, and the arrow shows the direction of increasing ω. In Figure 4.2 the Bode plot of our 7th order controller K_7 (the dashed-dotted lines) with C_{opt} in [6] (the solid lines). The corresponding sensitivities are shown in Figure 4.3. There is very little difference, the curves overlapping for the most part.

Example 4.2 A *loop shaping design.* In this example we consider the same delay system as in Example 4.1, but we follow the design philosophy in Kwakernaak [15] in the choice of the weights W_1, W_2 and V. This involves the steps

(i) Insert an integrator in G_f to achieve an integrating control action, and obtain a coprime factorization of $\frac{G_f}{s} = \frac{N}{D}$.

(ii) Choose $V = \frac{M}{D}$, where M is Hurwitz polynomial whose dominant zeros will be dominant poles desired in the closed loop system.

(iii) Choose $W_1 = 1$

(iv) Choose $W_2 = c(1 + rs)$, where $r \geq 0$ and $c > 0$.

We show below the results obtained for $\sigma = 1$, the delay $T = 0.37$. Now $G_3 = (0.0124s^2 - 0.1914s + 0.9814)/(s - 1)(0.0112s^2 + 0.1819s + 0.9833) =: N_2/D_3$. We choose $V = (0.0112s^4 + 0.2218s^3 + 1.2044s^2 + 1.0429s + 0.0492)/sD_3$, $W_1 = 1$, $W_2 = 0.7500s + 0.2000$. The achieved sensitivity is 1.1547. The performance of the 4th order suboptimal controller K_4 is shown in Figures 4.4-4.7.

Conclusions. In this paper we have shown that for the weighted mixed-sensitivity problem (1.1) a simple model reduction approach is a very effective design tool. It has the advantage that it works under very mild conditions on the weights (milder than in Özbay [19]) and for any transfer matrix which is the sum of an unstable rational part and a stable part which is approximable by rationals in the L_∞-norm. *No a priori factorization step* *needs to be performed*. The advantage over previous methods is that it gives *a priori* upper and lower bounds on the optimal performance and on the performance of suboptimal rational controllers. Moreover, the numerical tools are existing finite-dimensional ones available in MATLAB: you solve the problem for the reduced order model. Comparisons of the performance of optimal, low order controllers with the exact optimal infinite-dimensional controllers in [19] are very encouraging (see Example 4.1). In Example 4.2 it is illustrated how the loop-shaping design method of Kwakernaak [15] performs very well even for very large delays.

Acknowledgements. The first named author would like to take the opportunity to thank Professors John Lund and Ken Bowers and their staff for organizing such a successful and enjoyable conference. Both authors would like to acknowledge extremely useful discussions with Huib Kwakernaak and information from Hitay Özbay concerning unpublished work.

References

[1] R. Y. Chiang and M. G. Safonov. *Robust Control Toolbox, User's Guide*. Math Works, Inc., Aug 1992.

[2] R. F. Curtain. A synthesis of time and frequency domain methods for the control of infinite-dimensional systems: A system theoretic approach. In H. T. Banks, editor, *Control and Estimation in Distributed Parameter Systems, Frontiers in Applied Mathematics*, chapter Chap 5, pages 171–224. SIAM, Philadaphia, 1992.

[3] R. F. Curtain and K. Glover. Robust stabilization of infinite-dimensional systems by finite-dimensional controllers. *Systems and Control Letters*, 7:41–47, 1986.

[4] R. F. Curtain and G. Weiss. Well-posedness of triples of operators (in the sense of linear systems theory). In *Proc. 4th Int. Conf. on Control of Distributed Parameter Systems, Vorau, Italy 1988*, pages 41–59. Birkhäuser-Verlag, Basel, 1989. Int. Series of Numerical Math. Vol. 91.

[5] R. F. Curtain and Y. Zhou. A weighted mixed-sensitivity h_∞-control design for irrational transfer matrices. Technical Report W-9416, Department of Mathematics, University of Groningen, the Netherlands, 1994.

[6] Dale Enns, Hitay Özbay, and Allen Tannenbaum. Abstract model and controller design for an unstable aircraft. *Journal of Guidance, Control and Dynamics*, 15:498–508, 1992.

[7] D. S. Flamm and K. M. Crow. Numerical compensator of H^∞-optimal control for distributed parameter systems. submitted to IEEE Trans. Autom. Control, Feb 1994.

[8] D. S. Flamm and H. Yang. Optimal mixed sensitivity for general distributed plants. In *Proc. IEEE Conf on Decision and Control*, 1990.

[9] C. Foias, A. Tannenbaum, and G. Zames. Weighted sensitivity minimization for delay systems. *IEEE Trans. of Autom. Control*, AC-31:763–766, 1986.

[10] C. Foias, A. Tannenbaum, and G. Zames. On the H^∞-optimal sensitivity problem for systems with delays. *SIAM J. Control and Optimization*, 25:686–705, 1987.

[11] G. Glade, G. Högnäs, P. M. Mäkilä, and H. T. Toivonen. Approximation of delay system - a case study. *Int. J Control*, 53:369–390, 1991.

[12] K. Itô and K. Morris. An approximation theory of solutions to operator Riccati equations for H^∞-control. manuscript.

[13] K. Itô and H. T. Tran. Linear quadratic optimal control problem for linear systems with unbounded input and output operators: numerical approximations. *International Series of Numerical Mathematics*, 91:171–195, 1989.

[14] I. M. Krause. Comments on Grimble's comments on Stein's comments on rolloff of H^∞ optimal controllers. *IEEE Trans. Autom. Control*, AC-37:702–702, 1992.

[15] H. Kwakernaak. Robust control and H_∞-optimization – tutorial paper. *Automatica*, 29:255–274, 1993.

[16] C. McMillan and R. Triggiani. Min-max game theory and algebraic Riccati equations for boundary control problems with analytic semigroups: the stable case. In D. Elworthy, W. N. Everitt, and E. B. Lee, editors, *Differential Equations, Dynamical Systems and Control Science*. Marcel Dekker, 1994. Lecture Notes in Pure and Applied Mathematics.

[17] G. Meinsma. Unstable and nonproper weights in H_∞ control. manuscript.

[18] H. Özbay. Controller reduction in the two-block H_∞-optimal control design for distributed plants. *Int. J. Control*, pages 1291–1308, 1991.

[19] H. Özbay. Tutorial review: H^∞ optimal control design for a class of distributed parameter systems. *Int. J. Control*, 58:739–782, 1993.

[20] H. Özbay and A. Tannenbaum. A skew Toeplitz approach to H^∞ optimal control of multivariable distributed systems. *SIAM J. Control and Optimization*, 28:653–670, 1990.

[21] A. J. Pritchard and D. Salamon. The linear quadratic control problem for retarded systems with delays in control and observation. *IMA J. Control and Information*, 2:335–362, 1985.

[22] A. J. Pritchard and D. Salamon. The linear quadratic control problem for infinite-dimensional systems with unbounded input and output operators. *SIAM J. Contr. and Optimiz.*, 25:121–144, 1987.

[23] Bert van Keulen. *H_∞-control for distributed parameter systems: A state space approach.* Birkhäuser, Bosten, 1993.

[24] Bert van Keulen and Ruth F. Curtain. A state-space approach to the mixed-sensitivity minimization problem for delay systems. In *Proc. Workshop on Control of Partial Differential Equations, Italy*, New York, 1993. Lecture Notes in Pure and Applied Mathematics, Marcel Dekker.

[25] G. Zames. Feedback and optimal sensitivity: Model reference transformations, multiplicative seminorms, and approximate inverses. *IEEE Trans. Autom. Control*, AC-26:301–320, 1981.

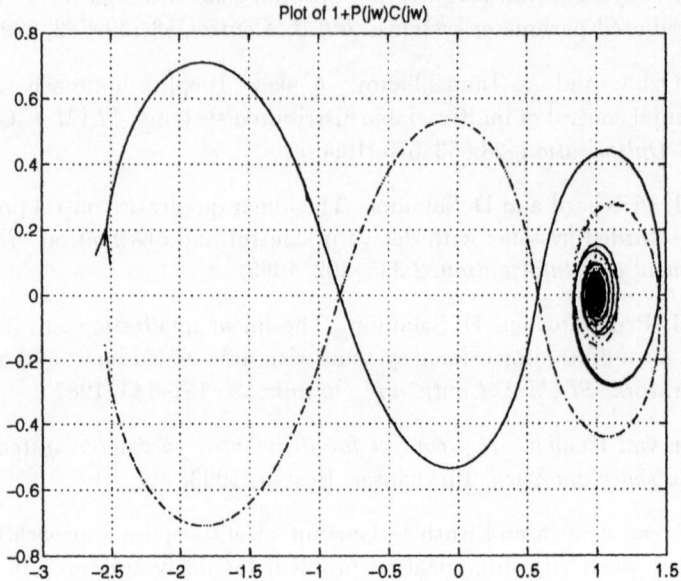

Figure 4.1: Comparison of Nyquist plots $1+G(j\omega)C_{opt}(j\omega)$ and $1+G(j\omega)K_7(j\omega)$

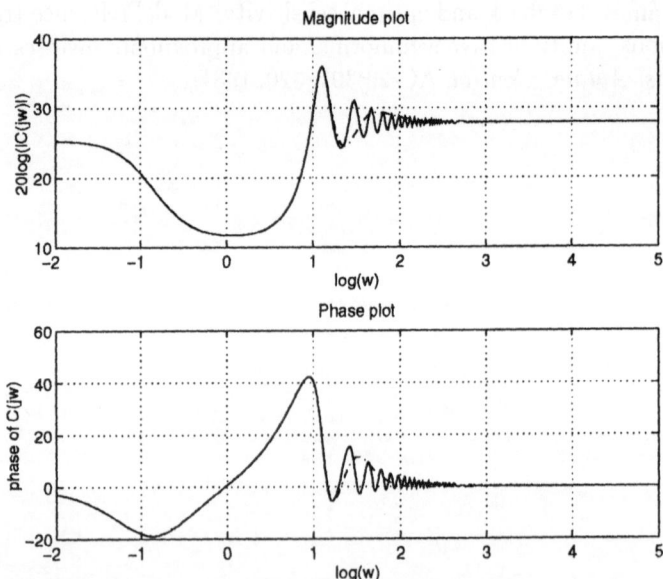

Figure 4.2: Comparison of Bode plots of C_{opt} and K_7

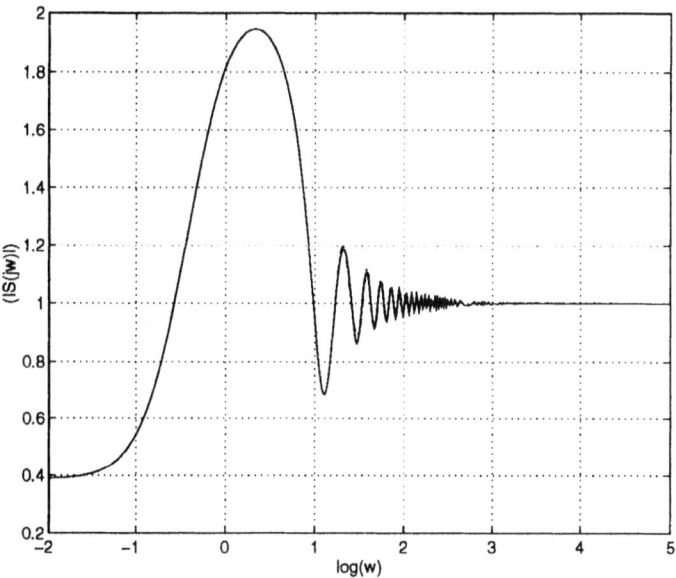

Figure 4.3: Comparison of $|1 + G(j\omega)C_{opt}(j\omega)|^{-1}$ and $|1 + G(j\omega)K_7(j\omega)|^{-1}$

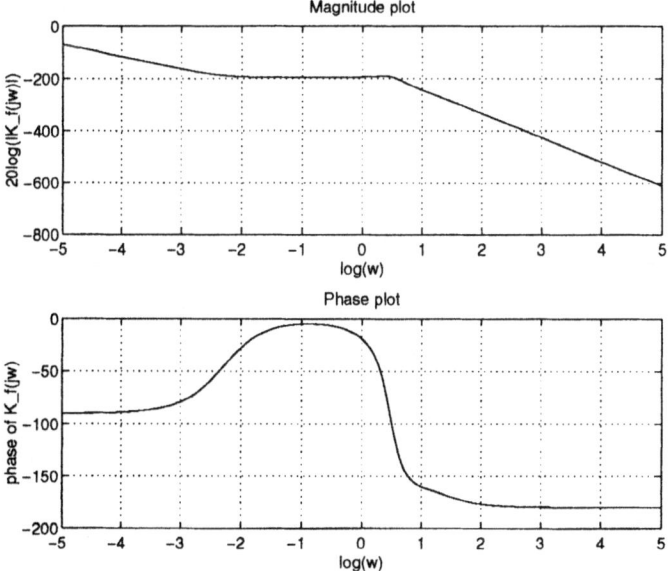

Figure 4.4: Bode plot of Controller K_4

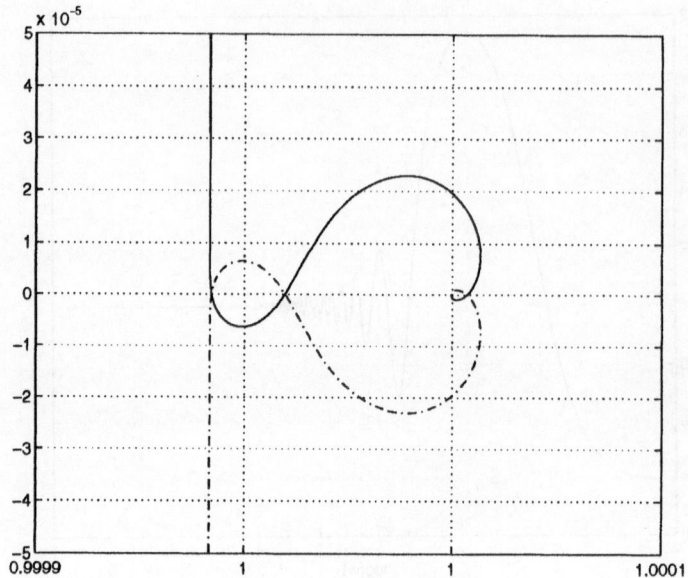

Figure 4.5: Nyquist plot of $1 + G(j\omega)K_4(j\omega)$

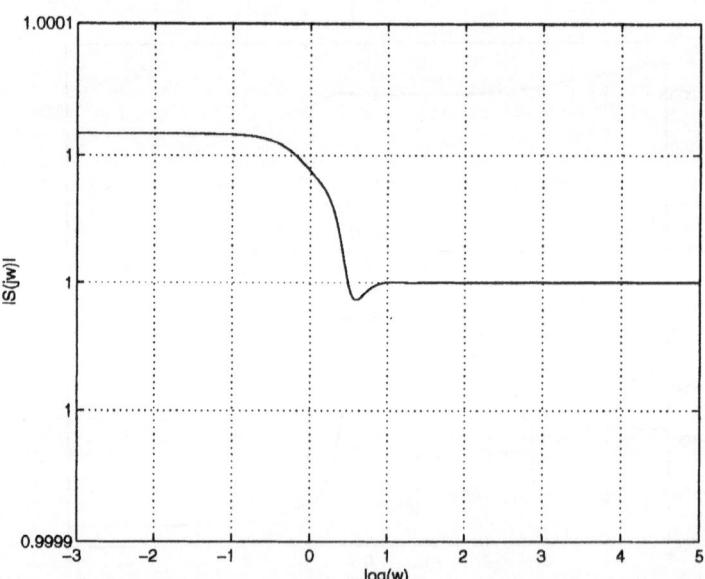

Figure 4.6: Sensitivity $|1 + G(j\omega)K_4(j\omega)|^{-1}$

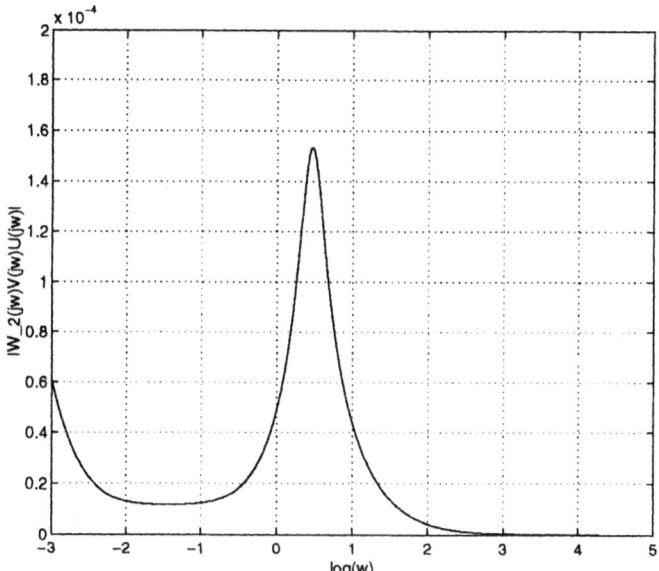

Figure 4.7: Robust stability test: $|W_2(j\omega)K_4(j\omega)(1 + G(j\omega)K_4(j\omega))^{-1}V(j\omega)|$

ACUITY OF OBSERVATION OF THE HEAT EQUATION ON A BOUNDED DOMAIN

Alisa DeStefano
Department of Mathematics
College of the Holy Cross
Worcester, MA 01610

Steven P. Kaliszewski
Department of Mathematics
University of Newcastle
Newcastle, NSW 2308
Australia

D. I. Wallace
Department of Mathematics
Dartmouth College
Hanover, NH 03755

1 Introduction

In [1], Gilliam, Li and Martin showed that the heat equation on a bounded domain in $\mathbb{R}^n$ is discretely observable under certain general conditions. Their sampling method was to sample at p points in the region, where p is the largest multiplicity of any eigenvalue in the corresponding eigenvalue problem, and to sample an infinite number of times. Of course, in practice one samples a finite number of times and then reconstructs an approximate solution to the equation. In this paper we investigate the accuracy of this process and show how the error in the estimate depends on the tail of the Fourier expansion of the initial condition. In some cases we can show that the size of the tail depends, in turn, on the smoothness of the initial condition.

Let Ω be a bounded domain in $\mathbb{R}^n$. Let $u(x,t)$ be a function on $\Omega \times \mathbb{R}_+$. The general heat equation for this situation is given by

$$u_t(x,t) = \Delta u(x,t) - q(x)u(x,t) \tag{1.1}$$

We will consider classical boundary conditions

$$\alpha(x)u(x,t) + \big(1 - \alpha(x)\big)u_\nu(x,t) = 0 \tag{1.2}$$

for $x \in \partial\Omega$, $0 \le \alpha(x) \le 1$, and u_ν the normal derivative to Ω, and initial condition

$$\lim_{t \to 0} u(x,t) = f(x).$$

With these boundary conditions the operator

$$A(x, \Delta) = \Delta - q(x)$$

is self-adjoint. The spectrum of A consists of a sequence of distinct eigenvalues $\{\lambda_j\}$ satisfying

$$0 < \lambda_1 < \lambda_2 < \ldots, \lambda_j \to \infty \quad \text{as } j \to \infty.$$

Let

$$\{\phi_{ij}\}, 1 \leq i \leq m_j$$

be a fixed orthonormal set of solutions to

$$A(x, \Delta)\phi = -\lambda_j \phi,$$

where m_j is the multiplicity of λ_j. We will further assume that

$$|m_j| \leq N \text{ for all } j.$$

With this notation, we can expand the initial condition into a Fourier series

$$f(x) = \sum_{j=1}^{\infty} \sum_{i=1}^{m_j} a_{ij}\phi_{ij}(x) \tag{1.3}$$

and the solution to the heat equation with the given boundary and initial conditions is

$$u(x, t) = \sum_{j=1}^{\infty} \sum_{i=1}^{m_j} a_{ij} e^{-\lambda_j t} \phi_{ij}(x) \tag{1.4}$$

To sample a solution one specifies N points in Ω, $x_1, \ldots, x_N$, and takes measurements

$$y_p(t) = u(x_p, t)$$

for a sequence of times $\{t_k\}_{k=1}^{\infty}$. This system, which includes the equation, the initial and boundary conditions, and the places and times of measurement and the manner of measurement, is called *discretely observable* if

$$y_p(t_k) = 0 \ \forall p = 1, 2, \ldots N, \forall k = 1, 2, \ldots$$

implies that

$$u(x, t) \equiv 0.$$

Let W_j be the matrix given by

$$W_j = \begin{pmatrix} \phi_{1j}(x_1) & \cdots & \phi_{m_j j}(x_1) \\ \vdots & \ddots & \vdots \\ \phi_{1j}(x_N) & \cdots & \phi_{m_j j}(x_N) \end{pmatrix}$$

Note that $m_j \leq N$. The main result of [1] is the following theorem.

Theorem 1.1 (Gilliam, Li, Martin, [1]) *With the equations and notation as above and assuming that*

$$\forall x \in \Omega, \ |\phi_{i,j}(x)| < \beta, \ \forall i, j$$

and

$$\sum_k e^{-\lambda_1 t_k} < \infty$$

then the system is discretely observable if and only if

$$\text{rank } W_j = m_j \ \forall j.$$

2   Approximation Error

For the remainder of this paper we will assume we are in a situation where Theorem 1.1 holds. In theory, once we have the samples of the solution, the solution is uniquely determined, or equivalently, the Fourier coefficients of the initial condition are uniquely determined. In practice, we can sample at only finitely many times $t_1, t_2, \ldots, t_H$. These samples then determine an approximate solution

$$\tilde{u}(x, t) = \sum_{j=1}^{H} \sum_{i=1}^{m_j} \tilde{a}_{ij} e^{-\lambda_j t} \phi_{ij}(x) \tag{2.5}$$

We will study how well this approximates the actual solution, i.e. the acuity of observation. Ultimately, we will measure the error in the Fourier coefficients since the solution $u(x, t)$ is completely determined by these coefficients. We compare the Fourier coefficients in the approximate solution to those of the actual solution. We measure the error by

$$\sum_{j=1}^{\infty} \sum_{i=1}^{m_j} |\tilde{a}_{ij} - a_{ij}|^2 \tag{2.6}$$

which is the ℓ^2 norm of the difference in the actual and approximate Fourier coefficients. We would like to find a bound on the error introduced and find conditions which make the error go to zero as the number of sample times increases. The total error consists of a head error

$$\sum_{j=1}^{H} \sum_{i=1}^{m_j} |\tilde{a}_{ij} - a_{ij}|^2 \tag{2.7}$$

and a tail error

$$\sum_{j=H+1}^{\infty} \sum_{i=1}^{m_j} |0 - a_{ij}|^2 = \sum_{j=H+1}^{\infty} \sum_{i=1}^{m_j} |a_{ij}|^2 \tag{2.8}$$

First, we will focus on the head error and show that it depends linearly on the tail error. Later, we will restrict our boundary conditions and find a bound for the decay of the tail error which depends on the smoothness of the initial condition.

3 Head Error

Before we can compute the head error, we will need to develop some notation. Note that the Fourier coefficients can be related to the samples of the solution by a matrix equation. To this end, suppose we have a solution

$$u(x,t) = \sum_{j=1}^{\infty} r_j(x)e^{-\lambda_j t}.$$

Here $r_j(x)$ is the projection of $u(x,0)$ onto the jth eigenspace of Δ, so

$$r_j(x) = \sum_{i=1}^{m_j} a_{ij}\phi_{ij}(x)$$

in the notation above. The samples are then given by

$$u(x_p, t_k) = \sum_j r_j(x_p)e^{-\lambda_j t_k}$$

for each p and each k. We combine these into the following matrix equation:

$$D = \begin{pmatrix} u(x_1,t_1) & \cdots & u(x_N,t_1) \\ \vdots & \ddots & \vdots \\ u(x_1,t_k) & \cdots & u(x_N,t_k) \\ \vdots & & \vdots \end{pmatrix}$$

$$= \begin{pmatrix} e^{-\lambda_1 t_1} & \cdots & e^{-\lambda_j t_1} & \cdots \\ \vdots & \ddots & \vdots & \\ e^{-\lambda_1 t_k} & \cdots & e^{-\lambda_j t_k} & \cdots \\ \vdots & & \vdots & \end{pmatrix} \begin{pmatrix} r_1(x_1) & \cdots & r_1(x_N) \\ \vdots & \ddots & \vdots \\ r_j(x_1) & & r_j(x_N) \\ \vdots & & \vdots \end{pmatrix} = M \cdot R.$$

Now to determine the head error in the Fourier coefficients, we follow a two step process. First, we divide up the matrices as follows

$$D = \left(\begin{array}{ccc} u(x_1 : t_1) & \cdots & u(x_N : t_1) \\ u(x_1 : t_2) & \cdots & u(x_N : t_2) \\ \vdots & \ddots & \vdots \\ u(x_1 : t_H) & \cdots & u(x_N : t_H) \\ \hline u(x_1 : t_{H+1}) & \cdots & u(x_N : t_{H+1}) \\ \vdots & & \vdots \end{array} \right) = \left(\frac{D_H}{D_T} \right)$$

and

$$M = \left(\begin{array}{ccc|cc} e^{-\lambda_1 t_1} & \cdots & e^{-\lambda_H t_1} & e^{-\lambda_{H+1} t_1} & \cdots \\ e^{-\lambda_1 t_2} & \cdots & e^{-\lambda_H t_2} & e^{-\lambda_{H+1} t_2} & \cdots \\ \vdots & \ddots & \vdots & & \\ e^{-\lambda_1 t_H} & \cdots & e^{-\lambda_H t_H} & e^{-\lambda_{H+1} t_H} & \cdots \\ \hline e^{-\lambda_1 t_{H+1}} & \cdots & e^{-\lambda_H t_{H+1}} & e^{-\lambda_{H+1} t_{H+1}} & \cdots \\ \vdots & & \vdots & \vdots & \end{array} \right) = \left(\begin{array}{c|c} M_H & N_H \\ \hline * & * \end{array} \right)$$

and

$$R = \left(\begin{array}{ccc} r_1(x_1) & \cdots & r_1(x_N) \\ r_2(x_1) & \cdots & r_2(x_N) \\ \vdots & \ddots & \vdots \\ r_H(x_1) & \cdots & r_H(x_N) \\ \hline r_{H+1}(x_1) & \cdots & r_{H+1}(x_N) \\ \vdots & & \vdots \end{array} \right) = \left(\frac{R_H}{R_T} \right)$$

Now using the first H sample times we actually have

$$D_H = M_H R_A \tag{3.9}$$

where R_A is of the same dimensions as R_H and R_A has entries

$$\tilde{r}_j(x) = \sum_{i=1}^{m_j} \tilde{a}_{ij} \phi_{ij}(x).$$

So the error introduced by R_A comes from two places, that is

$$R - \left(\frac{R_A}{0} \right) = \left(\frac{R_H}{R_T} \right) - \left(\frac{R_A}{0} \right) = \left(\frac{R_H - R_A}{0} \right) + \left(\frac{0}{R_T} \right).$$

In other words, there is an error in both the head, $R_H - R_A$, and the tail R_T of the approximation. Now

$$D_H = M_H R_H + N_H R_T$$

and

$$D_H = M_H R_A$$

so we have

$$M_H(R_A - R_H) - N_H R_T = 0$$

and hence

$$R_A - R_H = M_H^{-1} N_H R_T. \qquad (3.10)$$

That is, the error in the head is controlled by the size of the tail.

The second step is to determine the relationship between the head error in the Fourier coefficients and the head error $R_A - R_H$. Recall

$$R_H = \begin{pmatrix} r_1(x_1) & \cdots & r_1(x_N) \\ \vdots & \ddots & \vdots \\ r_H(x_1) & \cdots & r_H(x_N) \end{pmatrix}$$

and let $(R_H)_j = \big(r_j(x_1), \ldots, r_j(x_N)\big)$ and

$$\alpha_j = \begin{pmatrix} a_{1j} \\ \vdots \\ a_{m_j j} \end{pmatrix}.$$

We have that

$$r_j(x_p) = \sum_{i=1}^{m_j} a_{ij} \phi_{ij}(x_p)$$

for $p = 1, 2, \ldots, N$, and so

$$^t(R_H)_j = W_j \cdot \alpha_j$$

Now α_j is uniquely determined if rank $W_j = m_j$. Supposing this to be the case, let

$$\alpha_j = Y_j \cdot {}^t(R_H)_j$$

for Y_j a pseudoinverse of W_j.

Similarly, for the approximate problem we have

$$R_A = \begin{pmatrix} \tilde{r}_1(x_1) & \cdots & \tilde{r}_1(x_N) \\ \vdots & \ddots & \vdots \\ \tilde{r}_H(x_1) & \cdots & \tilde{r}_H(x_N) \end{pmatrix}$$

and we want to recover the $\tilde{a}_{ij}$ for each $\tilde{r}_j$. Let $(R_A)_j = \big(\tilde{r}_j(x_1), \ldots, \tilde{r}_j(x_N)\big)$ and

$$\tilde{\alpha}_j = \begin{pmatrix} \tilde{a}_{1j} \\ \vdots \\ \tilde{a}_{m_j j} \end{pmatrix}$$

We have that

$$\tilde{r}_j(x) = \sum_{i=1}^{m_j} \tilde{a}_{ij} \phi_{ij}(x)$$

and

$$^t(R_A)_j = W_j \cdot \tilde{\alpha}_j$$

so that

$$\tilde{\alpha}_j = Y_j \cdot {}^t(R_A)_j \qquad (3.11)$$

Using the notation just established, we see that

$$\sum_{i=1}^{m_j} |\tilde{a}_{ij} - a_{ij}|^2 = \|(\tilde{\alpha}_j - \alpha_j)\|^2 = \|Y_j {}^t((R_A)_j - (R_H)_j)\|^2 \quad (3.12)$$

$$\leq \|Y_j\|^2 \|^t((R_A)_j - (R_H)_j)\|^2$$

and we arrive at the following.

Lemma 3.1 *The head error in the Fourier coefficients is given by*

$$\sum_{j=1}^{H} \sum_{i=1}^{m_j} |\tilde{a}_{ij} - a_{ij}|^2 \leq \|R_A - R_H\|^2 \sup_{j \leq H} \|Y_j\|^2$$

Proof: Summing both sides of equation 3.12 over j (thinking of an ℓ^2 norm rather than an operator norm) yields

$$\sum_{j=1}^{H} \sum_{i=1}^{m_j} |\tilde{a}_{ij} - a_{ij}|^2 \leq \sum_{j=1}^{H} \|Y_j\|^2 \|^t((R_A)_j - (R_H)_j)\|^2$$

$$\leq \sup_{j \leq H} \|Y_j\|^2 \sum_{j=1}^{H} \sum_{k=1}^{N} |\tilde{a}_j(x_k) - a_j(x_k)|^2$$

$$= \sup_{j \leq H} \|Y_j\|^2 \|R_A - R_H\|^2$$

If we replace $R_A - R_H$ by $M_H^{-1} N_H R_T$ (see equation 3.10), we find that the error in the Fourier coefficients for the first H eigenspaces of $A(x, \Delta)$ in the approximate solution depends linearly on the tail of R. Formally,

Theorem 3.2 *With the hypothesis as in theorem 1.1 and notation as above, let*

$$u(x,t) = \sum_j e^{-\lambda_j t} \sum_{i=1}^{m_j} a_{ij} \phi_{ij}(x)$$

be a solution to the heat equation. If $\tilde{a}_{ij}$ are the Fourier coefficients of the approximate solution, then

$$\left(\sum_{j=1}^{H} \sum_{i=1}^{mj} |\tilde{a}_{ij} - a_{ij}|^2 \right)^{\frac{1}{2}} \leq \|M_H^{-1}\| \|N_H\| \|R_T\| \sup_{j \leq H} \|Y_j\|,$$

(where $\|R_T\|$ is the l^2 norm rather than operator norm), that is, the error in the Fourier coefficients for the first H eigenspaces of Δ in the approximate solution depends linearly on the tail of the Fourier expansion of the initial condition.

If we want an answer in terms of the a_{ij}, $j > H$, we can just use the same calculation to get

Corollary 3.3 *Equivalently,*

$$\left(\sum_{j=1}^{H} \sum_{i=1}^{mj} |\tilde{a}_{ij} - a_{ij}|^2 \right)^{\frac{1}{2}}$$

$$\leq \|M_H^{-1}\| \|N_H\| \sup_{j \leq H} \|Y_j\| \left(\sum_{j=H+1}^{\infty} \sum_{i=1}^{mj} |a_{ij}|^2 \right)^{\frac{1}{2}} \sup_{i > H} \|W_j\|$$

Note that $\|N_H\| \leq He^{-\lambda_H t_1}$ so that it goes to zero as $H \to \infty$. Conditioning problems arise from $\|M_H^{-1}\|$ and $\|Y_j\|$, the former being improved by a suitable choice of sampling times and the latter by a suitable choice of eigenfunctions and test points. If we can find an orthonormal basis of uniformly bounded functions then $\|W_j\|$ will be uniformly bounded. Note that if the bounded domain Ω were a symmetric space, then the asymptotics of

$$\lim_{H \to \infty} \sum_{j=H+1}^{\infty} \sum_{i=1}^{mj} |a_{ij}|^2$$

would be dictated by the Sobolev space to which the initial condition belonged (see Wallace and Wolf [2]). Also in certain cases we can interpret the decay of the tail of our Fourier series as a Sobolev condition.

4   Tail Error

For the remainder of this paper we will assume that the boundary conditions for the heat equation are given by $\alpha = 0$ or 1 in equation 1.2, that is, Dirichlet or von Neumann boundary conditions. Let B denote the differential operator A subject to these boundary conditions. If our initial condition is given by $f(x) = \sum_{i,j} a_{ij} \phi_{ij}(x)$, in terms of the eigenfunctions of the operator B, then Green's theorem tells us that the Fourier coefficients in

$$(Bf)(x) = \sum_{i,j} \phi_{ij}(x) < Bf, \phi_{ij} >$$

are given by

$$
\begin{aligned}
< Bf, \phi_{ij} > &= \int_{\Omega} (Bf)(x)\bar{\phi}_{ij}(x)dx \\
&= \int_{\Omega} f(x)(B\bar{\phi}_{ij})(x)dx + \\
&\quad \int_{\partial\Omega} \left(f(x)\bar{\phi}_{ij_\nu}(x) - f_\nu(x)\bar{\phi}_{ij}(x) \right) dx \\
&= < f, B\phi_{ij} > = -\lambda_j < f, \phi_{ij} >
\end{aligned}
$$

because the boundary conditions require either that $f = \bar{\phi}_{ij} = 0$ on $\partial\Omega$ or that $f_\nu = \bar{\phi}_{ij_\nu} = 0$ there.

To fix notation, let $\Delta^k(\Omega) = \{f : \Delta^k f \in L^2(\Omega)\}$. Then because $q(x)$ in equation 1.1 is assumed to be (Hölder) continuous on the compact domain $\bar{\Omega}$, $f \in \Delta^k(\Omega)$ implies $B^k f \in L^2(\Omega)$. This yields

$$
\begin{aligned}
\|B^k f\|_2^2 &= \sum_{i,j} | < B^k f, \phi_{ij} > |^2 \\
&- \sum_{i,j} \lambda_j^{2k} | < f, \phi_{ij} > |^2 \\
&= \sum_{i,j} \lambda_j^{2k} |a_{ij}|^2 < \infty,
\end{aligned}
$$

so $\lambda_j^{2k}|a_{ij}|^2 \to 0$ as $j \to \infty$, and we have proven

Lemma 4.1 *If $f = \sum_{ij} a_{ij}\phi_{ij}$ is the Fourier expansion in terms of the eigenfunctions of the operator B defined above and if $f \in \Delta^k(\Omega)$, then $|a_{ij}|^2 < \lambda_j^{-2k}$ for j sufficiently large.*

Therefore, we can interpret the rate of decay of the a_{ij} as a Sobolev norm for this situation.

Theorem 4.2 *For large H,*

$$
\begin{aligned}
\sum_{j=H+1}^{\infty} \sum_{i=1}^{m_j} |a_{ij}|^2 &< \sum_{j=H+1}^{\infty} \sum_{i=1}^{m_j} Pj^{-4k/n} \\
&\leq PN \sum_{j=H+1}^{\infty} j^{-4k/n} \\
&\leq \frac{PN}{4^{\frac{k}{n}} - 1} H^{\frac{-4k}{n}+1}
\end{aligned}
$$

Proof: Put $\mu_1 = \lambda_1, \mu_2 = \lambda_1, \ldots \mu_{m_1} = \lambda_1, \mu_{m_1+1} = \lambda_2, \ldots$. Then we have the estimate (see [1])

$$\lim_{j \to \infty} \frac{\mu_j}{j^{\frac{2}{n}}} = C = \text{constant}.$$

Thus $\frac{C}{2} j^{\frac{2}{n}} < \mu_j < 2C j^{\frac{2}{n}}$ for j sufficiently large. Since $m_j \leq N$ for all j, we have that $\mu_j \leq \lambda_j \leq \mu_{jN}$ for all j, so $\frac{C}{2} j^{2/n} < \lambda_j < 2CN^{2/n} j^{2/n}$ for j sufficiently large. Renaming the constants, we have $mj^{2/n} < \lambda_j < Mj^{2/n}$ for large j.

Now by the lemma, $|a_{ij}|^2 < \lambda_j^{-2k}$ for j large enough, so that eventually, $|a_{ij}|^2 < Pj^{-4k/n}$ where $P = 2CN^{\frac{2}{n}-2k}$. Thus, for large H,

$$\sum_{j=H+1}^{\infty} \sum_{i=1}^{m_j} |a_{ij}|^2 < \sum_{j=H+1}^{\infty} \sum_{i=1}^{m_j} Pj^{-4k/n}$$

$$\leq PN \sum_{j=H+1}^{\infty} j^{-4k/n}$$

$$\leq \frac{PN}{4\frac{k}{n} - 1} H^{\frac{-4k}{n}+1}.$$

5   Bounds on the Total Approximation Error

Now we can give a specific bound for the head error in the case of Dirichlet or Neumann boundary conditions.

Theorem 5.1 *Let $\alpha = 0$ or 1 in equation 1.2, and suppose the solution of equation 1.1 being sampled has initial conditions $f(x) = \lim_{t \to 0} u(x,t) = \sum_{i,j} a_{ij}\phi_{ij}(x)$ with $f \in \Delta^k(\Omega)$, then the head error in the Fourier coefficients is given by*

$$\left(\sum_{j=1}^{H} \sum_{i=1}^{m_j} |\tilde{a}_{ij} - a_{ij}|^2 \right)^{\frac{1}{2}} \leq \|M_H^{-1}\| \sup_{j \leq H} \|Y_j\| \sup_{j > H} \|W_j\| \|N_H\| \times$$

$$\left(\frac{PN}{4\frac{k}{n} - 1} H^{\frac{-4k}{n}+1} \right)^{\frac{1}{2}}$$

Proof: Using the previous theorem and Corollary 3.3 we have that for sufficiently large H,

$$
\left(\sum_{j=1}^{H} \sum_{i=1}^{m_j} |\tilde{a}_{ij} - a_{ij}|^2 \right)^{\frac{1}{2}} \leq \|M_H^{-1}\| \|N_H\| \sup_{j \leq H} \|Y_j\| \times
$$

$$
\left(\sum_{j > H} \sum_{i=1}^{m_j} |a_{ij}|^2 \right)^{\frac{1}{2}} \sup_{j > H} \|W_j\|
$$

$$
\leq \|M_H^{-1}\| \sup_{j \leq H} \|Y_j\| \sup_{j > H} \|W_j\| \|N_H\| \times
$$

$$
\left(\frac{PN}{4^{\frac{k}{n}} - 1} H^{\frac{-4k}{n} + 1} \right)^{\frac{1}{2}}
$$

Note: If we choose k large enough, we would expect that the growth in $\|M_H^{-1}\|$ will be dominated by the decay in the exponential term in the last inequality in the proof.

Corollary 5.2 *If the ϕ_{ij} are uniformly bounded by β, then $\|W_j\| \leq N\beta$, giving a statement of the theorem above which does not depend on the W_j.*

Finally, we have a bound for the total approximation error.

Theorem 5.3 *The total approximation error is given by*

$$
\sum_{j=1}^{\infty} \sum_{i=1}^{m_j} |\tilde{a}_{ij} - a_{ij}|^2 \leq \left(\|M_H^{-1}\|^2 \sup_{j \leq H} \|Y_j\|^2 \sup_{j > H} \|W_j\|^2 \|N_H\|^2 + 1 \right) \times
$$

$$
\left(\frac{PN}{4^{\frac{k}{n}} - 1} H^{\frac{-4k}{n} + 1} \right)
$$

Proof: Follows directly from the previous theorem and the fact that

$$
\sum_{j=1}^{\infty} \sum_{i=1}^{m_j} |\tilde{a}_{ij} - a_{ij}|^2 = \sum_{j=1}^{H} \sum_{i=1}^{m_j} |\tilde{a}_{ij} - a_{ij}|^2 + \sum_{j=H+1}^{\infty} \sum_{i=1}^{m_j} |a_{ij}|^2.
$$

A problem for the future is to determine bounds on $\|M_H^{-1}\|$ and $\|N_H\|$. The first is of exponential growth and the second is of exponential decay. If we compute these norms for specific eigenvalues and times, we may be able to prescribe a differentiability condition on the initial condition in order for the total approximation error to go to zero at a certain rate.

6 Conclusion

In the situation where the heat equation on a bounded domain in $\mathbb{R}^n$ is discretely observable, we have given bounds on the error in sampling at only finitely many points. Many of the theorems are similar in flavor to those in Wallace and Wolf [2], but differ in two respects. First, the observation points are mixed in space and time complicating the estimates somewhat. Second, placing the initial condition in a Sobolev space is not as straightforward as on a compact symmetric space. It would be interesting to know what sort of Sobolev lemma would hold for mixed boundary conditions, as that would affect observability.

Acknowledgement

The authors would like to thank Ken Bowers and John Lund for the opportunity to present this work at such a fruitful and enjoyable conference.

References

[1] D. S. GILLIAM, Z. LI and C. F. MARTIN, "Discrete Observability of the Heat Equation on Bounded Domains," *Internat. J. Control*, v. 48, 1988, pp. 755–780.

[2] D. I. WALLACE and J. A. WOLF, "Acuity of Observation for invariant evolution equations," *Proceedings of the Bozeman Conference, Bozeman, MT 1990*, 1991, pp. 325–350.

DESIGNING CONTROLLERS FOR INFINITE-DIMENSIONAL SYSTEMS WITH FINITE-DIMENSIONAL CONTROL SYSTEM DESIGN TOOLS *

M.A. Erickson,
RelMan, Inc.
444 Castro St., Suite 410
Mountain View, CA 94041

A.J. Laub
Dept. of Electrical and Computer Engineering
University of California
Santa Barbara, CA 93106-9560

1 Introduction

Practical implementation of a controller for an infinite-dimensional system generally requires that the controller be finite-dimensional. A design method is presented here that provides a link between infinite-dimensional system theory and finite-dimensional linear robust control system design tools that have been developed and refined over the last decade.

In particular, a design method is described that allows well known finite-dimensional H_∞ and μ-synthesis tools to be applied to single-input single-output hyperbolic systems that can be described in a separable Hilbert space $\mathcal{Z}$ by the abstract differential equation

$$\dot{z} = \tilde{A}z + \tilde{B}u, \quad y = \tilde{C}z, \quad z(0) = z_0,$$

where $\tilde{B} : \mathbb{R} \to \mathcal{Z}$ and $\tilde{C} : \mathcal{Z} \to \mathbb{R}$ are bounded linear operators and $\tilde{A}$ is a densely defined linear operator on $\mathcal{Z}$ with compact resolvent, eigenvalues $\{\tilde{\lambda}_i\}_{i=1}^{\infty}$, and eigenvectors $\{\tilde{\phi}_i\}_{i=1}^{\infty}$ which form a Riesz basis for $\mathcal{Z}$. Such systems are called *bounded spectral systems* in the sequel.

An important property of the design method is that it requires knowledge of only a small number of eigenvalues and eigenvectors of $\tilde{A}$. This is an important consideration that allows the method to be applied to models of physical systems that have complicated shapes, spatially variant mass density or stiffness, etc.

To demonstrate the utility of the method and illustrate its properties, consider a wave equation, which can be used to describe the motion of a stretched string on the interval $[0, 1]$, and written as the PDE

---

*This research was supported in part by the Air Force Office of Scientific Research under Grant No. F49620-94-1-0104DEF.

$$w_{tt}(x,t) = \frac{1}{\rho(x)}w_{xx}(x,t) - \gamma w_t(x,t) + \frac{b(x)}{\rho(x)}u(t),$$

$$y(t) = \int_0^1 c(x)w_x(x,t)dx, \qquad (1.1)$$

where $\rho(x)$ is the strictly positive mass density of the string, γ is a small positive viscous damping constant, and $b(x)$ and $c(x)$ are nonnegative smooth functions that describe sensor and actuator spatial properties. The initial conditions are $w(x,0) = w_0(x)$ and $w_x(x,0) = (w_0)_x(x)$, and the boundary conditions are $w(0,t) = w(1,t) = 0$. As formulated, the output $y(t)$ provides a measurement of the transverse velocity of the string.

To formulate (1.1) as a bounded spectral system, define the Hilbert space $\mathcal{Z} = H_0^1[0,1] \times L_2[0,1]$ with $v = [v_1 \quad v_2]^T \in \mathcal{Z}$ and the norm associated with the inner product $\langle v, w \rangle_{\mathcal{Z}} = \langle v_1, w_1 \rangle_{\tilde{H}_0^1} + \langle v_2, w_2 \rangle_{\rho}$, where

$$\langle v_1, w_1 \rangle_{\tilde{H}_0^1} = \int_0^1 (\bar{v}_1)_x(w_1)_x dx \quad \text{and} \quad \langle v_2, w_2 \rangle_{\rho} = \int_0^1 \bar{v}_2 w_2 \rho dx.$$

Next, define the operator A by

$$Av = \frac{1}{\rho}v_{xx} \text{ for } v \in D(A) = \{v : v_{xx} \in L^2[0,1] \text{ and } v \in H_0^1[0,1]\}.$$

A is densely defined on $L_2[0,1]$ and has compact-normal resolvent, negative simple eigenvalues $\{\lambda_i\}_{i=1}^\infty$, and corresponding eigenvector set $\{\phi_i\}_{i=1}^\infty$. Finally, define the operators $\tilde{A}$, $\tilde{B}$, and $\tilde{C}$ by

$$\tilde{A}v = \begin{bmatrix} 0 & I \\ A & -\gamma I \end{bmatrix} v, \qquad v \in D(\tilde{A}) = D(A) \times H_0^1[0,1],$$

$$\tilde{B}u = \begin{bmatrix} 0 \\ \frac{b(x)}{\rho(x)} \end{bmatrix} u, \qquad \tilde{C}v = \left\langle \frac{c(x)}{\rho(x)}, v_2(x) \right\rangle_{\rho},$$

and the formulation of the wave equation as a bounded spectral system is complete. This wave equation is simple and direct to formulate and is identical in block structure to wave equations with higher dimensional domain as well as to Euler-Bernoulli beam equations [11]. In fact, it is interesting to note that wave equation systems can be more challenging from a controller synthesis standpoint than beam equation systems. This is because the resonant frequencies of the modes of wave equation systems increase with increasing integer k, while beam equation system frequencies increase with k^2. The result is that wave equation system modes are more closely spaced than beam equation modes, making dynamic controller design more difficult.

The design method presented here is an example of the *approximate, then design* approach described by Balas in [2], and makes use of robust stability results described in [5, 6]. Its most attractive feature is that bounds on the error introduced by approximating an infinite-dimensional system are used to synthesize controllers that are guaranteed to meet stability and performance criteria when applied not only to the approximate system, but also to the original infinite-dimensional system.

2 Uncertainty and Robustness

The notion of robustness used here is that a robustly stabilizing controller should stabilize not just a single plant, but all plants within a given set. Similarly, robust performance means being able to meet some performance criterion for all plants in such a set. Robust controller synthesis can be accomplished by considering a nominal plant and a norm-bounded set of perturbations [10]. For the problem considered here, the nominal plant is a finite-dimensional approximation of the original infinite-dimensional plant and the perturbation set contains the infinite-dimensional subsystem that is discarded when forming the finite-dimensional approximation.

To more precisely describe the framework used here, consider a nominal plant represented by the transfer function $P(s) \in H_\infty$, a fixed stable weighting transfer function $W(s) \in RH_\infty$ with stable inverse, and a transfer function Δ that is a member of the set $B\Delta = \{\Delta : \Delta \in H_\infty, \|\Delta\|_\infty \leq 1\}$. Consider the additive perturbation structure shown in Figure 2.1. A standard formulation for a set of plants is then $\{P + \Delta W : \Delta \in B\Delta\}$. A bounded spectral system can be described as an element of this set by lumping a finite number of modes that qualitatively capture the behavior of the system into P. The part of the system not represented in P is then accounted for by selecting a weighting transfer function W with two properties. First, there exists a $\Delta \in B\Delta$ such that $P + \Delta W$ is the transfer function of the original infinite-dimensional system, and second, the infinite-dimensional subsystem represented by Δ is jointly stabilizable/detectable. A transfer function W and infinite-dimensional subsystem Δ that satisfy these respective criteria are called *admissible* with respect to the original infinite-dimensional system in question. Once the original infinite-dimensional system has been put in this form, a number of finite-dimensional LTI system analysis and controller synthesis tools may be applied to solve problems of interest. Representative robust stability and performance techniques are discussed in, e.g., [1, 9, 10, 13]. In particular, the D-K iteration approach to μ-synthesis implemented in the MATLAB package μ-Tools [1] can be used to design controllers that meet robust stability and robust performance criteria described in, e.g., [10, 12].

As an example, consider the robust stability problem for the feedback interconnection shown in Figure 2.2. Assume that a finite-dimensional feedback controller C stabilizes the finite-dimensional plant P. Then the nominal feedback loop and weight may be collapsed into the interconnection of $W(1 - CP)^{-1}C$ and Δ, yielding the feedback interconnection shown in Figure 2.3. The results presented in [6] yield a sufficient condition for the exponential stability of the complete closed-loop infinite-dimensional system. If the nominal feedback loop is exponentially stable and $\|W(1 - CP)^{-1}C\|_\infty < 1$, then the feedback controller will stabilize the plant with any $\Delta \in \mathcal{B}\Delta$ added. Since the admissible Δ is in $\mathcal{B}\Delta$, it can be concluded that the controller will exponentially stabilize the original infinite-dimensional system. This design method can be summarized as follows:

1. Find a finite-dimensional approximation of the original system and a frequency-domain bound on the approximation error.

2. Compute a stable rational transfer function W with stable inverse such that the magnitude of the frequency response of W overbounds the original frequency-domain error bound.

3. Compute C that stabilizes P so that $\|W(1 - CP)^{-1}C\|_\infty < 1$.

This method can be easily modified to accommodate, e.g., the robust performance criteria described in [10]. The following sections present readily computable formulas for frequency-domain error bounds that are not overly conservative. The derivation of these error bounds was originally inspired by [7], where it is shown that, for an exponentially stable bounded spectral system, an upper bound on the error between the infinite-dimensional frequency response $H(j\omega)$ and a truncated modal model frequency response $H_n(j\omega)$ formed by taking the first n terms of the series representation of $H(j\omega)$ is

$$\|H - H_n\|_\infty \leq \sum_{i=n+1}^{\infty} \frac{|C\phi_i|\,|B^*\psi_i|}{\operatorname{Re}\lambda_i}.$$

This bound can be used to calculate a constant (and hence admissible) W to set up the robust stability problem, but it has two disadvantages. First, it requires explicit knowledge of the real parts of all of the eigenvalues of the generator of the bounded spectral system. Second, since it is constant at all frequencies it is not well-suited to calculating bounds for lightly-damped hyperbolic systems. The improved bounds presented here require knowledge of only a finite number of eigenvalues and eigenvectors of the generator of the bounded spectral system along with an upper bound on the real parts of the rest of the eigenvalues. In the hyperbolic case, a frequency-dependent bound is derived in order to exploit the structure of the hyperbolic transfer function and reduce the conservatism of the bound.

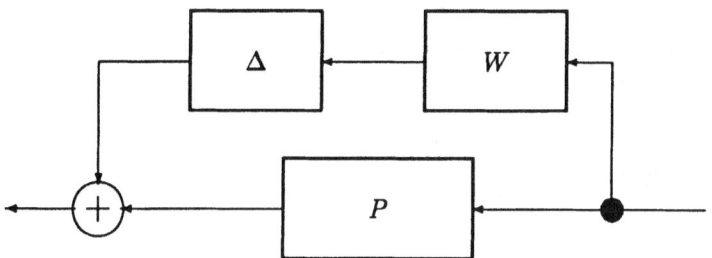

Figure 2.1: Additive uncertainty structure

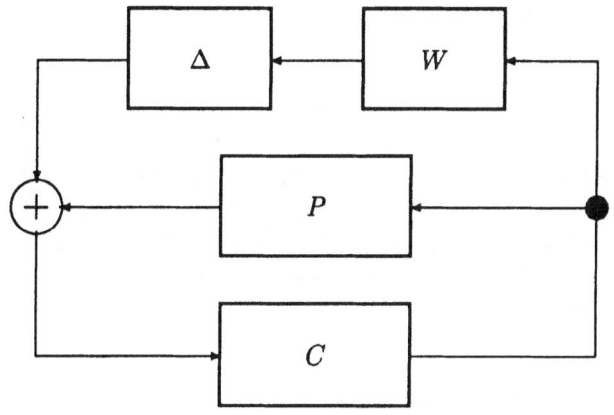

Figure 2.2: Uncertainty structure with feedback controller

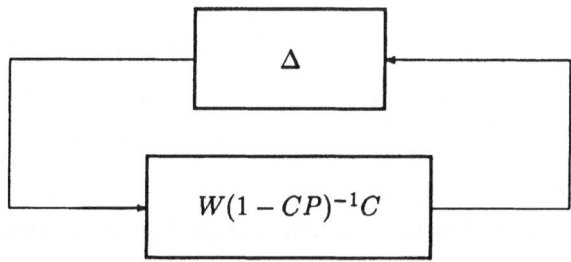

Figure 2.3: Collapsed uncertainty structure

3 Application to Hyperbolic Systems

Recall the wave equation system described in Section 1, and assume that only $n+1$ eigenvalues and n eigenvectors of $\tilde{A}$ are available. Consider its transfer function

$$H(s) = \sum_{i=1}^{\infty} \frac{s}{s^2 + \gamma s - \lambda_i} \left\langle \frac{c(x)}{\rho(x)}, \phi_i(x) \right\rangle_\rho \left\langle \frac{b(x)}{\rho(x)}, \phi_i(x) \right\rangle_\rho .$$

The truncated system with n modes retained is then

$$H_n(s) = \sum_{i=1}^{n} \frac{s}{s^2 + \gamma s - \lambda_i} \left\langle \frac{c(x)}{\rho(x)}, \phi_i(x) \right\rangle_\rho \left\langle \frac{b(x)}{\rho(x)}, \phi_i(x) \right\rangle_\rho ,$$

and the "error system," i.e., the part of the frequency response that is thrown away when H_n is used, is

$$H(s) - H_n(s) = \sum_{i=n+1}^{\infty} \frac{s}{s^2 + \gamma s - \lambda_i} \left\langle \frac{c(x)}{\rho(x)}, \phi_i(x) \right\rangle_\rho \left\langle \frac{b(x)}{\rho(x)}, \phi_i(x) \right\rangle_\rho .$$

Since, for a small γ, the terms that are thrown away each contribute a very large peak in the frequency response of $H(s)$, a constant bound on the error introduced by approximating $H(j\omega)$ with $H_n(j\omega)$ is extremely conservative. A much more useful frequency-dependent bound can be developed by considering more carefully the error incurred by throwing each modal term away. The result in the frequency region $[0, \omega_m]$ is the bound

$$|H(j\omega) - H_n(j\omega)| \le \left| \frac{j\omega}{-\omega^2 + j\gamma\omega - \lambda_{n+1}} \right| \times$$

$$\left(\left\| \frac{b}{\rho} \right\|_\rho^2 - \left\| \left(\frac{b}{\rho} \right)_n \right\|_\rho^2 \right)^{\frac{1}{2}} \left(\left\| \frac{c}{\rho} \right\|_\rho^2 - \left\| \left(\frac{c}{\rho} \right)_n \right\|_\rho^2 \right)^{\frac{1}{2}}, \qquad (3.1)$$

where ω_m is the frequency of the peak of the second-order rational function in (3.1) and $\left(\frac{b}{\rho} \right)_n$ and $\left(\frac{c}{\rho} \right)_n$ are the n-term truncated Fourier approximations of $\frac{b}{\rho}$ and $\frac{c}{\rho}$. This bound can be extended beyond the frequency range $[0, \omega_m]$ by taking its value at $\omega = \omega_m$ and extending that value to all frequencies $\omega \ge \omega_m$.

The bound for the position sensing case is similar to that for systems with velocity sensors. In the position sensing case the truncated system and error system transfer functions are

$$H_n(s) = \sum_{i=1}^{n} \frac{1}{s^2 + \gamma s - \lambda_i} \left\langle \frac{c(x)}{\rho(x)}, \phi_i(x) \right\rangle_\rho \left\langle \frac{b(x)}{\rho(x)}, \phi_i(x) \right\rangle_\rho$$

and

$$H(s) - H_n(s) = \sum_{i=n+1}^{\infty} \frac{1}{s^2 + \gamma s - \lambda_i} \left\langle \frac{c(x)}{\rho(x)}, \phi_i(x) \right\rangle_\rho \left\langle \frac{b(x)}{\rho(x)}, \phi_i(x) \right\rangle_\rho .$$

The final frequency-domain bound in the frequency region $[0, \omega_m]$ is

$$|H(j\omega) - H_n(j\omega)| \leq \left| \frac{1}{-\omega^2 + j\gamma\omega - \lambda_{n+1}} \right| \times$$

$$\left(\left\| \frac{b}{\rho} \right\|_\rho^2 - \left\| \left(\frac{b}{\rho} \right)_n \right\|_\rho^2 \right)^{\frac{1}{2}} \left(\left\| \frac{c}{\rho} \right\|_\rho^2 - \left\| \left(\frac{c}{\rho} \right)_n \right\|_\rho^2 \right)^{\frac{1}{2}} , \qquad (3.2)$$

where ω_m is defined in this case to be the frequency of the peak of the rational transfer function in (3.2). As with the position sensing case, this bound can be extended to higher frequencies by taking its value at ω_m and extending it to all frequencies $\omega \geq \omega_m$.

The bounds described here require knowledge of just n eigenvectors and $n + 1$ eigenvalues. As a result, systems with spatially variant parameters, whose eigenstructure must be computed numerically, are easily accommodated. The ease with which the bounds described here accommodate such systems sets them apart from similar bounds described in, e.g., [3, 4, 5, 7, 8]), which require knowledge of the tail of an infinite series of either eigenvalues or Hankel singular values in order to be computed.

The computation of the modal approximations and frequency-domain error bounds described here have been implemented as part of the PDE-Tools collection of MATLAB functions. This collection of tools is described in detail in [11]. Section 4 describes a complete design example using PDE-Tools and the bounds described here.

4   A Numerical Example

Consider the wave equation described in Section 1 with $\gamma = 0.01$ and the constant mass density $\rho = 1$. The eigenvalues of A are $\lambda_k = -\pi^2 k^2$ and corresponding orthonormal eigenvectors are $\phi_k(x) = \sqrt{2}\sin(k\pi x)$. It is easy to show that the eigenvalues of $\tilde{A}$ have real part -0.005. Let the input and output operators be defined by the concentrated functions

$$b(x) = \begin{cases} 10\, e^{16} e^{((x-0.73)^2 - 0.0625)^{-1}} & |x - 0.73| \leq 0.25 \\ 0 & |x - 0.73| > 0.25 \end{cases}$$

and

$$c(x) = \begin{cases} 10\, e^{16} e^{((x-0.67)^2 - 0.0625)^{-1}} & |x - 0.67| \leq 0.25 \\ 0 & |x - 0.67| > 0.25 \end{cases}$$

shown in Figure 4.1.

The block diagram shown in Figure 4.2 was developed to set up a robust-performance controller design problem as in [1, 10]. The 3-mode (6th-order) plant P and uncertainty bound described in Section 3 were calculated and are shown along with a 40th-order modal model in Figure 4.3. Note that the bound was calculated with only the first 4 eigenvalues and first 3 eigenvectors of $\tilde{A}$. A finite-dimensional LTI uncertainty weighting system W_1 was then developed to overbound the uncertainty bound. A performance weight W_2 was developed to specify disturbance rejection, i.e., better damping, for the first two modes. Since the plant P only contains information about the first three modes, the performance weight was designed to roll off quite steeply to avoid requiring damping of higher-frequency modes not contained in the plant model. The frequency responses of these weights are shown in Figure 4.4. The transfer functions are fairly high order to achieve the necessary rolloff.

A controller was then synthesized using H_∞-synthesis and D-K iteration functions in the μ-Tools toolbox. The resulting 48th-order controller was model-reduced using μ-Tools balanced truncation model reduction functions. The transfer function of the final 20th-order controller is shown along with the 48th-order controller transfer function in Figure 4.5. Note the very steep rolloff that corresponds to the steep sections of the uncertainty and performance weights shown in Figure 4.4. This controller meets the robust stability criterion described in Section 2 in that the transfer function from the "robustness input" to the "robustness output" in Figure 4.2 has H_∞ norm less than 1. Hence, if any LTI system with H_∞ norm less than 1 is connected between the "robustness input" and "robustness output," exponential stability of the resulting system is guaranteed. Since the uncertainty weight was chosen so that the truncated modes can be represented as such a system, exponential stability of the controller connected to the original infinite-dimensional system is thus guaranteed. In fact, the controller also yields *robust performance* in the sense that if any LTI system with H_∞ norm less than 1 is connected between the "robustness input" and "robustness output," the H_∞ norm of the transfer function from the "disturbance input" to the "performance output" is guaranteed to be less than 1 (for more information, see [1, 10]).

To test the final model-reduced controller, it was was connected to a 40th-order modal model. The resulting closed-loop transfer function is shown in Figure 4.6 along with the open-loop transfer function of the 40th-order modal model. As expected, the peaks associated with the first two modes are reduced substantially while the higher-frequency modes are essentially untouched.

The time-response of the closed-loop system consisting of the 40th-order modal model and the 20th-order controller was then computed. As in the output-feedback case, a square pulse with duration 0.01 seconds and height 10 was applied to the input, exciting a traveling wave on the string, and the resulting open-loop and closed-loop responses were computed. Figure 4.7 shows the open-loop and closed-loop positions of the string at time $t = 0.5$ seconds, and Figure 4.8 shows the open-loop and closed-loop positions of the string at time $t = 8.5$ seconds.

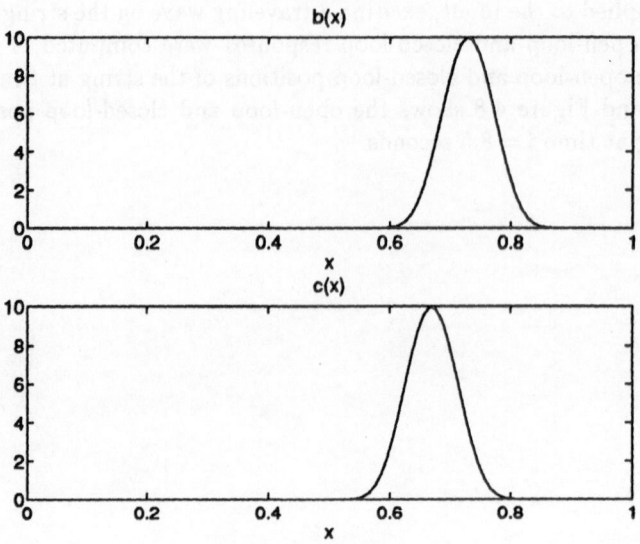

Figure 4.1: Input function $b(x)$ and output function $c(x)$

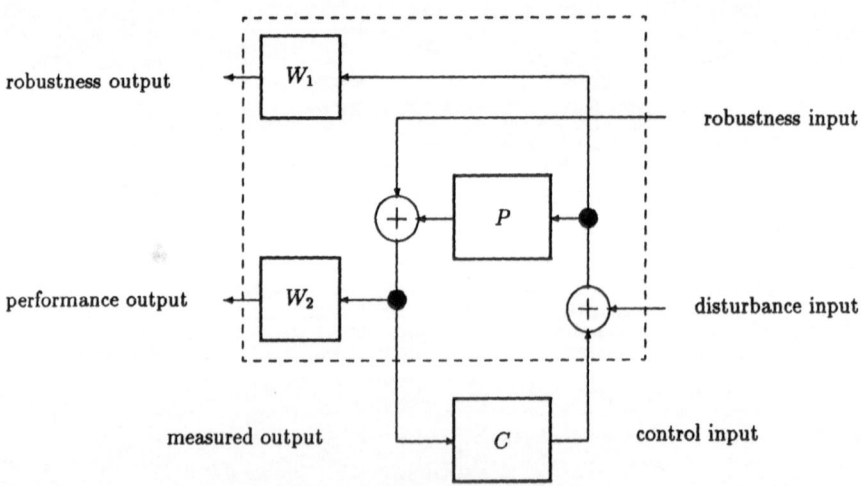

Figure 4.2: Synthesis block diagram

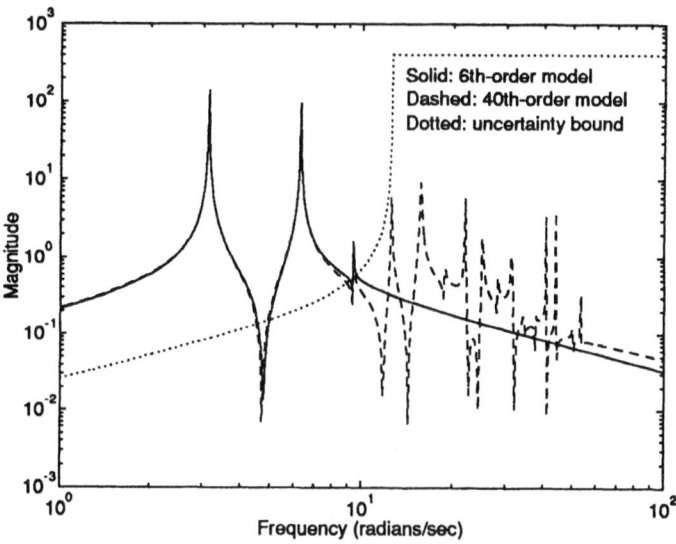

Figure 4.3: Open-loop transfer functions and uncertainty bound

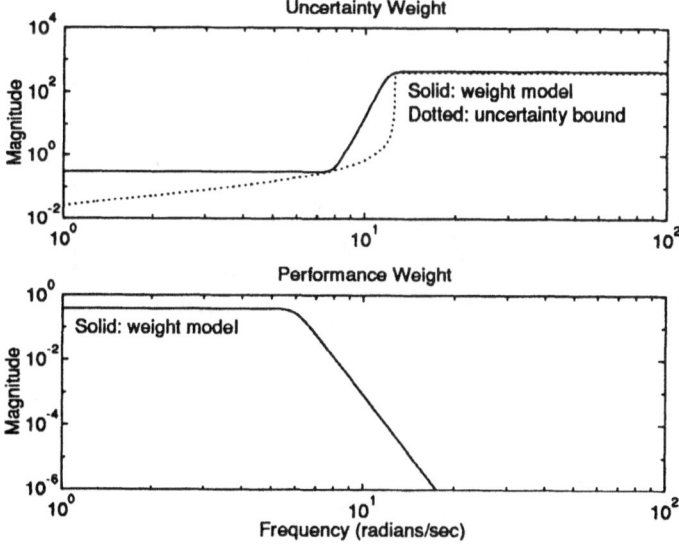

Figure 4.4: Uncertainty and performance weight transfer functions

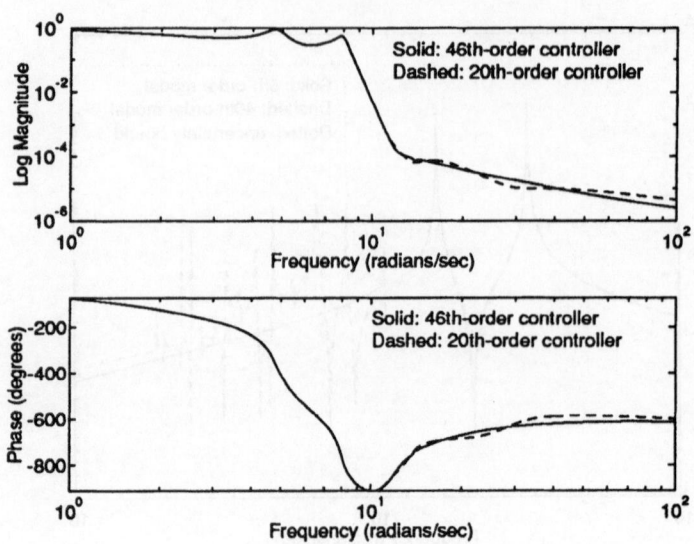

Figure 4.5: Controller and model-reduced controller transfer functions

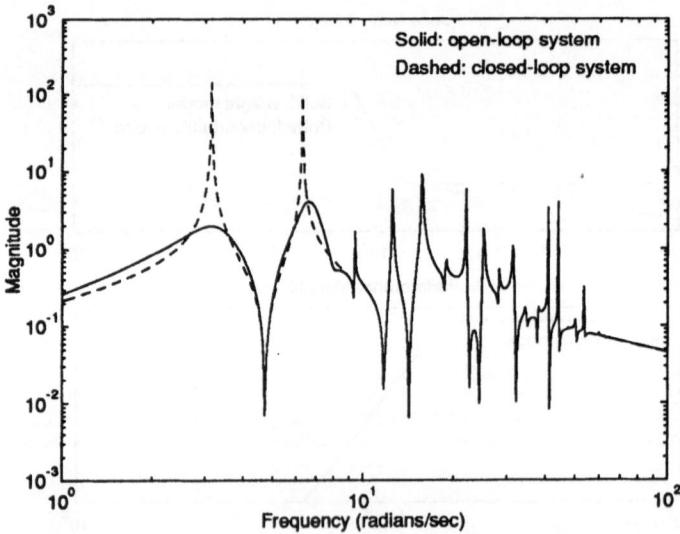

Figure 4.6: Open-loop and closed-loop transfer functions with 40th-order model

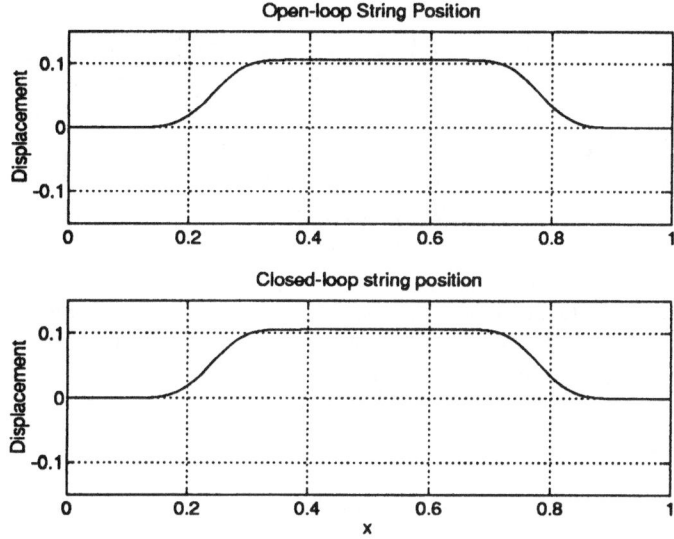

Figure 4.7: Open-loop and dynamic feedback closed-loop position at 0.5 sec

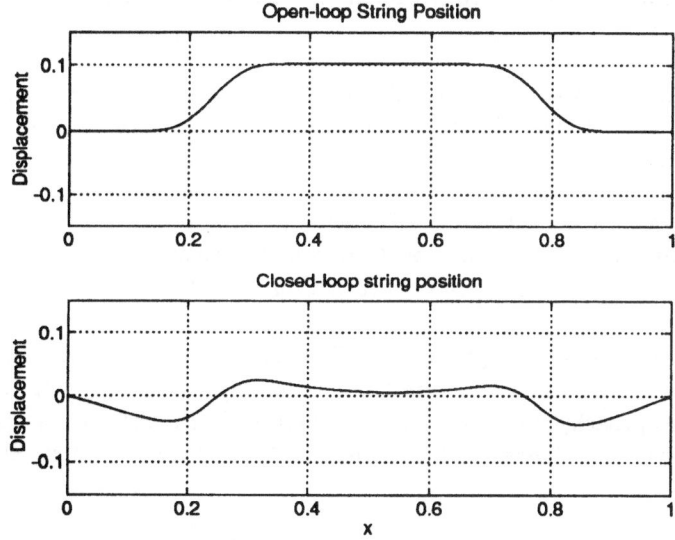

Figure 4.8: Open-loop and dynamic feedback closed-loop position at 8.5 sec

5 Conclusion

The control system design method described here is particularly useful for several reasons. First, it takes advantage of relatively mature finite-dimensional design tools that allow for the the solution of problems with multiple objectives, e.g., robust stability and performance in the presence of uncertainty. Second, relatively few quantities related to the original infinite-dimensional system must be calculated. In the example here, only four eigenvalues and three eigenvectors were needed to generate a controller with guaranteed stability and performance properties. The primary limitation of the results described here and in [11] is that they apply only to bounded spectral systems. Extension of this approach to broader classes of systems is an important area of future research.

References

[1] G. BALAS, J. DOYLE, K. GLOVER, A. PACKARD, and R. SMITH, μ-Analysis and Synthesis Toolbox (μ-Tools), The MathWorks, Inc., 1991.

[2] M.J. BALAS, "Toward A More Practical Control Theory for Distributed Parameter Systems," in Control and Dynamic Systems, C. Leondes (ed.), vol. 18, Academic Press, New York, 1982, pp. 361–421.

[3] Y. CHAIT, "Modeling Error Bounds for Flexible Structures with Application to Robust Control," AIAA J. Guid. Cont. Dyn., v. 14, 1991, pp. 665–667.

[4] Y. CHAIT, C.J. RADCLIFFE, and C.R. MACCLUER, "Frequency Domain Stability Criterion for Vibration Control of the Bernoulli-Euler Beam," ASME J. Dyn. Sys., Meas., Cont., v. 110, 1988, pp. 303–307.

[5] R.F. CURTAIN, "A Synthesis of Time and Frequency Domain Methods for the Control of Infinite-Dimensional Systems: A System Theoretic Approach," in Control and Estimation in Distributed Parameter Systems, H.T. Banks (ed.), Frontiers in Applied Mathematics, SIAM, 1992.

[6] R.F. CURTAIN, "Robust Controllers For Infinite-Dimensional Systems," in Analysis and Optimization of Systems: State and Frequency Domain Approaches for Infinite-Dimensional Systems, R.F. Curtain

(ed.), Springer-Verlag, 1993, Proceedings of the 10th International Conference, Sophia-Antipolis, France, June 9-12, 1992.

[7] R.F. CURTAIN and K. GLOVER, "Controller Design for Distributed Systems Based on Hankel-Norm Approximations," *IEEE Trans. Aut. Control*, v. AC-31, 1986, pp. 173–176.

[8] R.F. CURTAIN and K. GLOVER, "Robust Stabilization of Infinite-Dimensional Systems by Finite-Dimensional Controllers," *Sys. Control Lett.*, v. 7, 1986, pp. 41–47.

[9] C.A. DESOER and M. VIDYASAGAR, *Feedback Systems: Input-Output Properties*, Academic Press, New York, 1975.

[10] J.C. DOYLE, B.A. FRANCIS, and A.R. TANNENBAUM, *Feedback Control Theory*, Macmillan, New York, 1992.

[11] M.A. ERICKSON, *Computational Methods for Infinite-Dimensional Control Systems*, PhD thesis, University of California, Santa Barbara, June 1994.

[12] A.K. PACKARD, *What's New with μ: Structured Uncertainty in Multivariable Control*, PhD thesis, University of California, Berkeley, May 1988.

[13] M. VIDYASAGAR, *Control System Synthesis: A Factorization Approach*, MIT Press, Cambridge, MA, 1985.

PRESERVING EXPONENTIAL STABILITY UNDER APPROXIMATION FOR DISTRIBUTED PARAMETER SYSTEMS

Richard H. Fabiano
Department of Mathematics
Texas A&M University
College Station, TX 77843

1   Introduction

Consider a linear Cauchy problem

$$\begin{aligned}
\dot{z}(t) &= Az(t) + f(t) \\
z(0) &= z_0
\end{aligned} \tag{1.1}$$

on a Hilbert space Z, and assume that A is the infinitesimal generator of a C_0-semigroup $T(t)$ on Z. Assume also that $D(A) \subset V$, where $V \subset Z \subset V^*$, and the embeddings are dense and continuous. Consider also a generic approximation scheme for (1.1) that consists of finite dimensional subspaces $V^N \subset V$, associated orthogonal projections $P^N : Z \to V^N$, and operators $A^N : V^N \to V^N$. This defines a semidiscrete approximation of (1.1) given by

$$\begin{aligned}
\dot{z}^N(t) &= A^N z^N(t) + P^N f(t) \\
z^N(0) &= P^N z_0
\end{aligned} \tag{1.2}$$

Suppose that $T(t)$ is exponentially stable. That is, suppose that there exists $M \geq 1$ and $\omega > 0$ such that $\| T(t) \| \leq M e^{-\omega t}$ for all $t \geq 0$. It is natural to ask if the sequence of finite dimensional semigroups $T^N(t) = e^{tA^N}$ is uniformly exponentially stable. That is, does there exist $M_1 \geq 1$ and $\omega_1 > 0$ such that $\| T^N(t) \| \leq M_1 e^{-\omega_1 t}$ for all $t \geq 0$ for all sufficiently large N? A related question is the following - given finite dimensional spaces V^N, can one construct $A^N : V^N \to V^N$ so that $T^N(t) \to T(t)$ and $T^N(t)$ is uniformly exponentially stable? These issues are important for approximation of optimal control problems. In particular, an important sufficient condition for convergence of approximating feedback gains in the linear quadratic regulator problem is that the approximation schemes preserve the stability behavior of the original system (see [3], [7], [8], [11]).

In order to see how problems may arise and to motivate the use of an equivalent inner product, assume that there is a bilinear form $\sigma : V \times V \to \mathbb{C}$ such that

$$\langle Ax, y \rangle = \sigma(x, y) \qquad \forall x \in \text{dom}A, \ y \in V. \tag{1.3}$$

In this case, given finite dimensional spaces $V^N \subset V$, it is natural to define Galerkin approximations $A^N : V^N \to V^N$ by

$$\langle A^N x, y \rangle = \sigma(x, y) \qquad \forall x, y \in V^N. \tag{1.4}$$

Recall that the numerical range of an operator A (when (1.3) holds this is also the numerical range of the bilinear form σ) is the subset of the complex plane defined by $\Theta(A) = \{\langle Ax, x \rangle : x \in \text{dom} A, \| x \| = 1\}$. It is interesting to observe that for operators A^N defined via Galerkin projections as in (1.4), the point spectrum of A^N is contained in the numerical range of A. If it happens that $\Theta(A)$ is the entire left half plane, then with no further information (other than (1.4)), the most we can say about the eigenvalues of A^N is that they lie in the left half complex plane. Hence it is possible for the eigenvalues of A^N to be near to the imaginary axis. A worst case scenario occurs when $T(t)$ is exponentially stable, $\Theta(A)$ is the entire left half plane, and the eigenvalues of A^N approach the imaginary axis as $N \to \infty$. In this case $T^N(t)$ is not uniformly exponentially stable.

However, the numerical range is norm dependent, and hence inner product dependent. Thus we are motivated to seek an inner product which is equivalent to the original inner product (two inner products are equivalent if their compatible norms are equivalent) and which, for the case where $T(t)$ is exponentially stable, 'pushes' the numerical range into a half-plane of the form $\text{Re } \lambda \leq -\omega$ for some $\omega > 0$. We can then use this new inner product in (1.4) to define new Galerkin approximations with point spectra contained in this half-plane, and the corresponding semigroups will be uniformly exponentially stable. An important advantage of this approach compared to other methods (such as the Liapunov method, or semigroup results in which exponential stability follows from bounding the resolvent operator uniformly on the imaginary axis) is that a decay rate is obtained, both for the original semigroup and the approximating semigroups. Furthermore, uniform exponential stability follows *by construction*.

In Section 2 we consider an example of a wave equation with boundary damping in which the above mentioned worst case scenario occurs for Galerkin approximations using the standard energy inner product. We show how to construct a new inner product so as to obtain a uniformly stable approximation scheme. In Section 3, we show how to use an equivalent inner product in a different but related way. We construct an equivalent inner product for a scalar delay equation, but instead of using it to construct Galerkin approximations, we use the inner product to produce a simple proof that a known approximation scheme (the AVE scheme [1]) is uniformly exponentially stable. Previous proofs (see [10], [13]) of this result rely on delicate resolvent estimates.

2   A Weakly Damped Wave Equation

As a prototypical model for a weakly damped elastic system, we consider the following model of a 1-dimensional wave equation with boundary damping:

$$y_{tt}(t, \xi) = y_{\xi\xi}(t, \xi) \qquad 0 < \xi < 1 \qquad (2.1)$$

$$y(t, 0) = 0 \qquad y_\xi(t, 1) = -\alpha y_t(t, 1) \qquad \alpha > 0$$

$$y(0, \xi) = y_0(\xi) \qquad y_t(0, \xi) = v_0(\xi)$$

Here $y(t, \xi)$ represents displacement at time t and position ξ along an elastic string of length 1. The only energy dissipation occurs through the boundary via a velocity feedback term. By choosing a state $z(t) = (y(t, \cdot), y_t(t, \cdot))$, we can write (2.1) as the Cauchy problem

$$\dot{z}(t) = Az(t) \qquad (2.2)$$
$$z(0) = (y_0, v_0)$$

on the Hilbert space $Z = H_L^1(0, 1) \times L^2(0, 1)$, where $H_L^1(0, 1) = \{u \in H^1(0, 1) : u(0) = 0\}$. The energy norm on Z is given by

$$\| (u, v) \|^2 = \int_0^1 (|u'(\xi)|^2 + |v(\xi)|^2)d\xi, \qquad (2.3)$$

with the compatible energy inner product $\langle \cdot, \cdot \rangle$. The operator A in (2.2) is defined on the domain dom $A = \{(u, v) \in Z : u \in H^2(0, 1), v \in H_L^1(0, 1), u'(1) = -\alpha v(1)\}$ by $A(u, v) = (v, u'')$. If we set $V = H_L^1(0, 1) \times H_L^1(0, 1)$, then we can define a sesquilinear form $\sigma : V \times V \to \mathbb{C}$ by

$$\sigma((u, v), (f, g)) = \int_0^1 [v'(\xi)\overline{f'(\xi)} - u'(\xi)\overline{g'(\xi)}]d\xi - \alpha v(1)\overline{g(1)}. \qquad (2.4)$$

The operator A is related to σ by (1.3), and Galerkin approximations can be defined as in (1.4). It is known (see [4], [9]) that A generates an exponentially stable semigroup $T(t)$. However, we have the following result concerning the numerical range of A.

Theorem 2.1 *The numerical range of A contains the entire closed left half complex plane.*

Proof: Observe that for $z = (u, v) \in$ dom$A$, we have

$$\langle A(u, v), (u, v) \rangle = \int_0^1 (v'\overline{u}' + u''\overline{v})d\xi = 2i \operatorname{Im} \int_0^1 v'\overline{u}'d\xi - \alpha|v(1)|^2 \quad (2.5)$$

Given any integer M, define $v_M(\xi) = \frac{1}{2}(e^{2M\pi i\xi} - 1)$ and $u_M(\xi) = \int_0^\xi v(s)\,ds$. It is straightforward to check that $(u_M, v_M) \in$ dom$A$ and $\| (u_M, v_M) \| = 1$.

N	λ_{max}^N
8	-0.30123 + 26.246i
16	-0.08357 + 54.642i
32	-0.02148 + 110.452i
64	-0.00540 + 221.502i

Table 2.1: e^{tA^N} not uniformly stable

Further, from (2.5) we have $\langle A(u_M, v_M), (u_M, v_M)\rangle = M\pi i$. In a similar manner, for all positive integers $K > 1/\alpha^2$ define

$$v_K(x) = \begin{cases} 0 & \text{if } 0 \leq x \leq \frac{K-1}{K} \\ -\sqrt{K^2 x + (K - K^2)} & \text{if } \frac{K-1}{K} \leq x \leq 1 \end{cases}$$

$$w_K(x) = \begin{cases} 0 & \text{if } 0 \leq x \leq \frac{\alpha^2 K - 1}{\alpha^2 K} \\ \sqrt{\alpha^4 K^2 x + (\alpha^2 K - \alpha^4 K^2)} & \text{if } \frac{\alpha^2 K - 1}{\alpha^2 K} \leq x \leq 1 \end{cases}$$

If we set $u_K(x) = \int_0^x w_K(s)\, ds$, then $(u_K, v_K) \in \text{dom}A$, $\| (u_K, v_K) \| = 1$ and $\langle A(u_K, v_K), (u_K, v_K)\rangle = -\alpha K$. Since the numerical range is a convex set in the complex plane (see [12]), the result follows. □

Thus, as discussed in Section 1, there is a possibility that Galerkin approximations defined using the energy inner product will not be uniformly exponentially stable. In ([2]) Banks et al give a thorough investigation of several approximation schemes for this and related systems. They show that certain standard Galerkin finite element schemes are in fact not uniformly exponentially stable. We illustrate this in Table 2.1 by listing the eigenvalues (for the case $\alpha = 1$) of A^N with maximum real part for $N = 8, 16, 32, 64$. That is, $\text{Re}\lambda_{max}^N = \max_{\lambda \in \sigma(A^N)} \{\text{Re}\lambda\}$. The operator A^N is a Galerkin approximation defined as in (1.4), and $V^N = H_1^N \times H_1^N \subset V$, where H_1^N is the standard space of linear splines ('hat' functions). It is clear that this Galerkin approximation is not uniformly exponentially stable. In [2] Banks et al then make use of a discrete Liapunov function to construct a mixed finite element approximation scheme which is uniformly exponentially stable. The drawback of this approach is the unwieldiness of the discrete Liapunov function.

To complement the results in ([2]), we try the approach discussed in Section 1 with the following inner product:

$$\langle (u, v), (f, g)\rangle_e = \int_0^1 [u'\overline{f'} + v\overline{g}]d\xi + \int_0^1 a(\xi)[v(\xi)\overline{f'(\xi)} + u'(\xi)\overline{g(\xi)}]d\xi \quad (2.6)$$

It can be shown that $\langle \cdot, \cdot \rangle_e$ is an equivalent inner product for any $0 \leq a(\xi) \leq a_1 < 1$. A particularly nice choice for a is $a(\xi) = e^{\gamma\xi} - 1$ where

N	$\lambda^N_{e\,max}$
8	-0.67901 + 31.062i
16	-0.34795 + 68.934i
32	-0.34657 + 147.596i
64	-0.34657 + 308.361i

Table 2.2: $e^{tA^N_e}$ is uniformly stable

$\gamma = \ln\frac{(1+\alpha)^2}{1+\alpha^2}$. It then follows that $\operatorname{Re}\langle Az, z\rangle_e \leq -\frac{\gamma}{2}\parallel z \parallel^2_e$, so we get a decay rate $\omega = -\frac{\gamma}{2}$ for the semigroup $T(t)$ generated by A. In order to define uniformly stable Galerkin approximations, we need a bilinear form σ_e satisfying $\langle Ax, y\rangle_e = \sigma_e(x, y)$ for all $x \in \operatorname{dom}A$ and $y \in V$. This is the case for the bilinear form defined by

$$\sigma_e((u,v),(f,g)) = \int_0^1 [v'(\xi)\overline{f'(\xi)} - u'(\xi)\overline{g'(\xi)}]d\xi - \alpha v(1)\overline{g(1)}$$
$$+ \int_0^1 a(\xi)[u''(\xi)\overline{f'(\xi)} + v'(\xi)\overline{g'(\xi)}]d\xi$$
$$-a(1)[u'(1) + \alpha v(1)]\overline{f'(1)} \qquad (2.7)$$

Given an appropriate finite dimensional space $V^N \subset V$, we can define new Galerkin approximations $A^N_e : V^N \to V^N$ by

$$\langle A^N_e x, y\rangle_e = \sigma_e(x, y) \qquad \forall x = (u,v), y = (f,g) \in V^N. \qquad (2.8)$$

For any scheme defined by (2.8), the finite dimensional semigroups $T^N(t) = e^{tA^N_e}$ will be uniformly exponentially stable since the eigenvalues of A^N_e must have real part $\leq -\frac{\gamma}{2}$. Notice that we cannot use $V^N = H^N_1 \times H^N_1$ because of the u'' term in σ_e. Instead take $V^N = H^N_2 \times H^N_1$, where H^N_2 is a space of quadratic splines. Since the bilinear form σ_e requires one more derivative on u than v, it lends support to the speculation in [2] that mixed methods are somehow required in order to obtain uniform stability of approximations for weakly damped systems. In Table 2 we list the eigenvalues of A^N_e with maximum real part for $N = 8, 16, 32, 64$. That is, $\operatorname{Re}\lambda^N_{e\,max} = \max_{\lambda \in \sigma(A^N_e)} \{\operatorname{Re}\lambda\}$. The eigenvalues in Table 2 are for the case $\alpha = 1$, for which $\gamma = \ln\frac{(1+\alpha)^2}{1+\alpha^2} = \ln 2$ and $-\frac{\gamma}{2} = -.34657$. Thus the expected uniform stability behavior is observed. These eigenvalues are plotted in Figure 2.1.

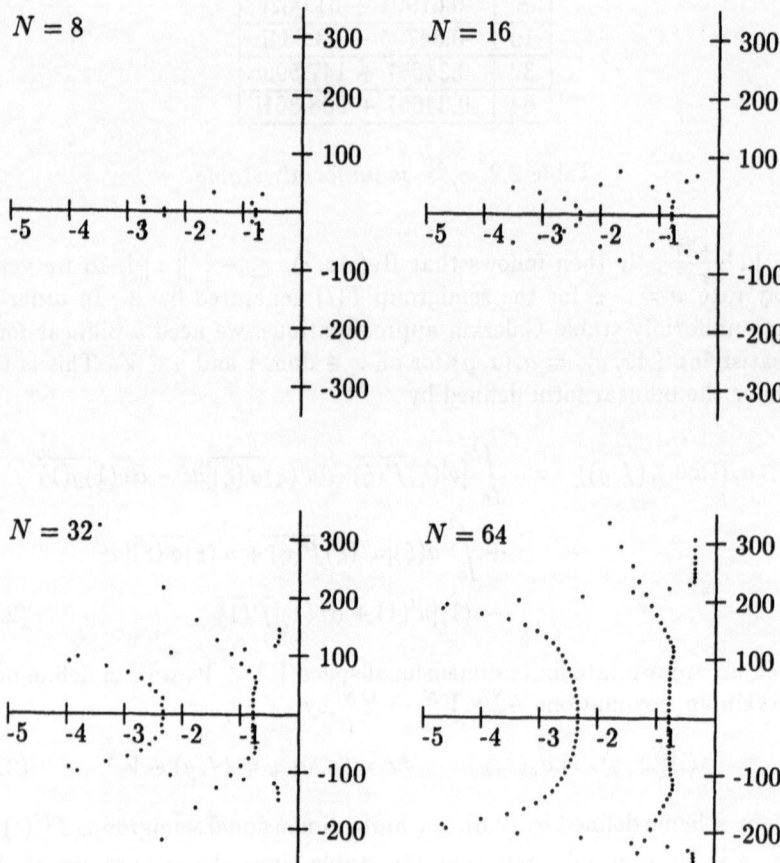

Figure 2.1: Eigenvalues of A_e^N

3 A Scalar Delay Equation

Consider the scalar delay equation

$$\dot{x}(t) = ax(t) + bx(t - r) \tag{3.1}$$

where a, b, and $r > 0$ are real. To reformulate (3.1) as an abstract Cauchy problem, consider the Hilbert space $Z = \mathbb{C} \times L^2(-r, 0; \mathbb{C})$ with the inner product

$$\langle (\eta, \phi), (\xi, \psi) \rangle = \eta\bar{\xi} + \int_{-r}^0 \phi(\theta)\overline{\psi(\theta)}\, d\theta, \tag{3.2}$$

and a compatible norm given by

$$\| (\eta, \phi) \|^2 = |\eta|^2 + \int_{-r}^0 |\phi(\theta)|^2\, d\theta. \tag{3.3}$$

Next define the operator $A : \operatorname{dom} A \subset Z \to Z$ by $\operatorname{dom} A = \{(\eta, \phi) \in Z : \phi \in H^1(-r, 0),\ \phi(0) = \eta\}$, and $A(\eta, \phi) = (a\eta + b\phi(-r), \frac{d}{d\theta}\phi)$. If the state at time $t \geq 0$ is defined to be $z(t) = (x(t), x_t)$, where $x_t(\theta) = x(t + \theta)$ for $-r \leq \theta \leq 0$, then (3.1) can be reformulated as an abstract Cauchy problem like (1.1). It is well known (see [1]) that A is the infinitesimal generator of a strongly continuous semigroup $T(t)$ of bounded linear operators on Z.

In recent work ([5], [6]) we showed how to use an equivalent inner product in order to construct uniformly stable spline based approximations for (3.1). In the present paper we will use the same inner product in a different way - to provide a simple proof that the well known averaging approximation scheme (AVE scheme, see [1]) for (3.1) is uniformly exponentially stable. This fact has been established elsewhere (see [10], [13]), and in both papers the proofs made use of delicate estimates of the approximating resolvent operators $(\lambda I - A^M)^{-1}$ along the imaginary axis. The proof given here uses a completely different approach. To be fair, we note that we only consider a scalar delay equation in this paper, whereas more general delay systems are considered in [10] and [13].

To proceed, let us briefly recall the AVE scheme (see [1] for details). For any positive integer M, partition $[-r, 0]$ into M equal subintervals with the mesh $\theta_j^M = -j\frac{r}{M}$, $j = 0, 1, \ldots, M$. Define the characteristic functions

$$\chi_j^M(\theta) = \begin{cases} 1 & \text{if } \theta_j^M \leq \theta \leq \theta_{j-1}^M \\ 0 & \text{elsewhere} \end{cases} \tag{3.4}$$

for $j = 1, 2, \ldots, M$. Set $B_0^M = (1, 0) \in Z$ and $B_j^M = (0, \chi_j^M)$ for $j = 1, 2, \ldots, M$, and define $Z^M \subset Z$ by $Z^M = \operatorname{span}\{B_j^M\}_{j=0}^M$. Define $A^M : Z^M \to Z^M$ by $A^M(\phi_0, \sum_1^M \phi_j\chi_j^M) = (a\phi_0 + b\phi_M, \sum_1^M \frac{M}{r}(\phi_{j-1} - \phi_j)\chi_j^M)$.

This defines the AVE approximation scheme, and next we define a useful equivalent inner product.

For any real γ we can endow Z with the following inner product

$$\langle(\eta,\phi),(\xi,\psi)\rangle_e = \eta\bar{\xi} + |b|e^{-\gamma r}\int_{-r}^{0} e^{-2\gamma\theta}\phi(\theta)\overline{\psi(\theta)}\,d\theta, \qquad (3.5)$$

and a compatible norm given by

$$\|(\eta,\phi)\|_e^2 = |\eta|^2 + |b|e^{-\gamma r}\int_{-r}^{0} e^{-2\gamma\theta}|\phi(\theta)|^2\,d\theta. \qquad (3.6)$$

It is clear that the inner products $\langle\cdot,\cdot\rangle$ and $\langle\cdot,\cdot\rangle_e$ are equivalent on Z. If we now choose γ to be the unique real root of the following equation (when $b > 0$, this is just the characteristic equation for (3.1))

$$a - \lambda + |b|e^{-\lambda r} = 0, \qquad (3.7)$$

then it follows (see [5]) that

$$\text{Re}\langle Az, z\rangle_e \leq \gamma\|z\|_e^2 \qquad (3.8)$$

for all $z \in \text{dom}A$. Our goal is to prove a similar inequality for the operators A^M. A few preliminary results are necessary. First, define

$$e_j = \int_{\theta_j^M}^{\theta_{j-1}^M} e^{-2\gamma\theta}\,d\theta = \frac{1}{2\gamma}\left(e^{-2\gamma\theta_j^M} - e^{-2\gamma\theta_{j-1}^M}\right),$$

for $j = 1, 2, \ldots, M+1$ (here we define $\theta_{M+1}^M = -(M+1)\frac{r}{M}$) and note that

$$\|(\phi_0, \sum_{j=1}^{M}\phi_j\chi_j^M)\|_e^2 = |\phi_0|^2 + |b|e^{-\gamma r}\sum_{j=1}^{M}e_j|\phi_j|^2. \qquad (3.9)$$

The next three lemmas are estimates for the e_j based on simple analysis of the exponential function.

Lemma 3.1 *Given $\epsilon > 0$ there exists $\tilde{M}$ such that* $-\frac{1}{2}\frac{M}{r}(e_j - e_{j+1}) \leq (\gamma+\epsilon)e_j$ *for $j = 1,\ldots,M$, for all $M \geq \tilde{M}$*

Proof: Consider the case $\gamma < 0$. The case $\gamma \geq 0$ is similar. Let sufficiently small $\epsilon > 0$ be given, and set $\delta = \frac{\gamma+\epsilon}{\gamma}$. Since $0 < \delta < 1$, we have $e^{-x}+\delta x \leq 1$ for all $x \in [0, -\ln\delta]$. It follows that (multiply by $1 - e^x$ and rearrange)

$$-\frac{1}{x} + \frac{2}{x}e^x - \frac{1}{x}e^{2x} \leq \delta(e^x - e^{2x}) \qquad \text{for all } x \in [0, -\ln\delta]. \qquad (3.10)$$

Set $x = -\frac{2\gamma r}{M}$ and observe that $x \in [0, -\ln \delta]$ for all sufficiently large M. With $x = -\frac{2\gamma r}{M}$ in (3.10), multiply by $\frac{1}{2}e^{-2\gamma\theta_{j+1}^M}$ to get

$$\frac{M}{2r}\left[\frac{1}{2\gamma}e^{-2\gamma\theta_{j+1}^M} - \frac{1}{\gamma}e^{-2\gamma\theta_j^M} + \frac{1}{2\gamma}e^{-2\gamma\theta_{j-1}^M}\right] \leq \frac{1}{2\gamma}\left(e^{-2\gamma\theta_j^M} - e^{-2\gamma\theta_{j-1}^M}\right)\gamma\delta$$

so that

$$-\frac{1}{2}\frac{M}{r}(e_j - e_{j+1}) \leq \gamma\delta e_j = (\gamma + \epsilon)e_j. \tag{3.11}$$

The result follows. $\qquad\square$

Lemma 3.2 *Given $\hat{\epsilon} > 0$ there exists $\tilde{M}$ such that $\frac{M}{r}e_1 < (1 + \hat{\epsilon})$ for all $M \geq \tilde{M}$.*

Proof: Again consider the case $\gamma < 0$, with the case $\gamma \geq 0$ similar. Let $\hat{\epsilon} > 0$ be given. Since $\lim_{x\downarrow 0}\frac{1 - e^{-x}}{x} = 1$, it follows that for all $x > 0$ sufficiently small, $\frac{1-e^{-x}}{x} \leq 1 + \hat{\epsilon}$. Set $x = -\frac{2\gamma r}{M}$ so that $x > 0$ is sufficiently small for all M sufficiently large. The result follows since $\frac{M}{r}e_1 = \frac{M}{r}\frac{1}{2\gamma}[e^{\frac{2\gamma r}{M}} - 1] = \frac{1-e^{-x}}{x}$. $\qquad\square$

Given $\epsilon > 0$, it follows from Lemma 3.2 by taking $\hat{\epsilon} = \left[\frac{1}{2}|b|c^{-\gamma r}\right]^{-1}\frac{\epsilon}{2}$ that

$$\frac{1}{2}|b|e^{-\gamma r}\frac{M}{r}e_1 < \frac{1}{2}|b|e^{-\gamma r} + \frac{\epsilon}{2} \qquad \text{for all } M \geq \tilde{M}. \tag{3.12}$$

Lemma 3.3 *Given $\hat{\epsilon} > 0$ there exists $\tilde{M}$ such that*

$$\frac{1}{2}|b|e^{\gamma r}\frac{r}{Me_{M+1}} \leq \frac{1}{2}|b|e^{-\gamma r}(1 + \hat{\epsilon})$$

for all $M \geq \tilde{M}$.

Proof: Again consider the case $\gamma < 0$, with the case $\gamma \geq 0$ similar. Let $\hat{\epsilon} > 0$ be given. Since $\lim_{x\downarrow 0}\frac{x}{1 - e^{-x}} = 1$, it follows that for all $x > 0$ sufficiently small, $\frac{x}{1-e^{-x}} \leq 1 + \hat{\epsilon}$. Set $x = -\frac{2\gamma r}{M}$ so that $x > 0$ is sufficiently small for all M sufficiently large. The result follows since $\frac{1}{2}|b|e^{\gamma r}\frac{r}{Me_{M+1}} = \frac{1}{2}|b|e^{-\gamma r}\frac{-2\gamma r/M}{1-e^{2\gamma r/M}} = \frac{1}{2}|b|e^{-\gamma r}\frac{x}{1-e^{-x}}$. $\qquad\square$

Given $\epsilon > 0$, it follows from Lemma 3.3 by taking $\hat{\epsilon} = \frac{\epsilon}{|b|e^{-\gamma r}}$ that

$$\frac{1}{2}|b|e^{\gamma r}\frac{r}{Me_{M+1}} \leq \frac{1}{2}|b|e^{-\gamma r} + \frac{\epsilon}{2} \tag{3.13}$$

Theorem 3.4 *Given $\epsilon > 0$, there exists $\tilde{M}$ such that*

$$\mathrm{Re}\langle A^M z, z\rangle_e \le (\gamma + \epsilon)\, \|z\|_e^2 \tag{3.14}$$

for all $z \in Z^M$ for all $M \ge \tilde{M}$.

Proof: Let $z = (\phi_0, \sum_1^M \phi_j \chi_j^M) \in Z^M$. Then

$$
\begin{aligned}
\mathrm{Re}\langle Az, z\rangle_e \;=\;& a|\phi_0|^2 + \mathrm{Re}\, b\phi_M\overline{\phi_0} + \mathrm{Re}\, |b|e^{-\gamma r}\sum_{j=1}^M \frac{M}{r}e_j(\phi_{j-1} - \phi_j)\overline{\phi_j} \\
\le\;& \left(a + \frac{1}{2}|b|e^{-\gamma r}\frac{M}{r}e_1\right)|\phi_0|^2 - \frac{1}{2}|b|e^{-\gamma r}\frac{M}{r}e_{M+1}|\phi_M|^2 \\
& + \mathrm{Re}\, b\phi_M\overline{\phi_0} - \frac{1}{2}|b|e^{-\gamma r}\frac{M}{r}\sum_{j=1}^M (e_j - e_{j+1})|\phi_j|^2
\end{aligned}
$$

where we used the fundamental inequality $\mathrm{Re}\, xy \le \frac{1}{2}|x|^2 + \frac{1}{2}|y|^2$.
Now $\mathrm{Re}\, b\phi_M\overline{\phi_0} \le \frac{1}{2}|b|e^{-\gamma r}\frac{M}{r}e_{M+1}|\phi_M|^2 + \frac{1}{2}|b|e^{\gamma r}\frac{r}{Me_{M+1}}|\phi_0|^2$, so we get

$$
\begin{aligned}
\mathrm{Re}\langle Az, z\rangle_e \;\le\;& \left(a + \frac{1}{2}|b|e^{-\gamma r}\frac{M}{r}e_1 + \frac{1}{2}|b|e^{\gamma r}\frac{r}{Me_{M+1}}\right)|\phi_0|^2 \\
& - \frac{1}{2}|b|e^{-\gamma r}\frac{M}{r}\sum_{j=1}^M (e_j - e_{j+1})|\phi_j|^2.
\end{aligned}
$$

Now apply Lemma 3.1 together with (3.12) and (3.13) to get

$$
\begin{aligned}
\mathrm{Re}\langle Az, z\rangle_e \;\le\;& \left(a + |b|e^{-\gamma r} + \epsilon\right)|\phi_0|^2 + (\gamma + \epsilon)|b|e^{-\gamma r}\sum_{j=1}^M e_j|\phi_j|^2 \\
\le\;& (\gamma + \epsilon)\,\|z\|_e^2
\end{aligned}
$$

$\square$

Recall that from (3.8) it follows that the semigroup $T(t)$ generated by A satisfies $\|T(t)\|_e \le e^{\gamma t}$. Theorem 1 shows that an inequality similar to (3.8) holds for the operators A^M. Thus, for any $\epsilon > 0$ the semigroups $T^M(t)$ generated by A^M satisfy $\|T^M(t)\|_e \le e^{(\gamma + \epsilon)t}$ for all sufficiently large M. This shows that the AVE scheme uniformly preserves the stability behavior of the original semigroup $T(t)$, and hence the semigroups $T^M(t)$ are uniformly exponentially stable whenever $T(t)$ is exponentially stable.

References

[1] H.T. BANKS and J.A. BURNS, "Hereditary control problems: numerical methods based on averaging approximations," *SIAM J. Control and Optimization*, v. 16, 1978, pp. 169–208.

[2] H.T. BANKS, K. ITO and C. WANG, "Exponentially stable approximations of weakly damped wave equations," *International Series in Numerical Mathematics*, Birkhäuser, 100, 1991, pp. 1–33.

[3] H.T. BANKS and K. KUNISCH, "The linear regulator problem for parabolic systems," *SIAM J. Control and Optimization*, v. 22, 1984, pp. 684–699.

[4] G. CHEN, "Energy decay estimates and exact boundary value controllability for the wave equation in a bounded domain," *J. Math. Pures Appl.*, v. 58, 1979, pp. 249–274.

[5] R.H. FABIANO, "Stability preserving spline approximations for scalar functional differential equations," *Computers and Mathematics with Applications*, to appear.

[6] R.H. FABIANO, "Uniformly stable spline approximations for scalar delay problems," Proceedings of Workshop on Optimal Control and Design, Blacksburg, VA, to appear.

[7] J.S. GIBSON, "Linear-quadratic optimal control of hereditary differential systems: infinite dimensional Riccati equations and numerical approximations," *SIAM J. Control and Optimization*, v. 21, 1983, pp. 95–139.

[8] J.S. GIBSON and A. ADAMIAN, "Approximation theory for linear quadratic Gaussian control of flexible structures, *SIAM J. Control and Optimization*, v. 29, 1991, pp. 1–37.

[9] J. LAGNESE, "Decay of solutions of wave equations in a bounded region with boundary dissipation," *J. Differential Equations*, v. 50, 1983, pp. 163–182.

[10] I. LASIECKA and A. MANITIUS, "Differentiability and convergence rates of approximating semigroups for retarded functional differential equations," *SIAM J. Numerical Analysis*, v. 25, 1988, pp. 883–907.

[11] I. LASIECKA and R. TRIGGIANI, *Differential and Algebraic Riccati Equations with Applications to Boundary/Point Control Problems: Continuous Theory and Approximation Theory*, Lecture Notes in Control and Information Sciences, v. 164, Springer-Verlag, 1991.

[12] T. KATO, *Perturbation Theory for Linear Operators*, Springer-Verlag, New York, 1976.

[13] D. SALAMON, "Structure and stability of finite dimensional approximations for functional differential equations," *SIAM J. Control and Optimization*, v. 23, 1985, pp. 928–951.

A COMPARISON OF ESTIMATION METHODS FOR HYDRAULIC CONDUCTIVITY FUNCTIONS FROM FIELD DATA

Ben G. Fitzpatrick* and Julius A. King[†‡]
Department of Mathematics and
Center for Research in Scientific Computation
North Carolina State University
Raleigh, NC 27695-8205

1 Introduction

Concern over the environmental impact of contaminated groundwater has motivated research aimed at developing improved, physically based models of transport processes in heterogeneous media. Large scale dispersion of contaminants is produced to a large degree by the spatial variability of the hydraulic properties of the media. Thus, in order to model transport processes, extensive field experiments providing three dimensional measurements of hydraulic properties are required. Estimation of the hydraulic parameters has been called the "Achilles' heel" of groundwater modeling because of the difficulty in inferring the distributed (and highly variable) hydraulic parameters from discrete data. We consider here two methods for estimating hydraulic conductivity. The first method is the commonly used kriging (after D.G. Krige [11]) method. Kriging is a minimum-mean-squared-error technique which is equivalent to optimal spatial linear estimation [4]. The second method is a regularized least squares, whose penalty term is suggested in [9] in the context of photon emission tomography.

Interpolation and approximation of functions from discrete data is a subject which has received a great deal of attention in the mathematics and statistics literature, and many different methods have been proposed, applied, and analyzed (see [14] and the discussion therein). In some cases, a specialized method is chosen corresponding the particular application, but in many other circumstances, simple, easy-to-apply methods are chosen and adjusted in various ways in order to fit data. Hydraulic data present many significant challenges, due to the scale and complexity of the properties to be interpolated as well as the sparsity and variability of the data.

The data to which we apply these techniques are from the MADE (MAcroDispersion Experiment) experiments at Columbus Air Force Base in Columbus, Mississippi. The data is comprised of measurements from 56 borehole flowmeters installed on the base. Measurements were at locations

*Supported in part by AFOSR grant F49620-93-1-0153
†Supported in part by AFOSR grant F49620-93-1-0153
‡Current Address: Stanley Holcombe and Associates, Atlanta, GA

scattered over a horizontal area approximately 200m by 350m, taken at six inch intervals in the z (depth) direction. Due to the varying levels of the land and underlying bedrock, samples vary from approximately a maximum of 62m - 50m above sea level to 59m - 56m. The papers [15, 16] describe in detail the MADE-1 experiment, and they also provide a great deal of analysis of the data. The MADE-2 experiment and data are described in the paper [3]. The heterogeneity of this particular site make this data an ideal testbed for approximation and interpolation methods in spatial statistics and general nonparametric regression.

The paper is organized as follows. In Section 2, we give some basic background on groundwater flow in order to motivate the interest in the estimation problem we consider. Kriging and regularized least squares are the topics of Sections 3 and 4, and in Section 5, we present results of our estimation. In Section 6 we give some remarks on future work and related ideas.

2   Groundwater flow and contaminant transport models

In this section we briefly summarize some commonly used models for solute transport in groundwater flow systems. These models are applicable in many situations, and complete descriptions may be found in [2, 8]. The transport of a single solute in groundwater is modeled by the equation

$$Rc_t + \nabla \cdot (vc) = \nabla \cdot (D\nabla c) - \frac{rc}{K_m + c},$$

on a region Ω in $\mathbb{R}^3$ which represents the site of interest. The coefficient R is called the retardation factor and models the process of adsorption (see [6]). The final term models biodegradation via Michaelis-Menten, or Monod, kinetics (see [6] and the references therein). The velocity of the groundwater is dependent on the hydraulic conductivity K and hydraulic head h through the empirical Darcys' Law

$$v = -K\nabla h,$$

which in steady state groundwater flow leads to the elliptic equation

$$\nabla \cdot K\nabla h = 0.$$

Another important feature of the model is the contaminant dispersion. The dispersion matrix D takes the form

$$D = d_0 I + D_1(v),$$

where $d_0 I$ models molecular diffusion and $D_1(v)$ models the hydrodynamic dispersion of the solute that is due to the pore system of the medium and collisions of solute particles with soil particles. Empirical evidence and statistical considerations (see [2]) suggest the following model for D_1:

$$(D_1(v))_{ij} = a_T |v| \delta_{ij} + (a_L - a_T) \frac{v_i v_j}{|v|}.$$

Here a_T is called the "transverse dispersivity," and a_L is called the "longitudinal dispersivity." In this dispersion model, the principal axis of dispersion is along the velocity vector (longitudinal), and there is less dispersion in directions orthogonal to v ($a_T < a_L$). Moreover, the dispersion is modeled as (roughly) proportional to speed, the dispersion being caused by collisions with soil particles. See [2] for a detailed exposition of dispersion.

Determination of the velocity field is crucial for prediction of contaminant transport; hence, estimation of the conductivity K is a major issue for transport modeling. In the following sections, we describe two methods for conductivity estimation.

3 Kriging Estimators

In a field situation, one may have observations $\{\tilde{K}_i\}_{i=1}^n$, which correspond to $K(x_i)$, where x_i denotes a known (three dimensional) coordinate at which observations are made. In order to apply differential equation models of groundwater flow and contaminant transport, one must have conductivity as a function of the spatial variable. Thus, we seek methods for estimating this function from the discrete, observed data. The method of kriging is based on best linear unbiased estimation for second order stationary spatial stochastic processes. Typically, in the hydrology literature, it is assumed that hydraulic conductivity is lognormally distributed, so that the logarithm of conductivity yields a Gaussian process.

We set $\tilde{Z}_i = \log(\tilde{K}_i)$, and we assume that these values are from a realization of a second order stationary process, Z (this logarithmic transformation of the data is commonly used: see [16]). We seek an estimator of the form

$$\hat{Z}(x) = \sum_{i=1}^n \lambda_i(x) \tilde{Z}_i,$$

where the weights $\{\lambda_i\}$ are chosen in such a way that Z is unbiased and that

$$J(\lambda) = (1/2) E |\hat{Z}(x) - Z(x)|^2$$

is minimized. Note that this minimization must be carried out for each x for which an estimator is desired. We will discuss this point further below. The kriging algorithm, which is explained quite well in [4], is given here for completeness.

To obtain an estimator which is unbiased, we require that $\sum_{i=1}^{n} \lambda_i(x) = 1$, for each x. In order to minimize J, we introduce the semivariogram $\gamma(h) = (1/2)E|Z(x+h) - Z(x)|^2$ (which includes the stationarity assumption in that the covariance only depends on the position difference). Using the constraint $\sum_{i=1}^{n} \lambda_i(x) = 1$, one can write J as

$$
\begin{aligned}
J(\lambda) &= E\left|\sum_{i=1}^{n} \lambda_i(x)Z(x_i) - Z(x)\right|^2 \\
&= \sum_{i=1}^{n}\sum_{j=1}^{n} \lambda_i\lambda_j\gamma(x_i - x_j) - 2\sum_{j=1}^{n} \lambda_j\gamma(x_j - x) + E|Z(x)|^2,
\end{aligned}
$$

which leads to a quadratic programming problem. The fact that we have a very simple linear constraint allows us to solve the quadratic program in a straightforward manner: differentiation leads to solving the linear system

$$
\begin{pmatrix} \Gamma & \vec{1} \\ \vec{1}' & 0 \end{pmatrix} \begin{pmatrix} \vec{\lambda}(x) \\ \mu \end{pmatrix} = \begin{pmatrix} \vec{b}(x) \\ 0 \end{pmatrix},
$$

where the matrix Γ is formed by $\Gamma_{ij} = \gamma(x_i - x_j)$, $\vec{1}$ denotes a (column) vector of n 1s and $\vec{1}'$ denotes its transpose, the vector $\vec{b}(x)$ is given by $b_i(x) = \gamma(x_j - x)$, and μ is the Lagrange mulitplier that arises from the unbiasedness constraint $\sum \lambda_i = 1$.

Two very important points become clear from the solution form. One is regarding implementation: while this system must be solved for x at which values of the process are desired, the matrix which is to be factored (inverted) is independent of the value of x at which the estimator is to be computed. Thus the matrix needs to be factored only once, an important fact for large n. The second important point to be made is that the optimal λ values are linear combinations of the functions $b_i(x) = \gamma(x_j - x)$; hence, the estimator derived from kriging is a linear combination of these functions. Viewed in this light, kriging resembles kernel based interpolation and estimation. Thus, the particular function γ used has a significant impact on the estimator.

A form of variogram which is commonly used in hydrology is

$$
\gamma(h) = \sigma^2\left(1 - \exp(-(h_1^2/\alpha_1^2 - h_2^2/\alpha_2^2 - h_3^2/\alpha_3^2)^{1/2})\right),
$$

in which $h = (h_1, h_2, h_3)$ denotes the distance vector, and the parameters $\sigma, \alpha_1, \alpha_2$, and α_3 must be estimated from the data. The typical approach to determining these variogram parameters is by comparing the functional form above with the nonparametric estimator

$$
\hat{\gamma}(h) = \frac{1}{2N_h} \sum_{i,j} |Z_i - Z_j|^2,
$$

where N_h denotes the number of observations whose positions are $h \pm \varepsilon$ apart, and the indices of summation denote those pairs. The tolerance ε is a parameter which provides some smoothing of $\hat{\gamma}$.

4 The penalized least squares approach

In seeking an estimation method that relaxes the stationarity assumptions of ordinary kriging, one can examine more general forms of kriging, such as universal kriging, in which one estimates a mean function in order to remove trends. Many such techniques still require very careful covariance modeling, however. Least squares or nonparametric regression methods, on the other hand, model the function to be estimated as deterministic, but unknown. When Bayesian techniques are applied, the function to be estimated is treated as random in the sense that a prior distribution reflects subjective uncertainty in the function's values. In particular, kriging can be viewed as a form of Bayesian estimation in which measurement noise can be neglected and the prior is for a second order stationary Gaussian process. However, from the nonparametric regression point of view, we need merely to consider a method for constraining the function which is consistent with our expectations of its behavior (e.g., regularity). We consider here a penalized least inspired by work in image reconstruction (see [9]).

The infinite dimensional form of the cost functional we minimize is given by

$$J_\beta(K) = \sum_{i=1}^{n} |K(x_i) - \tilde{K}_i|^2 + \beta U(K),$$

where the function U is given by

$$U(f) = \int_\Omega \Phi(|\nabla f(x)|),$$

with

$$\Phi(\xi) = 1 + \frac{-1}{1 + (\xi/\delta)^2},$$

δ and β being design parameters. The basic idea behind the nonquadratic weighting Φ is that one seeks a function that may have regions of rapid change but is primarily flat. Used in photon emission tomography, this penalty term has yielded excellent reconstructions (see [9] and the references therein). Were we using a Bayesian analysis, as is common in the statistical image reconstruction literature, we would be viewing $\exp(-\beta U(f))$ as a prior and the measurement noise as independent, identically distributed Gaussian random noise.

Analyzing this infinite dimensional minimization is is complicated in several ways. The boundedness and nonconvexity of the penalty make the use of standard regularization arguments invalid, and in this setting of the problem, we have no results concerning existence and uniqueness of minimizers of J_β, without additionally restricting J_β to a compact set (something one generally uses regularization to avoid).

The problem we solve numerically can be viewed as an approximation. We treat K as piecewise constant on a fixed grid, and we minimize

$$\hat{J}_\beta(K) = \sum_{i=1}^n |K(x_i) - \tilde{K}_i|^2 + \beta\hat{U}(K),$$

where

$$\hat{U}(f) = \sum_{j \in grid} \sum_{i \in nbhd(j)} \Phi(|f(x_i) - f(x_j)|),$$

where $x_i, i \in grid$ denotes the grid points, and $nbhd(i)$ denotes the indices of the points adjacent and diagonally adjacent to point i. This is exactly the form of penalty used in [9]. We remark here that approximation theoretic issues are rarely considered in much of the statistical literature on tomographic image reconstruction (see, e.g., [9, 13]). Analysis typically begins with a fixed pixel grid, which in turn makes for a finite dimensional problem of finding a piecewise constant function on the chosen pixel grid. The computational results we present below should be viewed in this light as well: "standard" theoretical results such as obtained in [1] for parameter estimation can not be applied using the penalty term we have chosen. However, the quality of computational results obtained (see below, as well as [9, 13] and there references therein) indicate that there is merit in this approach.

5   Conductivity Estimation for the MADE site

The MADE site on Columbus Air Force Base, Columbus, Mississippi, is an extremely heterogeneous aquifer composed of sandy gravel and gravelly sand, with some silt and clay. The soil layers are highly irregular. This site is certainly one of the most challenging from the point of view of quantitative characterization.

Figure 1 gives a "birds-eye" view of the borehole flowmeter locations (at which conductivity is measured). Measurements were taken vertically at each location from about 52m to about 62m above sea level, in 6in increments.

Horizontal slabs of 1 meter thickness centered on vertical planes at 54,55,56,57,58,59,60,61, and 62 meters above sea level were chosen for a vertical estimation grid. For kriging, a 132×47 grid evenly spaced through the parallelogram with corners at (76,155), (52,192), (216,373), and (192,410), which covered most of the borehole flowmeter sites, was used. The least squares method used a 43 × 12 subset of the kriging grid (with the same corner points). In the least squares methods, we estimated the conductivity functions independently for different vertical slabs, due to the computation time required by the IFFCO minimization routine (see [4]). This routine contains hueristic techniques for avoiding local minima; hence, the number of function evaluations is rather high. The values for δ and β used in the computations were 10^{-4} and 10^{-6}, respectively.

The kriging variogram used is that of Rehfeldt et. al. [16]:

$$\gamma(h) = \sigma^2 \left(1 - \exp(-(h_1^2/\alpha_1^2 - h_2^2/\alpha_2^2 - h_3^2/\alpha_3^2)^{1/2})\right),$$

with $\sigma^2 = 4.5, \alpha_1 = 11.9, \alpha_2 = 12.2$, and $\alpha_3 = 1.6$ estimated from the data using the sample variance, as well as the sample variogram given in Section 3 above.

The figures below illustrate the results of the computations. Spatial dimensions are all in meters, while conductivity values are in cm/sec. Depth or z measurements are in meters above sea level. In Figures 2 and 3, we see level curves of conductivity at the various vertical slices, for the kriging and penalized least squares estimates. Figures 4 through 9 display estimates surfaces (as black circles) and data points within the 1 meter thick slab (as white circles). In Figures 4 and 5, we show the kriging and penalized least squares estimates for the depth plane $z = 58$, Figures 6 and 7 have a similar view for the plane $z = 60$, and Figures 8 and 9, for $z = 62$. The differences in the two estimates are striking, particularly the relative flatness of the least squares method.

6   Conclusions

We have presented two methods for estimating spatially varying functions from discrete data. The underlying assumptions in the kriging method are that the data are observations of a realization of a second order stationary process. While trend analysis can provide some relaxation of the constant mean requirement, one encounters the difficulty of choosing appropriate forms for the trend, in itself a challenging problem.

The least squares estimates tended to be less oscillatory, which is to be expected since they need not be interpolatory (as is kriging). In continuing efforts we are examining the effects of the parameters β and δ on the results.

While we have no sort of convergence proof for the finite dimensional approximations, we feel that the results indicate the utility of this approach. In future studies, we plan to examine using the least squares estimators for trend removal: kriging can then be used to understand "high frequency, low amplitude" variations in the data.

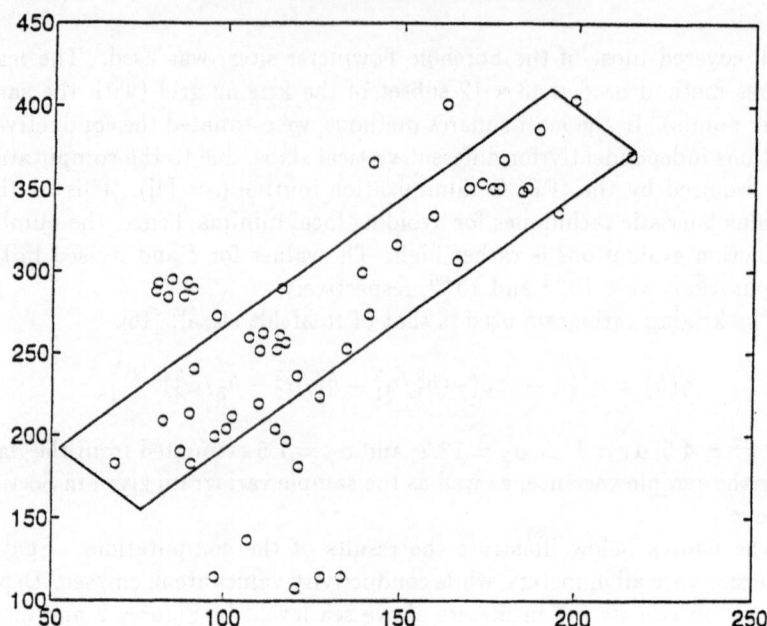

Figure 5.1: Borehole Flowmeter Positions and Estimation Region.

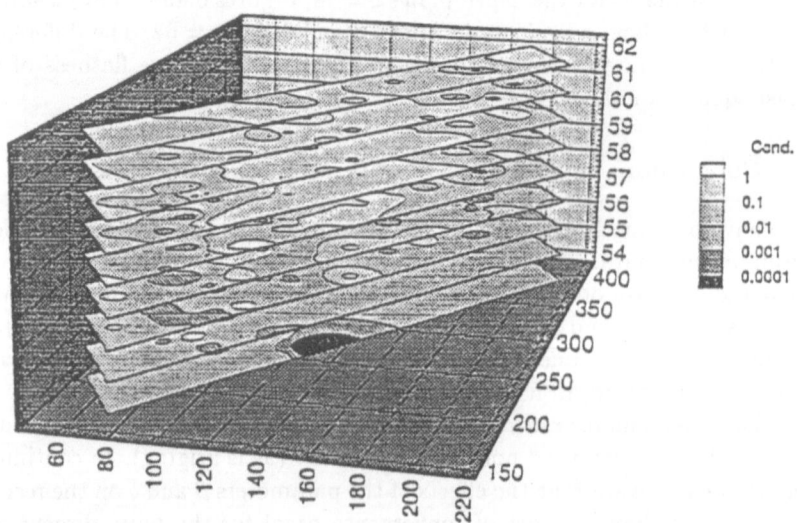

Figure 5.2: Three dimensional view of kriged estimator.

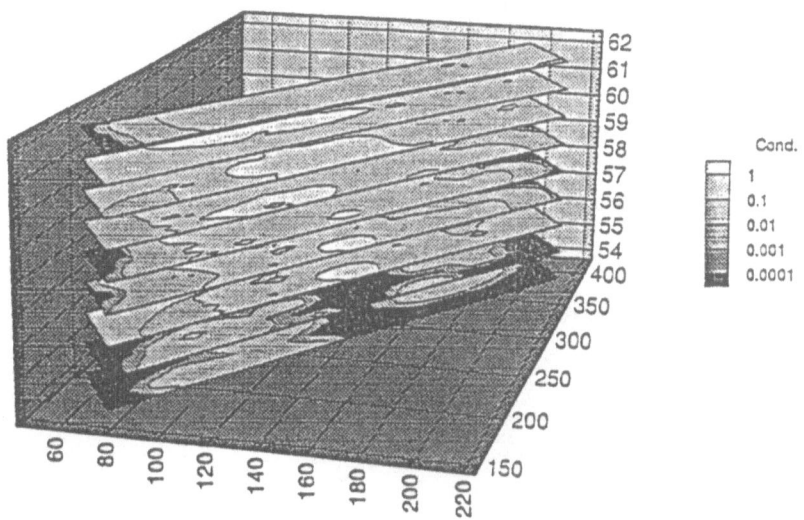

Figure 5.3: Three dimensional view of least squares estimator.

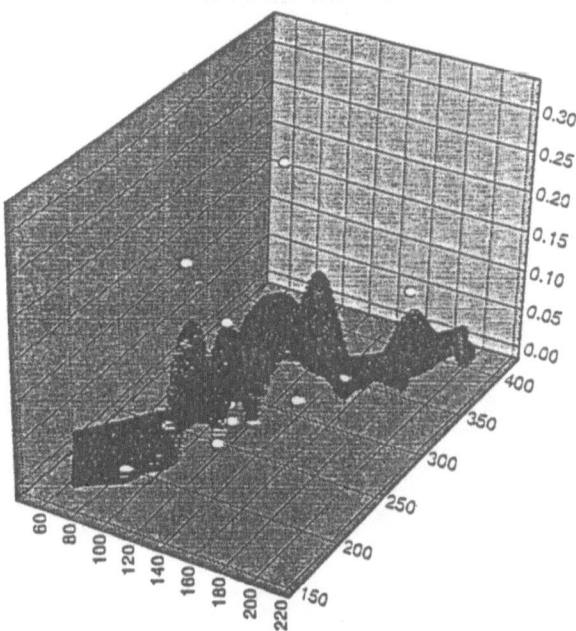

Figure 5.4: Kriged estimator for the vertical slice at $z = 58m$. Black circles give the kriged surface, white circles are data values.

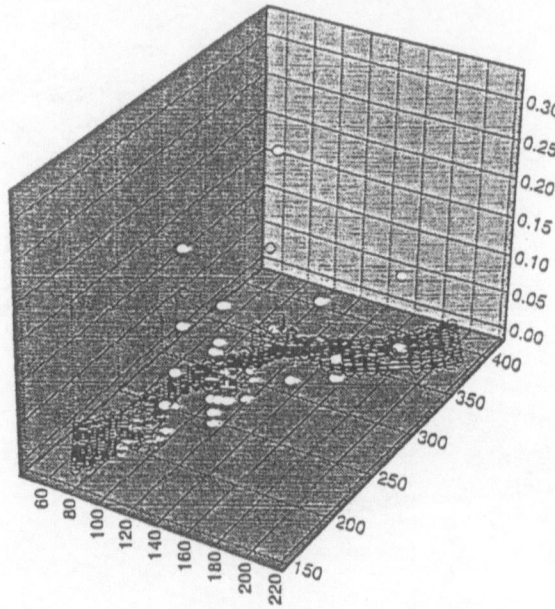

Figure 5.5: Least squares estimator for the vertical slice at $z = 58m$. Black circles give the least squares surface, white circles are data values.

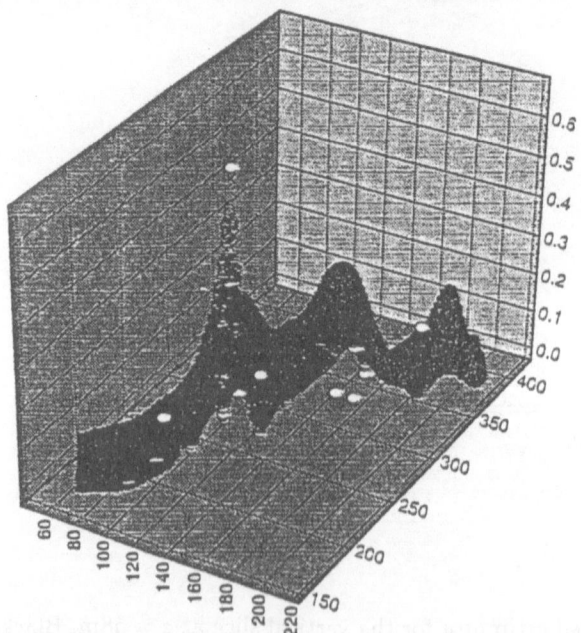

Figure 5.6: Kriged estimator for the vertical slice at $z = 60m$.

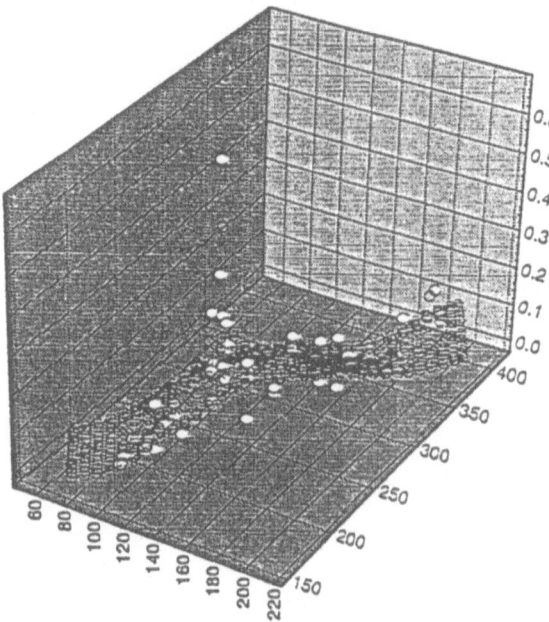

Figure 5.7: Least squares estimator for the vertical slice at $z = 60m$.

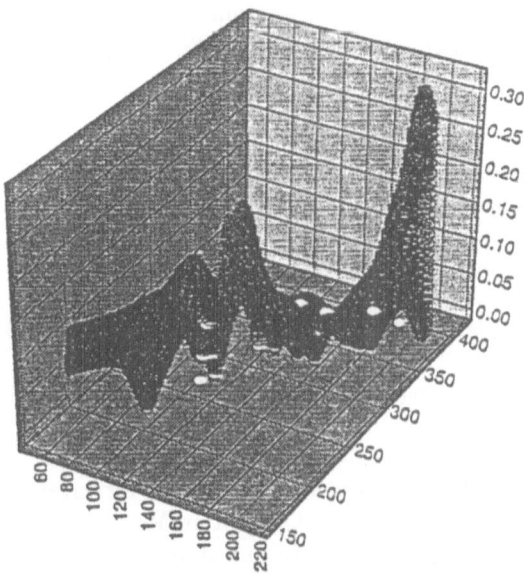

Figure 5.8: Kriged estimator for the vertical slice at $z = 62m$.

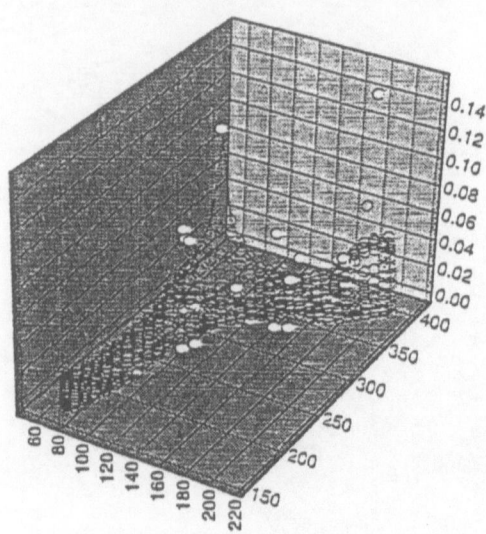

Figure 5.9: Least squares estimator for the vertical slice at $z = 62m$.

References

[1] Banks, H. T., and K. Kunisch, *Estimation Techniques for Distributed Parameter Systems*, Birkhäuser, Boston, 1989.

[2] Bear, *Dynamics of Fluids in Porous Media*, Elsevier, New York, 1972.

[3] Boggs, J. M., L. M. Beard, S. E. Long, M. P. McGee, W. G. MacIntyre, C. P. Antworth, and T. B. Stauffer, Database for the Second Macrodispersion Experiment (MADE-2), Electric Power Research Institute Technical Report TR-102072, EPRI, Palo Alto, February, 1993.

[4] Cressie, N. A. C. *Statistics for Spatial Data*, Wiley, New York, 1991.

[5] deMarsily, G. *Quantitative Hydrogeology*, Academic Press, San Diego, 1986.

[6] Fitzpatrick, B. G. "Analysis and Approximation for Inverse Problems in Contaminant Transport and Biodegradation Models," Center for Research in Scientific Computation Technical Report CRSC-TR-94-12, North Carolina State University, Raleigh, 1994.

[7] Fitzpatrick, B. G., and M. A. Jeffris. "Parameter Estimation in Groundwater Flow Models with Distributed and Pointwise Observations," Cen-

ter for Research in Scientific Computation Technical Report CRSC-TR94-17, North Carolina State University, Raleigh, 1994.

[8] Freeze, R. A., and J. A. Cherry. *Groundwater,* Prentice-Hall, Engiewood Cliffs, 1978.

[9] Geman, S., K. M. Manbeck, and D. E. McClure. "A Comprehensive Statistical Model for Single-Photon Emission Tomography," Chapter 5 of *Markov Random Fields: Theory and Applications,* R. Chellapa and A. Jain, editors, Academic Press, Boston, 1993.

[10] Gilmore, P. "Implicit Filtering for constrained Optimization, Users' Guide," Center for Research in Scientific Computation Technical Report CRSC-TR93-7, North Carolina State University, Raleigh, 1993.

[11] Journel, A. G., and G. T. Huijbregts. *Mining Geostatistics,* Academic Press, New York, 1978.

[12] Kitanidis, P. K., and E. Vomvoris. "A Geostatistical Approach to the Inverse Probiem in Groundwater Modeling (Steady State) and One Dimensional Simulations," *Water Resources Research,* **19** (3): 677-690, 1983.

[13] Künsch, H. R. "Robust Priors for Smoothing and Image Restoration," *Ann. Inst. Stat. Math.,* **46** (1), pp. 1-19, 1994.

[14] Laslett, G. M. "Kriging and Splines: An Empirical Comparison of Their Predictive Performance in Some Applications" (with discussion), *J. Amer. Stat. Assoc.,* **89** (426), pp. 391-409, 1994.

[15] Rehfeldt, K. R., L. W. Gelhar, S. C. Young, E. E. Adams, L. M. Beard, and J. M. Boggs. "Fieid Study of Dispersion in a Heterogeneous Aquifer: 1. Oveview and Site Description," *Water Resources Research,* **28** (12): 3281-3291, 1992.

[16] Rehfeldt, K. R., L. W. Gelhar. and J. M. Boggs. "Field Study of Dispersion in a Heterogeneous Aquifer: 3. Geostatistical Analysis of Hydraulic Conductivity," *Water Resources Research,* **28** (12): 3309-3324, 1992.

AN ADAPTIVE STENCIL FINITE DIFFERENCING SCHEME FOR LINEAR FIRST ORDER HYPERBOLIC SYSTEMS—A PRELIMINARY REPORT

Robert H. Hoar, * and C. R. Vogel †
Department of Mathematical Sciences
Montana State University
Bozeman, Montana 59717

1 Introduction

The simulation of wave propagation through highly heterogeneous media is important in many applications, e.g., seismic and ultrasound modeling and imaging. In these applications, material parameters may vary widely over short length scales. In practice, these rapid variations are often modeled with hyperbolic partial differential equations having discontinuous parameters.

To solve these equations, asymptotic techniques may sometimes be applied. They can effectively handle discontinuities, but they are difficult to implement if the geometry of the discontinuities becomes too complex. Numerical methods—primarily finite difference methods—are a widely used alternative. While relatively easy to implement, these methods perform poorly when applied to discontinuous coefficient equations. Even for equations with smoothly varying coefficients, standard finite difference methods suffer from numerical dispersion, i.e., the propagation of differing frequency components at different speeds. This is manifested in the spread of waveforms and the introduction of spurious oscillations [3].

In this paper, we present an adaptive stencil scheme to deal with these difficulties. This scheme combines elements of the Courant-Isaacson-Rees (CIR) method [2], the essentially non-oscillatory (ENO) scheme [5], and the Strang splitting [7]. While the method is adaptive in nature, it does not seek to increase accuracy or resolution by refining the computational grid. Instead, it uses a fixed regular grid and adapts the finite difference stencil from point to point. This strategy eliminates spurious oscillations, reduces the spreading of waveforms, and allows numerical propagation of waveforms across discontinuities in spatially dependent parameters. The adaptive nature of the mesh selection strategy makes the method nonlinear—in spite of the fact that it is designed for linear hyperbolic systems in one or more spatial dimensions.

To illustrate this method, we describe in detail its implementation for the scalar wave equation. Section 2 deals with the case of one space dimension, while section 3 covers the 2-D case. Section 2.1 reviews the conversion

---

*Supported in part by a DOE-EPSCoR Graduate Fellowship.
†Supported in part by the NSF under Grant DMS-9303222.

of the 1-D scalar wave equation to first order system form and the implementation of the CIR method for this linear system. The CIR method exhibits only first order accuracy. In section 2.2, we utilize the ENO criterion to develop a stencil selection strategy which has higher order spatial accuracy. Section 2.3 presents an approach for increasing temporal accuracy. Section 2.4 deals with issues of computational complexity. In section 3.1, we review the conversion of the 2-D scalar wave equation to first order system form. Section 3.2 describes the implementation of Strang splitting for this system. This allows us to propagate solutions in two (or more) space dimensions using a product of 1-D propagators. Numerical results are presented in section 4. A one dimensional example with non-smooth initial data is presented, as well as a pair of two dimensional problems that have discontinuous parameters (wave speeds). The numerical implementation of boundary conditions is also discussed in this final section.

2   The One Dimensional Case

In this section, we outline how to apply the adaptive stencil finite difference method to the one dimensional acoustic wave equation. First, we review the CIR method; then we extend it to gain higher order methods.

2.1   The CIR Method

The CIR method of Courant, Isaacson and Rees [2] was first presented in 1952. The adaptive stencil method presented here may be viewed as an extension of CIR.

The one dimensional hyperbolic acoustic wave equation

$$\partial_t^2 u - c^2 \partial_x^2 u = 0 \tag{2.1}$$

is equivalent to the first order system

$$\partial_t \vec{v} + A \partial_x \vec{v} = \vec{0}, \tag{2.2}$$

where

$$\vec{v} = \begin{bmatrix} \partial_x u \\ \partial_t u \end{bmatrix}, \quad A = \begin{bmatrix} 0 & -1 \\ -c^2 & 0 \end{bmatrix}. \tag{2.3}$$

Let E be the matrix of eigenvectors and $\Lambda = diag\{\lambda_1, \lambda_2\}$ be the diagonal matrix of corresponding eigenvalues of A so that $A = E\Lambda E^{-1}$. Then (2.2) becomes

$$E^{-1} \partial_t \vec{v} + \Lambda E^{-1} \partial_x \vec{v} = 0.$$

Letting $\vec{w} = E^{-1}\vec{v}$, we have (assuming c is a function of space but not time)

$$\partial_t \vec{w} + \Lambda \partial_x \vec{w} = \Lambda (\partial_x E^{-1}) E \vec{w}. \tag{2.4}$$

The left hand side is decoupled, and with the spatial and temporal discretizations $\vec{w}_j^m \equiv [w_{1j}^m \; w_{2j}^m]^T$, where $w_{ij}^m = w_i(m\Delta t, j\Delta x)$ and $m \geq 0$, $j = \ldots, -1, 0, 1, \ldots$, the CIR[1] method yields

$$\frac{\vec{w}_j^{m+1} - \vec{w}_j^m}{\Delta t} + \left[\begin{array}{c} \lambda_1 \, \Delta_x(\lambda_1) w_{1j}^m \\ \lambda_2 \, \Delta_x(\lambda_2) w_{2j}^m \end{array} \right] = \Lambda(\partial_x E^{-1}) E \vec{w}_j^m, \qquad (2.5)$$

where we define for a spatial grid function $f = \{f(j\Delta x)\} = \{f_j\}$,

$$\Delta_x(\lambda) f_j = \left\{ \begin{array}{ll} \dfrac{f_j - f_{j-1}}{\Delta x}, & \text{if } \lambda > 0 \quad \bullet \; \circ \\[2ex] \dfrac{f_{j+1} - f_j}{\Delta x}, & \text{if } \lambda < 0 \quad \circ \; \bullet \end{array} \right. \qquad (2.6)$$

Hence, upwinding is used in the selection of the spatial difference stencils. At any given time step, one can recover $\vec{v}$ by computing $\vec{v} = E\vec{w}$.

The symbols to the right in (2.6) indicate which stencil is being used. The solid dot $\bullet$ represents the point at which the spatial derivative is being approximated, and the circles $\circ$ represent other points that are to be used. Relative position indicates whether upwinding is to the left or to the right. For example, $\circ \; \bullet$ indicates that upwinding is to the left, and the first order finite difference approximation that requires f at x_j and x_{j-1} is taken.

The local truncation error [4] for the CIR method is $O(\Delta x) + O(\Delta t)$. Since $|\lambda_i| = c$, the CFL stability condition reduces to

$$\max_x c(x) \frac{\Delta t}{\Delta x} \leq 1.$$

Given stability, CIR is first order accurate in both space and time. Hence, one obtains first order accurate approximations to the components of $\vec{v}$. One can then integrate the approximation of u_t with respect to t and obtain a second order accurate approximation to the solution u of (2.1).

Note that the stencil *adapts* to the sign of the λ_i. For this reason, CIR can be viewed as an adaptive stencil finite difference method. Also for this reason, CIR deals with numerical dispersion better than standard methods based directly on difference approximations to the second derivatives in problem (2.1) (see [8, pp. 80-83]).

An important goal of this paper is to obtain higher order spatial truncation error while retaining the desirable numerical dispersion characteristics of the CIR method.

---

[1]The CIR approximation assumes c is constant, so that the right hand side of (2.4) is the zero vector.

2.2   Higher Order Stencil Selection

Clearly $O(\Delta x^n)$ approximations to the derivative $\partial_x w_i$ at the point x_j may be obtained using $n+1$ distinct, consecutive points [1]. These need not be centered at x_j. In fact, there are $n+1$ possible stencils containing x_j that yield an $O(\Delta x^n)$ truncation error.

For $n=1$, there are two stencil choices. With the CIR method, the sign of the λ_i's is used to select the appropriate stencil. For $n>1$, an additional criterion is needed to select the stencil. In order to minimize undesirable effects of numerical dispersion, we adopt the ENO criterion, which was originally developed by Harten and Osher [5, 6] for scalar conservation laws.

To determine which $(n+1)$ spatial grid points to use in the finite difference stencil, we start with the point x_j and gradually build up the stencil that has certain properties based on the local smoothness of the data. The necessary decisions and resulting stencils are displayed in the tree in Figure 2.1. Again, the solid dot • refers to the point x_j, and the circles ◦ refer to other points that are used in the stencil.

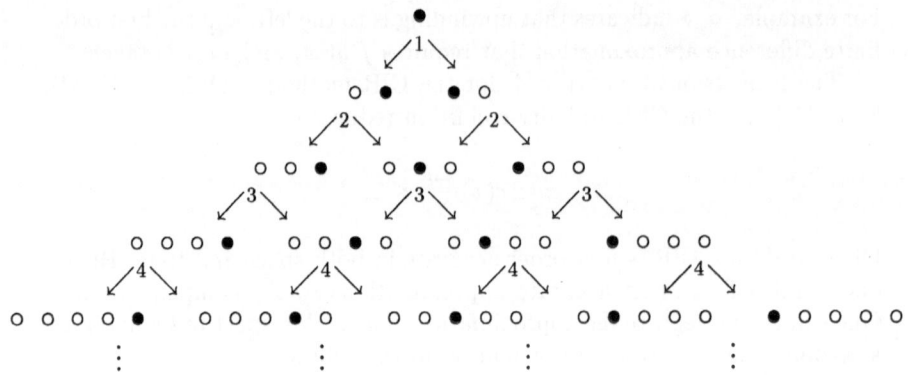

Figure 2.1: Tree of possible finite difference stencils.

The $O(\Delta x^n)$ stencils are found on the $(n+1)^{st}$ level of the tree, thus n decisions must be made in order to reach this level. The first decision adds the point that is in the upwind direction which is determined by the sign of λ_i, as in the CIR method, c.f. (2.6). Take the arrow to the left at level 1 if $\lambda_i > 0$, otherwise take the arrow to the right. This ensures the use of an upwind scheme at each point [2]. The remaining $n-1$ decisions are aimed at minimizing the spurious oscillations that often occur when using a fixed finite difference stencil [3].

Each finite difference stencil in Figure 2.1 corresponds to the derivative of a polynomial interpolant. Harten, Osher, and others have shown that choosing polynomial interpolants with small (in magnitude) higher derivatives will reduce the spurious oscillations [5, 6]. This selection criterion yields what are known as essentially non-oscillatory (ENO) methods. Given a vector f of length N, let $D(f)$ be the vector of length $N-1$ whose j^{th} entry $D(f)_j$ is $f_{j+1} - f_j$ and define $D^n(f) = D(D^{n-1}(f))$ for $n > 1$. Consider the decision that must be made at level k. The arrow to the left is taken if

$$|D^k(f)_i| < |D^k(f)_{i+1}|, \tag{2.7}$$

otherwise the one on the right is taken. This choice ensures that the method will choose the stencil on the left at decision k if the leading coefficient of the degree k polynomial interpolating the data represented on the left is smaller in magnitude than the leading coefficient for the data on the right, which is the ENO criterion.

The value of i in (2.7) will depend on the results of the previous decisions. If l is the number of previous decisions that resulted in taking a path to the left (i.e., the number of o 's to the left of $\bullet$ in the current stencil), then $i = j - l$. Hence, the resulting method, even though applied to a linear system, is nonlinear.

2.3 Temporal Accuracy

Expanding the solution to

$$\partial_t \vec{v} + A\partial_x \vec{v} = 0 \tag{2.8}$$

in a Taylor series about t,

$$\vec{v}(x, t + \Delta t) = (1 + \Delta t \partial_t + \frac{1}{2}\Delta t^2 \partial_t^2 + \ldots)\vec{v}(x, t). \tag{2.9}$$

Using (2.8) to replace the temporal derivatives with spatial derivatives, we obtain

$$\vec{v}(x, t + \Delta t) = (1 - \Delta t A\partial_x + \frac{1}{2}\Delta t^2 (A\partial_x)^2 - \ldots)\vec{v}(x, t). \tag{2.10}$$

Taking the first 2 terms on the right hand side, diagonalizing A, and employing the spatial derivative approximation found in level 2 of the tree gives the CIR method. One can derive higher order methods simply by taking additional terms in the right hand side of (2.10) (increasing temporal accuracy), and by continuing to lower levels in the tree (increasing spatial accuracy).

2.4   Computational Complexity

Note that second order temporal accuracy requires computation of

$$(A\partial_x)^2\vec{v} = A\partial_x(A\partial_x\vec{v}),$$

and hence, an additional application of the approximation to $A\partial_x$. Similarly, n applications of $A\partial_x$ are required for n^{th} order temporal accuracy.

There are $n+1$, n^{th} order stencils for each f_j to choose from, and each of these stencils have a value $D^n(f)_i$ associated with it. Although this appears to imply that $N(n+1)$ of the $D^n(f)_i$ must be computed in order to determine the appropriate stencil at the N points f_k, the trees (and therefore the $D^n(f)_i$) of neighboring points overlap. So, in fact, the one vector $D^n(f)$ contains all of the needed information for the decisions that need to be made on level n of the tree. And, since the vectors $D^n(f)$ decrease in length as n increases, the *added* amount of work to go to the next level of the tree actually decreases, even though the number of possible stencil choices increases.

For w_k near a boundary, the tree in Figure 2.1 must be truncated. How these trees are truncated will affect the realized boundary conditions, and will be discussed in section 4.2.

3   Solving Multi-Dimensional Problems

Clearly, the method described above can be extended to any first order linear hyperbolic system in one space dimension. Here we present a scheme that utilizes the splitting techniques of Strang [7] to solve linear hyperbolic systems in more than one space dimension.

3.1   System form of 2-D Scalar Wave Equation

To demonstrate this approach, consider the two dimensional acoustic wave equation

$$\partial_t^2 u - c^2(\partial_x^2 u + \partial_y^2 u) = 0.$$

This is equivalent to the first order system

$$\partial_t\vec{v} + A\partial_x\vec{v} + B\partial_y\vec{v} = \vec{0}, \tag{3.11}$$

where

$$\vec{v} = \begin{bmatrix} \partial_x u \\ \partial_y u \\ \partial_t u \end{bmatrix}, \quad A = \begin{bmatrix} 0 & 0 & -1 \\ 0 & 0 & 0 \\ -c^2 & 0 & 0 \end{bmatrix}, \quad B = \begin{bmatrix} 0 & 0 & 0 \\ 0 & 0 & -1 \\ 0 & -c^2 & 0 \end{bmatrix}.$$

Unfortunately, A and B do not commute and cannot be simultaneously diagonalized, so the above approach must be modified.

3.2   Strang Splitting

Note that (2.10) can be expressed as

$$\vec{v}(x, t + \Delta t) = e^{-\Delta t A \partial_x} \vec{v}(x, t),$$

where

$$e^{-\Delta t A \partial_x} = I - \Delta t A \partial_x + \frac{\Delta t^2}{2}(A \partial_x)^2 - \cdots$$

defines the propagator for the operator $A \partial_x$. In the two dimensional case,

$$\vec{v}(x, y, t + \Delta t) = e^{-\Delta t (A \partial_x + B \partial_y)} \vec{v}(x, y, t).$$

Strang's idea [7] was to approximate the two dimensional propagators by a product of one dimensional propagators.

One can show

$$e^{-\Delta t (A \partial_x + B \partial_y)} = e^{-\Delta t A \partial_x} e^{-\Delta t B \partial_y} + \frac{\Delta t^2}{2}(AB - BA) + O(\Delta t^3), \quad (3.12)$$

so that a first order accurate approximation is based on

$$\vec{v}(x, y, t^{m+1}) = e^{-\Delta t A \partial_x}(e^{-\Delta t B \partial_y} \vec{v}(x, y, t^m)). \quad (3.13)$$

One can also derive a second order approximation, utilizing the fact that

$$e^{-\Delta t (A \partial_x + B \partial_y)} = e^{-\frac{1}{2}\Delta t A \partial_x} e^{-\Delta t B \partial_y} e^{-\frac{1}{2}\Delta t A \partial_x} + O(\Delta t^3). \quad (3.14)$$

Applications of one dimensional propagators require numerical solutions to systems of the form

$$\partial_t \vec{v} + C \partial_s \vec{v} = 0, \quad (3.15)$$

where C is either A or B corresponding to $s = x$ or $s = y$, respectively. $\vec{v}$ is a function of x, y, and t, but one of x or y is treated as a parameter. For example, to implement (3.13), one first propagates data $\vec{v}^m(x_i, y_j)$ from t^m to $t^m + \Delta t$ by numerically solving N_x linear systems

$$\partial_t \vec{v} + B \partial_y \vec{v} = 0,$$

one for each x_i. Call the resulting approximation $\vec{v}^{intermediate}(x_i, y_j)$. This is then taken as initial data, and one solves the N_y systems

$$\partial_t \vec{v} + A \partial_x \vec{v} = 0,$$

one for each y_j, to obtain $\vec{v}^{m+1}$.

4   Numerical Results

This section contains a number of examples that point out the attributes of
the adaptive stencil method described above. The methods used to obtain
the results in this section employ the Strang splitting (3.14) and, unless
otherwise stated, an $O(\Delta x^2)$ spatial stencil (level 3 in the tree).

4.1   A one-dimensional problem with non-smooth initial data

The first example is the one dimensional initial value problem

$$\partial_t^2 u - c^2 \partial_x^2 u = 0 \tag{4.16}$$
$$u(x,0) = f(x), \qquad \partial_t u(x,0) = 0,$$

where $f(x)$ is the piecewise linear function on the left in Figure 4.1 and
$c = 1$. The derivative $f'(x)$ is shown on the right.

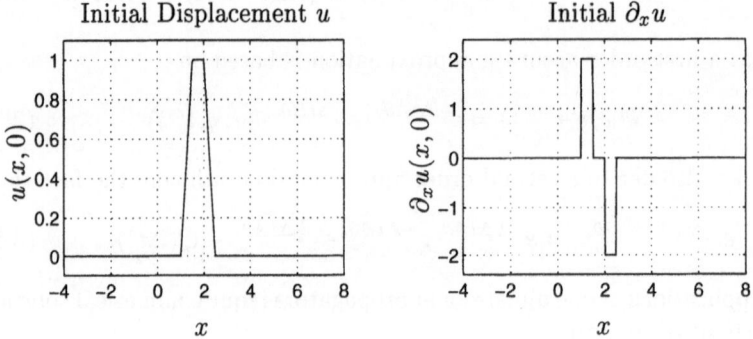

Figure 4.1: Initial data for 1-D homogeneous acoustic initial value problem.

Recalling the first order system formulation in (2.3), the numerical
method seeks to approximate the derivatives of u. The solution of (4.16) is

$$u(x,t) = \frac{1}{2}f(x-t) + \frac{1}{2}f(x+t), \tag{4.17}$$

and so

$$\partial_x u(x,t) = \frac{1}{2}f'(x-t) + \frac{1}{2}f'(x+t), \tag{4.18}$$

i.e., $\partial_x u$ is the sum of translates of the piecewise constant function $\frac{d}{dx}f(x)$
shown on the right in Figure 4.1. In Figure 4.2, we present the approxima-
tion of $\partial_x u$ at various times.

The approximations to the spatial derivative in the above example are

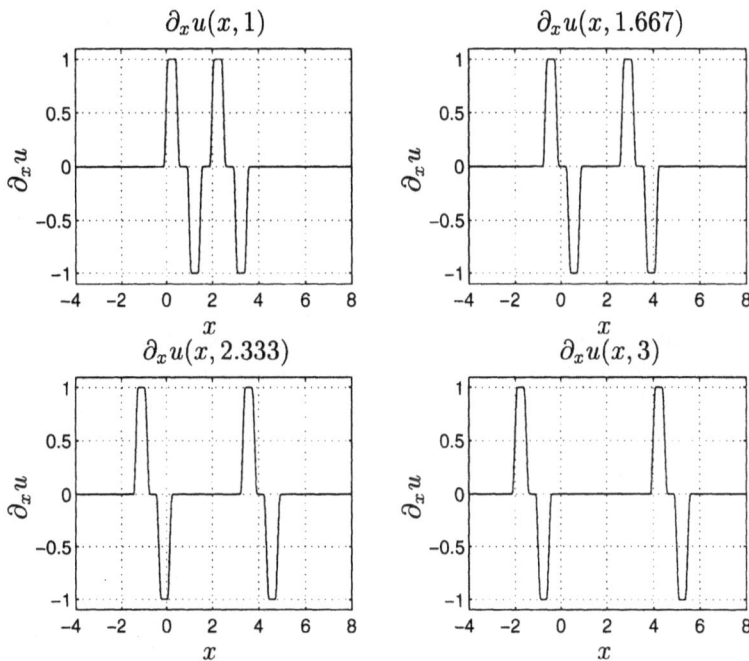

Figure 4.2: Results for the 1-D homogeneous acoustic initial value problem at times $t = 1, \frac{5}{3}, \frac{7}{3}$, and 3.

$O(\Delta x^2)$. In Figure 4.3 we present results obtained using approximations that are $O(\Delta x^n)$ for $n = 1, 2$ and 3. Observe that we have zoomed in on the subinterval $3 \leq x \leq 6$ at time $t = 3$. Note that the approximations improve as n increases, and that none exhibit spurious oscillations.

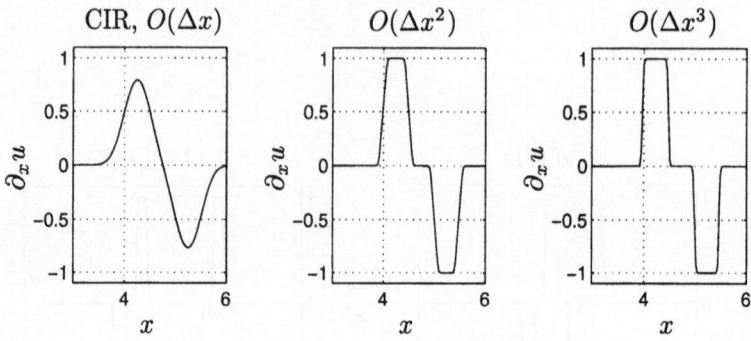

Figure 4.3: Effects of degree on ENO routine.

As a final note to this first example, we present a fixed stencil finite difference approximation to the second order problem. We used standard second order finite difference approximations to the derivatives in (4.16) yielding,

$$\frac{u_j^{m+1} - 2u_j^m + u_j^{m-1}}{\Delta t^2} - \frac{u_{j+1}^m - 2u_j^m + u_{j-1}^m}{\Delta x^2} = 0. \qquad (4.19)$$

We then solved for u_j^{m+1} in terms of known data. We use the same Δt and Δx as in the previous approximations to (4.16), and the CFL condition was met. On the left in Figure 4.4 is the approximation of u at time $t = 3$, on the right is the approximate derivative of $u(x, 3)$. Note the spread in the waveform and introduction of spurious oscillations in the approximation of u. Taking the derivative $\partial_x u$ greatly magnifies these oscillations.

4.2 Boundary Conditions

The adaptive stencil selection procedure outlined above cannot be applied near the boundary of the computational domain. Here we discuss modifications required to accommodate various boundary conditions.

The free surface boundary conditions $\left(\dfrac{\partial u}{\partial n} = 0 \right)$ are easily handled since the components of ∇u (rather than u itself) are computed at each time step. For instance, in the 1-D case, we can *set* $\partial_x u$ equal to zero at the end points

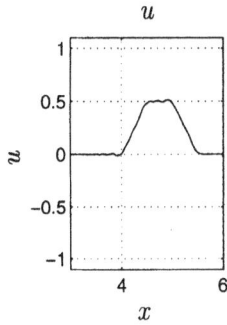

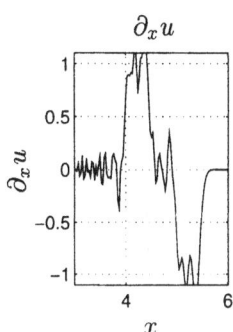

Figure 4.4: Standard finite difference approximation.

at each time step. Standard methods (c.f., (4.19)) require additional, often quite complicated, difference approximations near the endpoints.

Absorbing boundary conditions allow a wave to seemingly pass through a computational boundary with no reflections. The first order wave equation is easily absorbed by using stencils that *look inward* at the computational boundaries. Suppose we wish to approximate the derivative of $\vec{f} = (f_1, \ldots, f_N)$ with the procedure outlined above. We can force the routine to select these inward looking stencils near the boundary by setting $f_0 = f_{N+1} = \infty$, or some relatively large value. The selection scheme described by the tree in Figure 2.1 will *not* select these new end nodes at any level, resulting in absorption.

Figure 4.5 is another series of approximations of $\partial_x u$ for the one dimensional initial value problem (4.16). These show that the wave is absorbed into the computational boundary (located at the ends of the window).

4.3   A two-dimensional problem in heterogeneous media

We now consider

$$\partial_t^2 u - c^2 \left(\partial_x^2 u + \partial_y^2 u \right) = f(t)\delta(x - x_0)\delta(z), \qquad (4.20)$$

where $c(x, z)$ is the piecewise constant function shown on the left in Figure 4.6 and $f(t)$ is the Ricker wavelet commonly used in geophysical problems [3], shown on the right in Figure 4.6.

The method above requires values for $\partial_z c$ and $\partial_x c$. Since c is not differentiable, we use approximations to the distributional derivatives. The support of the distributional derivative that corresponds to $\partial_z c$ consists of the horizontal interfaces and is the short vertical interface for $\partial_x c$.

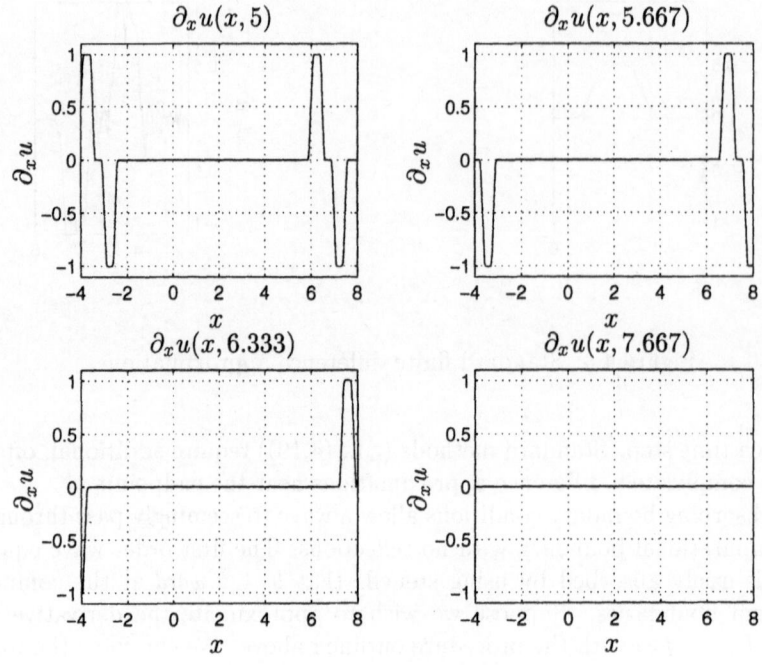

Figure 4.5: Absorbing the one-dimensional wave.

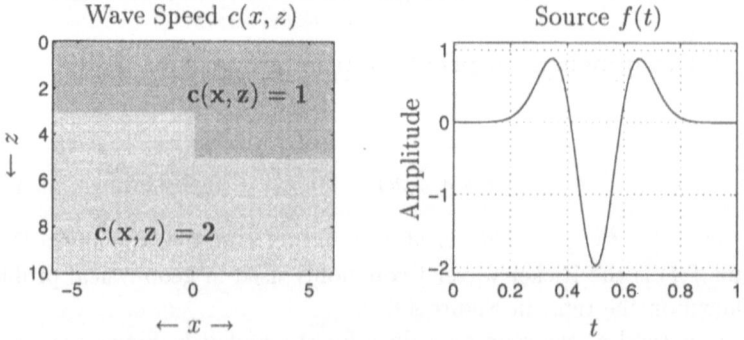

Figure 4.6: Wave Speed and Forcing Function.

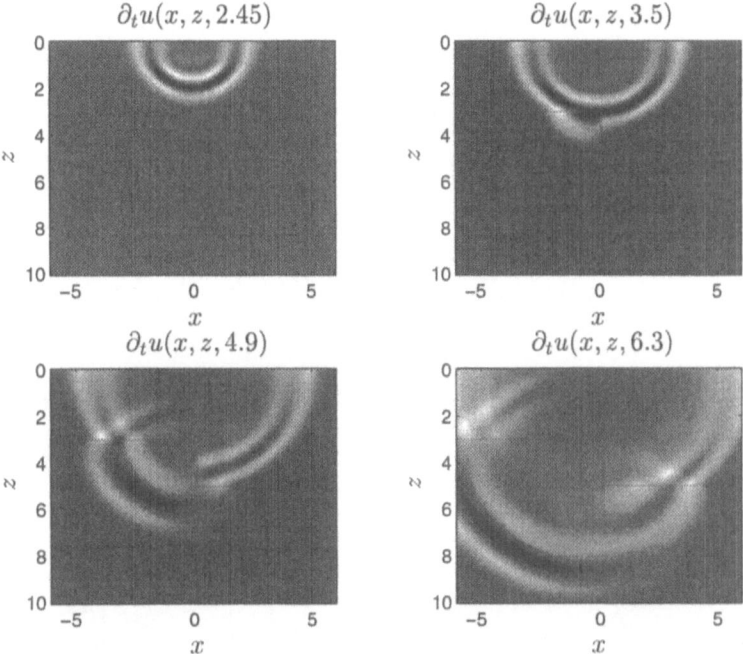

Figure 4.7: $\partial_t u$ at Various Times t.

So that the reflections may be easily seen, absorbing boundary conditions were used on all four sides of the computational boundary in the numerical results shown in Figure 4.7.

In the first frame a wave propagates from a point source at the top center of the media. In the second frame we see the wave soon after it reaches the left interface. Note both the returning reflection and the higher speed transmitted wave. In the third frame, the main wave meets the right horizontal interface causing a reflection which can be seen in the fourth and final frame. The effect of the absorbing boundary conditions can also be seen in this final frame.

4.4 A two-dimensional problem with a slanted interface.

Again consider (4.20), but with the wave speed and forcing function given in Figure 4.8.

In the first frame a wave propagates from a point source at the top center of the media. In the second frame we see the wave soon after it

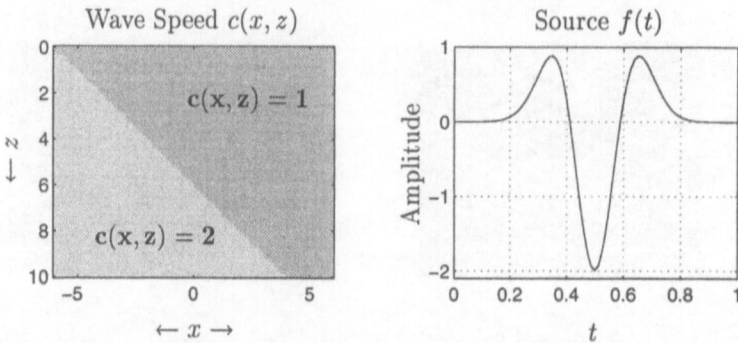

Figure 4.8: Wave Speed and Forcing Function for the Slanted Interface Problem.

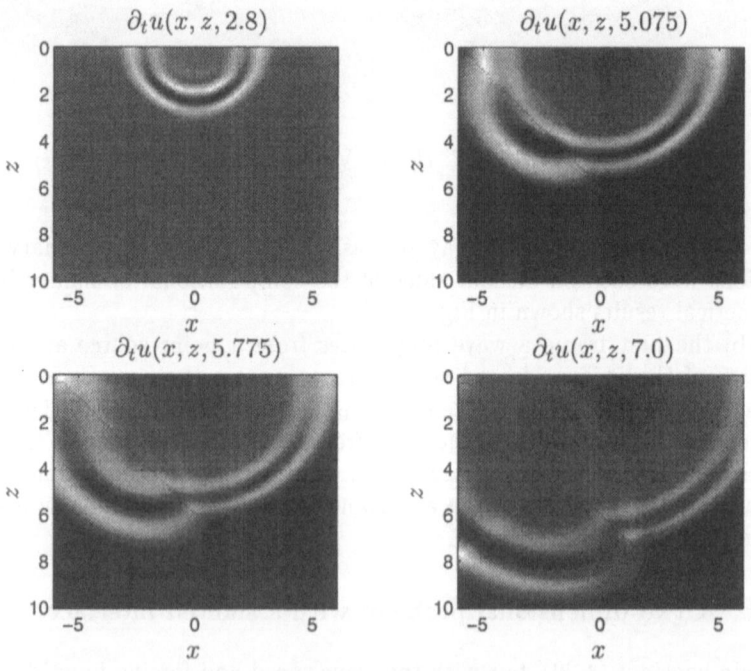

Figure 4.9: $\partial_t u$ at Various Times t for the Slanted Interface Problem.

reaches the slanted interface. In the third frame, part of the transmitted wave reaches the computational boundary and is absorbed, and the reflected wave becomes apparent. In the final frame, both the transmitted and the reflected wave are being absorbed.

In this test case, there seem to be no difficulties associated with discontinuities that are not oriented along the computational grid.

References

[1] R. L. Burden and J. D. Faires. *Numerical Analysis*. Prindle, Weber & Schmidt, Boston, 3rd edition, 1985.

[2] R. Courant, E. Isaacson, and M. Rees. On the solution of nonlinear hyperbolic differential equations by finite differences. *Comm. Pure Appl. Math.*, 5:243–255, 1952.

[3] M. A. Dablain. The application of high-order differencing to the scalar wave equation. *Geophysics*, 51:54–66, 1985.

[4] A. Hall and T. A. Porsching. *Numerical Analysis of Partial Differential Equations*. Prentice-Hall, New Jersey, 1st edition, 1990.

[5] A. Harten, S. Osher, B. Engquist, and S. R. Chakravarthy. Some results on uniformly high-order accurate essentially nonoscillatory schemes. *Applied Numerical Mathematics*, 2:347–377, 1986.

[6] S. Osher. Fronts propagating with curvature dependent speed: Algorithms based on Hamilton-Jacobi formulations. Report No. 87-66, ICASE, September 1987.

[7] G. Strang. On the construction and comparison of difference schemes. *SIAM J. Numer. Anal.*, 5(3):506–517, September 1968.

[8] R. Vichnevetsky and J. B. Bowles. *Fourier analysis of Numerical approximations of Hyperbolic Equations*. SIAM, Philadelphia, 1st edition, 1982.

EIGENVALUE APPROXIMATIONS FOR STURM–LIOUVILLE DIFFERENTIAL EQUATIONS WITH MIXED BOUNDARY CONDITIONS

Mary Jarratt*

Department of Mathematics and Computer Science
Boise State University
Boise, Idaho 83725

1 Introduction

The Sinc-Galerkin method in [1] was used to approximate the eigenvalues of Sturm-Liouville differential equations with Dirichlet boundary conditions on an interval (a, b). The discrete system of the method led to a symmetric generalized eigenvalue problem with the entries of the matrices point evaluations of known functions. The eigenvalues of this system are approximations to the eigenvalues of the differential equation (for either regular or singular differential equations) accurate to within $O\left(\exp(-k\sqrt{N})\right)$ where k is a positive constant and N is related to the number of basis functions used.

In this paper we consider a Sinc–Galerkin scheme on Sturm–Liouville differential equations with mixed boundary conditions at one or both endpoints of the interval where the problem is based. Choosing the basis functions for the Galerkin scheme is important to maintain the symmetry of the system as well as to use the information already known from the Dirichlet case. For the appropriate choice of basis functions, what arises is a symmetric "bordered" matrix system; that is, the matrix system of the new problem is built with the matrices from the Dirichlet problem bordered by vectors incorporating the changes in the boundary conditions. Symmetry is maintained and the elements of the matrices remain relatively simple to find. In fact, the majority of the entries are again point evaluations of known functions with only the eight "corners" of the system found by integration (For regular Sturm-Liouville problems, these integrations can be done very accurately with a numerical quadrature).

The organization of the paper is as follows. Section 2 summarizes the Sinc-Galerkin method for Sturm–Liouville differential equations with Dirichlet boundary conditions on the interval (a, b) (finite or infinite) and states the error bound theorem. Sections 3 (finite interval case) and 4 (infinite interval case) show the process of building (from the basic system of Section 2) the matrix systems for the Sturm-Liouville differential equation with mixed boundary conditions. Section 5 gives examples that show the

*Supported by Idaho State Board of Education Grant # S93–059

implementation and accuracy of the method (It appears in these examples that the exponential accuracy of the Dirichlet case is maintained.). The final section gives the statement and proof of the error bound theorem for the method described in Sections 3 and 4.

2 Sinc–Galerkin Method

In [1] and [3], the Sinc-Galerkin method (equivalent to the sinc–collocation method because of properties of the sinc function) was used to approximate

$$Lu = -u''(x) + q(x)u(x) = \lambda\rho(x)u(x), \ a < x < b \ \Big\} \tag{2.1}$$
$$u(a) = u(b) = 0$$

The function q is assumed to be nonnegative and ρ is assumed to be positive. The interval on which the problem is based can be either finite or infinite.

The sinc function is defined by

$$\text{sinc}(t) = \frac{\sin(\pi t)}{\pi t}, \ \ t \in (-\infty, \infty).$$

For a function f defined on the whole real line and $h > 0$, the Whittaker Cardinal expansion of f (when the series converges) is given by

$$C(f, h)(t) = \sum_{k=-\infty}^{\infty} f(kh)S_k(t) \tag{2.2}$$

where

$$S_k(t) = \text{sinc}\left(\frac{t - kh}{h}\right).$$

(See [5] and [6] for properties of (2.2).)

Of interest in this setting is a class of functions $B(D_S)$ that are approximated by (2.2) and characterized by the following definition:

Definition 2.1 *Let* $B(D_S)$, *be the class of functions* f *analytic in* D_S *where, for* $d > 0$

$$D_S = \left\{t + is : |s| < d \le \frac{\pi}{2}\right\}$$

that satisfy

$$\int_{-d}^{d} |f(t + is)| ds \to 0 \ as \ t \to \pm\infty$$

and

$$N_2(f) \equiv \max\left\{\lim_{s \to d^-}\left(\int_{-\infty}^{\infty} |f\left(|t + is|^2\right) \ dt\right)^{1/2}\right\} < \infty.$$

Since (2.1) is defined on an arbitrary interval (a, b), the following definition allows the interval to be mapped to $(-\infty, \infty)$ and all results dealing with the Whittaker Cardinal expansion on R^1 can be carried over to (a, b).

Definition 2.2 *Let D_d be a simply connected domain in the complex plane with boundary points $a \neq b$. Let φ be a conformal mapping from D_d onto D_S with $\varphi(a) = -\infty$ and $\varphi(b) = \infty$. Let ψ be the inverse mapping of φ and define $\Gamma = \{\psi(t) : -\infty < t < \infty\}$ and $x_k = \psi(kh)$ for $k = 0, \pm 1, \pm 2, \ldots$.*

With the substitution

$$f = (\sqrt{\varphi'}u) \circ \psi(t)$$

(2.1) is transformed to the whole real line, yet retains the same eigenvalues. Let f and f'' be approximated by

$$f_m \equiv C_{M,N}(t) = \sum_{k=-M}^{N} f(kh)S_k(t),$$

and

$$f_m'' \equiv \frac{d^2 C_{M,N}}{dt^2} = \sum_{k=-M}^{N} f(kh)\frac{d^2}{dt^2}(S_k(t))$$

where $m = M + N + 1$.

To find the unknowns $f(kh), k = -M, \ldots, N$, use the Galerkin scheme defined by

$$0 = (Lf_m - \lambda \rho f_m, S_k \circ \varphi) = \int_a^b (Lf_m - \lambda \rho f_m)(S_k \circ \varphi)(1/\sqrt{\varphi'})\,dt; \quad (2.3)$$

that is, orthogonalize the residual with basis functions $S_k \circ \varphi$, $k = -M, \ldots, N$ and weight function $1/\sqrt{\varphi'}$ (using collocation at the nodes $t_j = jh$, $j = -M, \ldots, N$ yields the same results). The matrix system that results is the generalized eigenvalue problem

$$A\vec{z} = \mu B\vec{z} \quad (2.4)$$

with $\vec{z} = D(\sqrt{\varphi'})\vec{u}$.

In the notation of the original problem

$$A = -\frac{1}{h^2}I^{(2)} + D\left(-\left(\frac{1}{\sqrt{\varphi'}}\right)'' \frac{1}{(\varphi')^{3/2}} + \frac{q}{(\varphi')^2}\right) \quad (2.5)$$

and

$$B = D\left(\frac{\rho}{(\varphi')^2}\right) \quad (2.6)$$

where

$$(I^{(2)})_{ij} = \begin{cases} \frac{-\pi^2}{3} & i = j \\ \frac{-2(-1)^{i-j}}{(i-j)^2} & i \neq j \end{cases}$$

and diagonal matrices are denoted $D(\cdot)$ with

$$(D(f))_{ij} = \begin{cases} f(t_i) & i = j \\ 0 & i \neq j \end{cases}$$

Notice that since $I^{(2)}$ is symmetric, A is a symmetric matrix. The error in approximating a true eigenvalue of (2.1) by (2.4) is given in the following theorem:

Theorem 2.3 *Let λ_0 and u_0 be an eigenpair of (2.1). Assume $\sqrt{\varphi'} u \circ \psi \in B(D_S)$ and that positive constants α, β, and c exist so that*

$$|\sqrt{\varphi'} u| \leq c \begin{cases} \exp(-\alpha|\varphi(x)|), & x \in \Gamma_a \\ \exp(-\beta|\varphi(x)|), & x \in \Gamma_b \end{cases}$$

where $\Gamma_a = \{\psi(t) : t \in (-\infty, 0]\}$ and $\Gamma_b = \{\psi(t) : t \in (0, \infty)\}$. If there is a constant $\delta > 0$ so that $|\gamma_q(x)| \geq \delta^{-1}$ where

$$\gamma_q(x) = -\left(\frac{1}{\sqrt{\varphi'}}\right)'' \frac{1}{(\varphi')^{3/2}} + \frac{q}{(\varphi')^2}$$

then for $h = (\pi d/\alpha M)^{1/2}$ and $N = \left[\left|\frac{\alpha}{\beta} M\right|\right]$, there is an eigenvalue μ_p of (2.4) satisfying

$$|\mu_p - \lambda_0| \leq k(\delta\lambda_0)^{1/2} M^{1/2} \exp(-(\pi d\alpha M)^{1/2}).$$

The mappings that have been studied in connection with various intervals are found in [1], [2] and [3]. For (a, b) finite, $\varphi(x) = \ln\left(\frac{x-a}{b-x}\right)$. On $(0, \infty)$, two choices of mapping are discussed in [1]: $\varphi(x) = \ln(x)$ and $\varphi(x) = \ln(\sinh(x))$. For the case of the whole real line, the identity mapping, $\varphi(x) = x$, is considered in detail in [3].

3　Mixed Boundary Conditions: Finite Interval

Now consider the boundary value problem on a finite interval (a, b) given by

$$\left.\begin{array}{l} Lu = -u''(x) + q(x)u(x) = \lambda\rho(x)u(x) \\ k_1 u'(a) - k_2 u(a) = 0 \\ K_1 u'(b) + K_2 u(b) = 0 \end{array}\right\} \tag{3.1}$$

where $q \geq 0$ and $\rho > 0$, as before, and k_1, k_2, K_1, and K_2 are nonnegative constants. Let an approximate solution to (3.1) be

$$u_n(x) = c_{-M-1}\omega_a(x) + \sum_{j=-M}^{N} c_j \frac{S_j \circ \varphi(x)}{\varphi'(x)} + c_{N+1}\omega_b(x) \qquad (3.2)$$

with $n = M + N + 3$ and the c_j's unknown constants. The conformal mapping, $\varphi(x)$, of (a, b) onto $(-\infty, \infty)$ is the mapping in Section 2. The function $\omega_a(x)$ is dependent on k_1 and k_2 and $\omega_b(x)$ is dependent on K_1 and K_2. They are chosen so that u_n satisfies the boundary conditions in (3.1). The functions are

$$\omega_a(x) = -(k_2(b-a) + 2k_1)\left(\frac{b-x}{b-a}\right)^3 + (k_2(b-a) + 3k_1)\left(\frac{b-x}{b-a}\right)^2 \quad (3.3)$$

and

$$\omega_b(x) = -(K_2(b-a) + 2K_1)\left(\frac{x-a}{b-a}\right)^3 + (K_2(b-a) + 3K_1)\left(\frac{x-a}{b-a}\right)^2. \quad (3.4)$$

When $k_1 = 0$, the first boundary condition simplifies to $u(0) = 0$ and there is no need for ω_a; hence, let $c_{-M-1} = 0$ in (3.2). Similarly if $K_1 = 0$, let $c_{N+1} = 0$.

Looking at this problem from a Galerkin standpoint, the coefficients c_j are determined by orthogonalizing the residual with respect to the set of basis functions, $\xi_p(x)$, $p = -M - 1, -M, \ldots, N, N + 1$, with

$$\xi_p(x) = \begin{cases} \omega_a(x) & p = -M - 1 \\ (S_p \circ \varphi(x))/\sqrt{\varphi'(x)} & p = -M, \ldots, N \\ \omega_b(x) & p = N + 1 \end{cases}$$

and weight function 1; that is, from the $M + N + 3$ conditions given by

$$0 = (Lu_n - \lambda\rho u_n, \xi_p) = \int_a^b (-u_n'' + qu_n - \lambda\rho u_n)\xi_p \, dx \qquad (3.5)$$

solve for the c_j's. For this set of basis functions, define the $(M + N + 3) \times (M + N + 3)$ system associated with (3.5) by

$$\mathcal{A}\vec{c} = \mu \mathcal{B}\vec{c}. \qquad (3.6)$$

The following discussion will explicitly give the forms of $\mathcal{A}, B$, and $\vec{c}$ and reasons for the choice of basis functions.

Rewrite the inner product in (3.5) as

$$-\int_a^b u_n''\xi_p \, dx = \int_a^b (-qu_n + \lambda\rho u_n)\xi_p \, dx \qquad (3.7)$$

and let $p = -M, \ldots, N$. Consider first

$$\int_a^b (-qu_n + \lambda \rho u_n)\xi_p \, dx = \int_a^b (-q + \lambda \rho) \left(\frac{S_p \circ \varphi}{\sqrt{\varphi'}} \right) u_n \, dx.$$

Using the sinc quadrature rule on the above integral, applying the properties of the sinc function [5], and substituting (3.2) in for u_n yields

$$\int_a^b (-q + \lambda \rho) \left(\frac{S_p \circ \varphi}{\sqrt{\varphi'}} \right) u_n \, dx = h \sum_{j=-\infty}^{\infty} \left(\frac{-q_j + \lambda \rho_j}{(\varphi_j')^{3/2}} \right) (S_p \circ \varphi)_j (u_n)_j$$

$$= h \left(\frac{-q_p + \lambda \rho_p}{(\varphi_p')^{3/2}} \right) \left(c_{-M-1}(\omega_a)_p + \frac{c_p}{\varphi_p'} + c_{N+1}(\omega_b)_p \right)$$

where $f_p = f(x_p)$.

For the integral on the left-hand side of (3.7), two integrations by parts are done. Then letting the boundary terms $(u_n' \xi_p - u_n \xi_p')|_0^1 = BT(u_n)$,

$$\int_a^b u_n'' \xi_p \, dx = \int_a^b u_n \xi_p'' \, dx + BT(u_n)$$

$$= c_{-M-1} \int_a^b \omega_a \xi_p'' \, dx + c_{N+1} \int_a^b \omega_b \xi_p'' \, dx \qquad (3.8)$$

$$+ h \sum_{j=-M}^{N} c_j \left(\frac{1}{(\varphi_j)^{3/2}} \right) (\xi_p'')_j + BT(u_n)$$

(using again the sinc quadrature and properties of the sinc function). Following the same procedure for the integrals on the right-hand side of (3.8),

$$\int_a^b \omega_i \xi_p'' \, dx = \int_a^b \omega_i'' \xi_p \, dx + BT(\omega_i)$$

$$= \frac{h(\omega_i'')_p}{(\varphi_p')^{3/2}} + BT(\omega_i), \ i = a, b.$$

With u_n written as in (3.2), a short computation shows that $BT(u_n) + BT(\omega_a) + BT(\omega_b) = 0$ (one reason for the choice of ω_a, ω_b, and ξ_p).

Also from (3.5)

$$(\xi_p'')_j = (S_p \circ \varphi)_j \left(\frac{1}{\sqrt{\varphi'}} \right)_j'' + (S_p'' \circ \varphi)_j \frac{(\varphi_j')^2}{\sqrt{\varphi_j'}}$$

$$+(S_p' \circ \varphi)_j \left\{ 2\varphi_j' \left(\frac{1}{\sqrt{\varphi'}} \right)_j' + \frac{1}{\sqrt{\varphi_j'}} \varphi_j'' \right\} \qquad (3.9)$$

$$= \delta_{pj}^{(0)} \left(\frac{1}{\sqrt{\varphi'}} \right)_j'' + \frac{1}{h^2} \delta_{pj}^{(2)} (\varphi_j')^{3/2}$$

where

$$\delta_{pj}^{(k)} = h^k \left(\frac{d^k}{dx^k} (S_p) \circ \varphi(x) \right) \bigg|_{x=x_j} \qquad \text{and} \quad k = 0, 2.$$

Notice that the bracketed term in (3.9) is zero. Because of this, the matrix system that eventually results is symmetric (another reason for the choice of ξ_p).

Finally, then, for $p = -M, \ldots, N$, (3.7) becomes

$$-\int_a^b u_n'' \xi_p \, dx = -h \sum_{j=-M}^{N} c_j \frac{1}{(\varphi_j')^2} \left\{ \delta_{pj}^{(0)} \left(\frac{1}{\sqrt{\varphi'}} \right)_j'' + \frac{1}{h^2} \delta_{pj}^{(2)} (\varphi_j')^{3/2} \right\}$$

$$-h \frac{(\omega_a'')_p}{(\varphi_p')^{3/2}} c_{-M-1} - h \frac{(\omega_b'')_p}{(\varphi_p')^{3/2}} c_{N+1} \qquad (3.10)$$

$$= h \left(\frac{-q_p + \lambda \rho_p}{(\varphi_p')^{3/2}} \right) \left\{ c_{-M-1}(\omega_a)_p + \frac{c_p}{\varphi_p'} + c_{N+1}(\omega_b)_p \right\}$$

$$= \int_a^b (-qu_n + \lambda \rho u_n) \xi_p \, dx.$$

The matrix form of (3.10) can be written as a $(M+N+1) \times (M+N+3)$ partitioned system

$$[\, \vec{\alpha}_a \mid A \mid \vec{\alpha}_b \,] \vec{c} = \lambda [\, \vec{\gamma}_a \mid B \mid \vec{\gamma}_b \,] \vec{c}$$

where A and B are (2.5) and (2.6) from the Dirichlet problem (both $(M+N+1) \times (M+N+1)$). For $i = a$ and b

$$(\vec{\alpha}_i)_j = \frac{((\omega_i'')_j - q_j(\omega_i)_j)}{(\varphi_j')^{3/2}} \qquad (3.11)$$

$$(\vec{\gamma}_i)_j = \frac{\rho_j(\omega_i)_j}{(\varphi_j')^{3/2}} \qquad (3.12)$$

and

$$
\vec{c} =
\begin{bmatrix}
c_{-M-1} \\
c_{-M}/\sqrt{\varphi'_{-M}} \\
\vdots \\
c_N/\sqrt{\varphi'_N} \\
c_{N+1}
\end{bmatrix}
$$

where $\vec{\alpha}_i$ and $\vec{\gamma}_i$ are $(M+N+1) \times 1$ and $\vec{c}$ is $(M+N+3) \times 1$.

To complete filling the matrices $\mathcal{A}$ and $\mathcal{B}$, consider again the inner product (3.5) for $p = -M-1$ and $N+1$.

$$
\int_a^b (-qu_n + \lambda \rho u_n)\xi_{-M-1}\, dx = c_{-M-1}\int_a^b (-q+\lambda\rho)\omega_a^2\, dx
$$
$$
+ h\sum_{j=-M}^N c_j \left(\frac{-q_j + \lambda\rho_j}{(\varphi'_j)^2}\right)(\omega_a)_j
$$
$$
+ c_{N+1}\int_a^b (-q+\lambda\rho)\omega_a\omega_b\, dx
$$

and

$$
\int_a^b (-qu_n + \lambda\rho u_n)\xi_{N+1}\, dx = c_{-M-1}\int_a^b (-q+\lambda\rho)\omega_a\omega_b\, dx
$$
$$
+ h\sum_{j=-M}^N c_j \left(\frac{-q_j + \lambda\rho_j}{(\varphi'_j)^2}\right)(\omega_b)_j
$$
$$
+ c_{N+1}\int_a^b (-q+\lambda\rho)\omega_b^2\, dx.
$$

Also

$$
\int_a^b u_n''\xi_{-M-1}\, dx = \int_a^b u_n \xi''_{-M-1}\, dx
$$
$$
= c_{-M-1}\int_a^b \omega_a\omega_a''\, dx + c_{N+1}\int_a^b \omega_b\omega_a''\, dx
$$
$$
+ h\sum_{j=-M}^N c_j \frac{(\omega_a'')_j}{(\varphi'_j)^2}.
$$

Similarly

$$
\int_a^b u_n''\xi_{N+1}\, dx = c_{-M-1}\int_a^b \omega_a\omega_b''\, dx + c_{N+1}\int_a^b \omega_b\omega_b''\, dx
$$

$$+ h \sum_{j=-M}^{N} c_j \frac{(\omega_b'')_j}{(\varphi_j')^2}.$$

The choice of ω_a and ω_b as the basis functions for $p = -M-1$ and $p = N+1$ and the boundary conditions from the given differential equation zero out the boundary terms in the two integrations by parts above. The vector forms for these cases (taking into account $\vec{c}$) are $\vec{\alpha}_i$ and $\vec{\gamma}_i$ for $i = a$ and b, as in (3.11) and (3.12). Letting $(i, j = a, b)$

$$k_{ij} = \frac{1}{h} \int_a^b (-\omega_i'' \omega_j + q \omega_i \omega_j) \, dx \tag{3.13}$$

and

$$s_{ij} = \frac{1}{h} \int_a^b \rho \omega_i \omega_j \, dx, \tag{3.14}$$

the system becomes $\mathcal{A}\vec{c} = \mu \mathcal{B}\vec{c}$ with

$$\mathcal{A} = \left[\begin{array}{c|c|c} k_{aa} & \vec{\alpha}_a^T & k_{ab} \\ \hline \vec{\alpha}_a & A & \vec{\alpha}_b \\ \hline k_{ba} & \vec{\alpha}_b^T & k_{bb} \end{array} \right] \tag{3.15}$$

and

$$\mathcal{B} = \left[\begin{array}{c|c|c} s_{aa} & \vec{\gamma}_a^T & s_{ab} \\ \hline \vec{\gamma}_a & B & \vec{\gamma}_b \\ \hline s_{ba} & \vec{\gamma}_b^T & s_{bb} \end{array} \right].$$

With a short computation it is easy to show that $k_{ab} = k_{ba}$ and clearly, $s_{ab} = s_{ba}$. Therefore $\mathcal{A}$ and $\mathcal{B}$ are both symmetric. All but eight of the entries of $\mathcal{A}$ and $\mathcal{B}$ are point evaluations of known functions. The corner entries are integrals and can be calculated by some means dependent on the form of q and ρ.

4   Mixed Boundary Conditions: Semi–infinite Interval

When the boundary value problem is defined on the semi–infinite interval $(0, \infty)$, the differential equation and boundary conditions are given by

$$\left. \begin{array}{l} Lu = -u''(x) + q(x)u(x) = \lambda \rho(x)u(x) \\ k_1 u'(0) - k_2 u(0) = 0 \\ u \text{ bounded at } \infty \end{array} \right\} \tag{4.1}$$

where k_1 and k_2 are nonnegative constants. With only the mixed condition at the finite endpoint of the interval, the approximate solution to the differential equation is given by

$$u_n(x) = c_{-M-1}\omega_0(x) + \sum_{j=-M}^{N} c_j \frac{S_j \circ \varphi(x)}{\varphi'(x)} \tag{4.2}$$

with $n = M + N + 2$ and

$$\omega_0(x) = -(k_2 + 2k_1)\left(\frac{1}{1+x}\right)^3 + (k_2 + 3k_1)\left(\frac{1}{1+x}\right)^2 \tag{4.3}$$

As before, ω_0 was chosen so that u_n satisfies the boundary conditions.

For the Dirichlet problem on $(0, \infty)$, two choices were available for the mapping: $\varphi(x) = \ln(x)$ or $\varphi(x) = \ln(\sinh(x))$. Both mappings give the expected exponential error in that setting; however, given a particular problem, one mapping may give the desired accuracy of the eigenvalue approximation with a smaller matrix system than the other mapping. In the setting of the mixed boundary conditions, for $\varphi(x) = \ln(x)$, u_n in (4.2) is not bounded at ∞. With $\varphi(x) = \ln(\sinh(x))$, the boundedness condition at ∞ is satisfied (as well as the boundary condition at $x = 0$) so that will be the mapping for this problem.

When applying the Galerkin scheme to (4.1), the basis functions are $\xi_p(x)$, $p = -M - 1, -M, \ldots, N$, with

$$\xi_p(x) = \begin{cases} \omega_0(x) & p = -M - 1 \\ (S_p \circ \varphi(x))/\sqrt{\varphi'(x)} & p = -M, \ldots, N \end{cases}$$

and weight function 1. The $(M+N+2)$ x $(M+N+2)$ matrix system that arises is given by

$$A\vec{c} = \mu B\vec{c}. \tag{4.4}$$

The procedure to find the entries of A and B is the same as for the finite interval case. In fact, this procedure yields a matrix system that is exactly the upper left corner of the finite interval system. That is,

$$A = \left[\begin{array}{c|c} k_{00} & \vec{\alpha}^T \\ \hline \vec{\alpha} & A \end{array} \right] \tag{4.5}$$

and

$$B = \left[\begin{array}{c|c} s_{00} & \vec{\gamma}^T \\ \hline \vec{\gamma} & B \end{array} \right]$$

for k_{00}, s_{00}, $\vec{\alpha}$, $\vec{\gamma}$, A, B, and $\vec{c}$ as defined in Section 3 (using ω_0 in (4.3) in the definitions of k_{00}, s_{00}, $\vec{\alpha}$, and $\vec{\gamma}$). Of course, the lack of a mixed boundary condition at the right–hand endpoint of the interval accounts for the lack of the bottom row and right–hand column in the A and B of this case that are found in the matrices of the finite interval case.

5   Numerical Examples

This section includes two examples that illustrate the Sinc–Galerkin method of Section 3. The differential equation for both examples will be

$$-u''(x) = -\lambda u(x).$$

The boundary conditions will determine the eigenvalues, λ. The integrals given by (3.13) and (3.14) were approximated by the 'quad8' routine in Matlab with accuracy 10^{-16}. For both examples, these integrals could have been done exactly, but the author chose to include the routine in her program to make it fit the general setting where the integrals might not be able to be done exactly.

The tables following each example give the errors in the approximations of the first four eigenvalues. The number of basis functions used in the approximation, u_n (from (3.2)) is given by $n = 2M + 3$ (the matrix systems are $(2M + 3)$ x $(2M + 3)$ since a symmetric sum $(M = N)$ was used in (3.2). In both cases, notice that the first eigenvalues are approximated more closely than the latter eigenvalues. This same phenomena can be seen in the Dirichlet case [1] where the eigenvalue itself is a part of the error bound. It is the feeling of this author that the error bound for the mixed case will also have this property.

Example 5.1:
$$u'(0) = 0, \quad u'(\pi) = 0$$

The true eigenvalues using these boundary conditions are $\lambda_n = n^2$ for n a nonnegative integer. To find ω_0 and ω_π (in (3.3) and (3.4)) in the approximate solution u_n, note that $k_1 = 1$, $k_2 = 0$, $K_1 = 1$, and $K_2 = 0$. Therefore

$$\omega_0(x) = -2 \left(\frac{\pi - x}{\pi} \right)^3 + 3 \left(\frac{\pi - x}{\pi} \right)^2$$

and

$$\omega_\pi(x) = -2 \left(\frac{x}{\pi} \right)^3 + 3 \left(\frac{x}{\pi} \right)^2.$$

The conformal mapping for the interval $(0, \pi)$ is $\varphi(x) = \ln \left(\frac{x}{\pi - x} \right)$. The nodes that will be used are $x_k = \frac{\pi e^{kh}}{e^{kh} + 1}$ (from Definition 2.2). Also notice that $q(x) = 0$ and $\rho(x) = 1$.

λ_p	M =	20	30	40
0		1.1×10^{-16}	4.9×10^{-16}	4.3×10^{-16}
1		3.1×10^{-7}	4.0×10^{-9}	8.3×10^{-11}
4		3.2×10^{-4}	6.4×10^{-6}	2.0×10^{-7}
9		1.6×10^{-2}	6.5×10^{-4}	3.2×10^{-5}

Table 5.1: Eigenvalues and Errors in Approximations of Eigenvalues for Example 5.1

λ_p	M =	20	30	40
0.74017		8.1×10^{-11}	6.0×10^{-13}	2.4×10^{-14}
11.73486		6.3×10^{-6}	8.5×10^{-8}	1.8×10^{-9}
41.43880		3.8×10^{-3}	7.9×10^{-5}	2.5×10^{-6}
90.80821		1.8×10^{-1}	7.1×10^{-3}	3.6×10^{-4}

Table 5.2: Eigenvalues and Errors in Approximations of Eigenvalues for Example 5.2

Example 5.2:
$$u'(0) - u(0) = 0, \quad u'(1) = 0$$

The true eigenvalues using these boundary conditions are the roots of $\tan(\sqrt{\lambda}) - 1/\sqrt{\lambda} = 0$. The first 4 roots were calculated using the 'fzero' command in Matlab with tolerance 10^{-16} (only 5 decimal places are noted in the table). Because $k_1 = 1$, $k_2 = 1$, $K_1 = 1$, and $K_2 = 0$,

$$\omega_0(x) = -3(1-x)^3 + 4(1-x)^2$$

and

$$\omega_1(x) = -2x^3 + 3x^2.$$

The conformal mapping for the interval $(0,1)$ is $\varphi(x) = \ln\left(\dfrac{x}{1-x}\right)$ with nodes $x_k = \dfrac{e^{kh}}{e^{kh}+1}$. Again, $q(x) = 0$ and $\rho(x) = 1$.

For both examples it is interesting to note that $\mathcal{B}$, for each of the three sizes given, is essentially positive semi-definite with condition numbers extremely large (10^{12} to 10^{17}). In example 2, $\mathcal{A}$, for each size, is nicely positive definite with condition number on the order of 10 to 100. However, in example 1 $\mathcal{A}$, for each size, is essentially positive semi-definite (condition numbers from 10^{12} to 10^{16}). Even with the indications from the condition numbers that there could be problems in approximating the eigenvalues using these matrices, the eigenvalues have been approximated very accurately.

6   Error Bound

In this section we consider the error bound for the Sinc–Galerkin method as described in Sections 3 and 4. We begin with some notation and results from matrix and Sturm–Liouville theory that will aid in the proof of the error bound theorem.

In applying the Sinc–Galerkin method to the differential equation (3.1) or (4.1), the matrix system that arises is

$$\overline{L(u_n)_G} \equiv \mathcal{A}\vec{c} = \mu\mathcal{B}\vec{c}$$

where $\mathcal{A}$ and $\mathcal{B}$ are as defined in (3.15) and (4.5).

Let u_t and λ_t be a true eigenpair of (2.1) with u_t normalized, that is,

$$\int_a^b u_t^2(x)\rho(x)dx = 1. \tag{6.1}$$

When u_t is put into the differential equation and evaluated at the points $x = jh$, $j = -M - l, \ldots, N + 1$, the system that arises is given by

$$\overline{Lu_t} = \lambda_t \mathcal{D}u_t$$

where

$$\mathcal{D} = \begin{bmatrix} \rho_{-M-1} & 0 & \cdots & & & 0 \\ 0 & \rho_{-M}\sqrt{\varphi'_{-M}} & 0 & \cdots & & \vdots \\ \vdots & & \ddots & & & \\ & & & & \rho_N\sqrt{\varphi'_N} & 0 \\ 0 & & \cdots & & 0 & \rho_{N+1} \end{bmatrix}$$

and

$$\vec{u}_t = \begin{bmatrix} (u_t)_{-M-1} \\ (u_t)_{-M}/\sqrt{\varphi'_{-M}} \\ \vdots \\ (u_t)_N/\sqrt{\varphi'_N} \\ (u_t)_{N+1} \end{bmatrix}.$$

Notice that since $\rho > 0$ by definition and $\sqrt{\varphi'_j} > 0$ for all $j = -M - 1, \ldots, N + 1$, $\mathcal{D}$ is positive definite.

Now let

$$\overline{\Delta u_t} = \overline{L(u_t)}_G - \overline{L u_t} = (\mathcal{A} - \lambda_t \mathcal{D})\vec{u}_t \tag{6.2}$$

Because $\mathcal{A}$ and $\mathcal{D}$ are symmetric and $\mathcal{D}$ is positive definite, there exist independent eigenvectors $\vec{z}_i$ and eigenvalues μ_i $(i = -M - 1, \ldots, N + 1)$ so that

$$Z^T \mathcal{A} Z = D(\mu_i) \text{ and } Z^T \mathcal{D} Z = I \tag{6.3}$$

This implies that

$$\mathcal{A}\vec{z}_i = \mu_i \mathcal{D}\vec{z}_i.$$

Because the set of eigenvectors is independent, there exist constants β_i so that

$$\vec{u}_t = \sum_{i=-M-1}^{N+1} \beta_i \vec{z}_i \tag{6.4}$$

The following lemma gives a bound on the two–norm of an eigenvector $\vec{z}_i$ in terms of its corresponding eigenvalue μ_i and the smallest eigenvalue of $\mathcal{A}$.

Lemma 6.1 *Let $\mathcal{A}$ be a positive definite matrix with eigenvalues $\{\alpha_i\}_{i=-M-1}^{N+1}$ and corresponding eigenvectors $\{\vec{z}_i\}_{-M-1}^{N+1}$. Also let the eigenvalues of $\mathcal{A}$ be ordered so that $0 < \alpha_{-M-1} \le \alpha_{-M} \le \cdots \le \alpha_{N+1}$. Then*

$$\| \vec{z}_j \|_2^2 \le \frac{\mu_j}{\alpha_{-M-1}}. \tag{6.5}$$

Proof: Using the Rayleigh quotient and $\vec{z}_j^T \mathcal{A}\vec{z}_j = \mu_j$ from (6.3), we have

$$\begin{aligned}
\alpha_{-M-1} = \min(\alpha_j) &= \min_{\vec{x}^T \vec{x}} \left(\frac{\vec{x}^T \mathcal{A}\vec{x}}{\vec{x}^T \vec{x}} \right) \\
&\le \frac{\vec{z}_j^T \mathcal{A}\vec{z}_j}{\vec{z}_j^T \vec{z}_j} \\
&= \frac{\mu_j}{\| \vec{z}_j \|_2^2}
\end{aligned}$$

The conclusion follows from this.

We now state and prove the error bound theorem for the Sinc–Galerkin method.

Theorem 6.2 *Let λ_t and u_t be an eigenpair of (3.1) or (4.1). Assume the hypotheses of Lemma 5.1. Also assume that $(u_t \circ \psi)\psi' \in B(D_S)$ and there exist constants c, α, and β so that*

$$(u_t \circ \psi)\psi' \leq c \begin{cases} \exp(-\alpha|x|), & x \in (-\infty, 0) \\ \exp(-\beta|x|), & x \in [0, \infty) \end{cases}$$

Then for $h = (\pi d/\alpha M)^{1/2}$ and $N = \left[\left| \frac{\alpha}{\beta} M \right| \right]$, there is an eigenvalue μ_p of (3.6) or (4.4) satisfying

$$|\mu_p - \lambda_t| \leq \sqrt{\frac{2\lambda_t}{\alpha_{-M-1}}} \, \| \overline{\Delta u_t} \|_2 \, \sqrt{2h(M + N + 3)}.$$

Proof: Putting (6.4) into (6.2) yields

$$\begin{aligned} \overline{\Delta u_t} &= (\mathcal{A} - \lambda_t \mathcal{D}) \sum_{i=-M-1}^{N+1} \beta_i \vec{z}_i \\ &= \sum_{i=-M-1}^{N+1} (\beta_i \mathcal{A} \vec{z}_i - \lambda_t \beta_i \mathcal{D} \vec{z}_i) \\ &= \sum_{i=-M-1}^{N+1} \beta_i (\mu_i - \lambda_t) \mathcal{D} \vec{z}_i \end{aligned}$$

Taking an inner product of this with $\vec{z}_p$, $(p = -M - 1, \ldots, N + 1)$ gives

$$\begin{aligned} \vec{z}_p^T \overline{\Delta u_t} &= \mathcal{D} \vec{z}_p^T \sum_{i=-M-1}^{N+1} \beta_i (\mu_i - \lambda_t) \mathcal{D} \vec{z}_i \\ &= \sum_{i=-M-1}^{N+1} \beta_i (\mu_i - \lambda_t) \vec{z}_p^T \mathcal{D} \vec{z}_i \\ &= \beta_p (\mu_p - \lambda_t) \end{aligned} \qquad (6.6)$$

Applying $\mathcal{D}$ to both sides of (6.4) gives

$$\mathcal{D} \vec{u}_t = \sum_{i=-M-1}^{N+1} \beta_i \mathcal{D} \vec{z}_i$$

and taking an inner product of this with $\vec{u}_t$ (recalling that $Z^T \mathcal{D} Z = I$) gives

$$\vec{u}_t^T \mathcal{D} \vec{u}_t = \vec{u}_t^T \sum_{i=-M-1}^{N+1} \beta_i \mathcal{D} \vec{z}_i$$

$$= \sum_{i=-M-1}^{N+1} \beta_i \vec{z}_i \sum_{i=-M-1}^{N+1} \beta_i \mathcal{D}\vec{z}_i$$

$$= \sum_{i=-M-1}^{N+1} \beta_i^2$$

$$\leq \beta_p^2(M + N + 3) \qquad (6.7)$$

for

$$|\beta_p| = \max_{-M-1 \leq i \leq N+1} |\beta_i|.$$

Applying the quadrature rule in [5] to (6.1) yields

$$1 = h \sum_{i=-M-1}^{N+1} u_t^2(ih)\rho(ih) + \varepsilon(u_t, M, N) = h\vec{u}_t^T \mathcal{D}\vec{u}_t + \varepsilon(u_t, M, N).$$

The assumptions on u_t guarantee that as M and N go to infinity, ε goes to zero. Therefore, for sufficiently large M and N, $|\varepsilon| \leq \frac{1}{2}$ which implies that

$$h\vec{u}_t^T \mathcal{D}\vec{u}_t = 1 - \varepsilon \geq 1 - \frac{1}{2} = \frac{1}{2}$$

so

$$\frac{1}{2h} \leq \vec{u}_t^T \mathcal{D}\vec{u}_t \leq \beta_p^2(M + N + 3).$$

Therefore

$$|\beta_p| \geq \frac{1}{\sqrt{2h(M + N + 3)}}.$$

Assume that $|\mu_p - \lambda_t| \leq \lambda_t$ (the case of $|\mu_p - \lambda_t| > \lambda_t$ will discussed later). Then

$$\mu_p = \mu_p - \lambda_t + \lambda_t \leq |\mu_p - \lambda_t| + \lambda_t \leq 2\lambda_t. \qquad (6.8)$$

Combining this with (6.5) gives

$$\| \vec{z}_p \|_2^2 \leq \frac{2\lambda_t}{\alpha_{-M-1}} \qquad (6.9)$$

Let θ_p be the angle between vectors $\vec{z}_p$ and $\overline{\Delta u_t}$. Taking absolute values and expanding the inner product on the left–hand side of (6.6) and using (6.7) and (6.9) gives

$$|\mu_p - \lambda_t| = \frac{|\vec{z}_p^T \overline{\Delta u_t}|}{|\beta_p|}$$

$$\begin{aligned}
&= \frac{\| \vec{z_p} \|_2 \| \overline{\Delta u_t} \|_2 \, |\cos(\theta_p)|}{|\beta_p|} \\
&\leq \sqrt{\frac{2\lambda_t}{\alpha_{-M-1}}} \, \| \overline{\Delta u_t} \|_2 \, \sqrt{2h(M + N + 3)}.
\end{aligned} \qquad (6.10)$$

The generalized eigenvalue problem can be a more difficult problem to solve than the standard eigenvalue problem because of the hypersensitivity of the eigenvalues to perturbations [4]. However, given that $\mathcal{A}$ and $\mathcal{B}$ are restricted to the class of positive definite matrices, these hypersensitivity problems will not arise. Because of this, along with the inequality (6.5) in Lemma 5.1, the hypotheses that $\mathcal{A}$ and $\mathcal{B}$ are positive definite matrices are included in the theorem. From the example section, though, we can see that without these hypotheses there is still the possibility of getting good results from the method.

It is still an open question to give a usable bound for $\| \overline{\Delta u_t} \|_2$. It appears from the examples that this term can be bounded exponentially as was the case in the Dirichlet setting [1].

In (6.8) we used the assumption that $|\mu_p - \lambda_t| \leq \lambda_t$. Consider the case when $|\mu_p - \lambda_t| > \lambda_t$. If indeed the error in (6.10) can be bounded exponentially, then as M and $N \to \infty$, the error goes to zero; and therefore μ_p must approach λ_t. Hence, as M and N increase, the original assumption $|\mu_p - \lambda_t| \leq \lambda_t$ will be valid and the error bound, as given, will apply.

References

[1] N. EGGERT, M. JARRATT, and J. LUND, "Sinc Function Computation of the Eigenvalues of Sturm-Liouville Problems," *J. of Computational Physics,* 69, 1987, 209-229.

[2] M. JARRATT, "Eigenvalue Approximations on the Entire Real Line," *Computation and Control,* Birkhäuser, 1989, 133-144.

[3] M. JARRATT, J. LUND, and K. BOWERS, "Galerkin Schemes and the Sinc-Galerkin Method for Singular Sturm-Liouville Problems," *J. of Computational Physics,* 89 , 1990, 41-62.

[4] B. PARLETT, *The Symmetric Eigenvalue Problem,* Prentice–Hall, 1980.

[5] F. STENGER, "Numerical Methods Based on the Whittaker Cardinal, or Sinc Functions," *SIAM Review,* 23, 1981, 165-224.

[6] F. STENGER, *Numerical Methods Based on Sinc and Analytic Functions*, Springer–Verlag, 1993.

EXISTENCE OF FUNCTIONAL GAINS FOR PARABOLIC CONTROL SYSTEMS

Belinda B. King, *
Department of Mathematics
Oregon State University
Corvallis, Oregon 97331

1 Introduction

This paper is concerned with some fundamental issues regarding the existence and smoothness of functional gains for LQR feedback control of parabolic systems. It is well known, under suitable stabilizability conditions, a solution to the LQR problem exists in the form of a bounded linear feedback operator K. The corresponding optimal control is given in feedback form

$$u(t) = -Kx(t).$$

Here, $K : X \to U$ maps the state space X into the control space U. Typically, X is an infinite dimensional Hilbert space and often, U is the finite dimensional space $\mathbb{R}^m$ for some m. In this case the Riesz Representation Theorem provides an explicit representation of K. If the control space U is also an infinite dimensional Hilbert space, then the problem becomes much more difficult. However, provided that the algebraic Riccati equation has sufficiently smooth solutions, the corresponding feedback gain operator will have explicit representations.

We consider a specific problem governed by the one dimensional heat equation and allow rather general input operators. In particular, we do not require that the input operator be Hilbert-Schmidt (or even bounded) and we allow for non-compact weighting operators. We show that it is the specific structure of the parabolic problem that provides the smoothness needed to prove the existence of functional gains.

This analysis is motivated by the need for specific representations for Riccati operators that can be used in the development of computational schemes for problems where the input and output operators are not Hilbert-Schmidt. This situation occurs in many boundary control problems and in certain distributed control problems associated with optimal sensor/actuator placement.

---

*This research was supported in part by the Air Force Office of Scientific Research under grant F49620-93-1-0280 while the author was a visiting scientist at the Air Force Center for Optimal Design and Control, Virginia Polytechnic Institute and State University, Blacksburg, VA 24061–0531, and by the National Science Foundation under grant DMS-9409506.

In [7], Lupi, Chun, and Turner considered a distributed parameter LQR problem for an Euler-Bernoulli beam model. The approach in [7] is of interest in that they make no prior assumptions regarding the form of the controls/actuators in an effort to make decisions about where actuators and sensors are best placed. They assumed that the input operator was the identity. Miller and van Schoor [9] considered the problem of constructing kernels for integral representations of feedback control laws obtained by solving LQR control problems for various beam models. They used these kernels to shape and design area averaging polyvinylidene fluoride sensors (a type of piezoelectric film). These sensors enable the real-time implementation of full state feedback for the infinite dimensional system governed by the Euler-Bernoulli equation (also, see [9]). As in [7], Miller and van Schoor assumed the existence of an integral representation for the feedback control law and then proceed to "approximate" these kernels by using finite element models. In both papers, fundamental mathematical questions concerning the existence of integral representations and the smoothness of the corresponding integral kernels are not considered. These issues are important in the development and analysis of rigorous numerical approximations. Also, the ability to accurately compute these kernels is an essential component in the study of actuator/sensor placement.

In recent papers by Rosen [10, 11] it was shown that, under suitable assumptions on the system input, output and weighting operators, the Riccati operator is Hilbert-Schmidt. This observation made it possible to develop an approximation theory in the space of Hilbert-Schmidt operators that can be used to analyze the convergence of numerical approximations.

The papers noted above represent two basic approaches to the problem. Rosen [10] considered the problem for control systems where the generator of the semigroup was strongly coercive and developed a theory for this restricted class of systems. In [3] the approach was to consider general dynamical systems (not necessarily analytic semigroups), and then require that certain system operators (input, weighting, etc.) be nuclear. The problem we consider here lies between these two approaches, although it is more in the spirit of Rosen [10]. The results below apply to control problems with unbounded input operators.

2   The Problem Description

Consider the control system

$$\dot{x}(t) = Ax(t) + Bu(t), \quad x(0) = x_0 \tag{2.1}$$

and cost function

$$J(u) = \frac{1}{2} \int_0^\infty \left[\langle Qx(t), x(t) \rangle + \langle u(t), u(t) \rangle \right] dt, \tag{2.2}$$

where the state weighting matrix Q is self-adjoint and non-negative definite. Although it is possible to consider more general systems or the case with Q unbounded and B bounded, we limit our discussion to systems satisfying the following standing hypothesis

(H) The spaces X and U are separable Hilbert spaces and;

  (i) The linear operator A is the generator of a C_0-semigroup $S(t)$ on X and there exist $M > 0, \omega > 0$ such that $\|S(t)\| \leq Me^{-\omega t}$.

 (ii) The operator $Q : X \to X$ is bounded.

(iii) The (possibly unbounded) linear operator B maps U into $[\mathcal{D}(A^*)]'$. Moreover, there exists γ with $0 \leq \gamma < 1$, such that $A^{-\gamma}B \in \mathcal{L}(U, X)$.

When the optimal control exists, it is given in feedback form

$$u_{opt}(t) = -B^* \Pi x_{opt}(t) = -K x_{opt}(t), \qquad (2.3)$$

where Π is the non-negative definite solution to the algebraic Riccati equation (ARE)

$$\langle \Pi x, Az \rangle_X + \langle Ax, \Pi z \rangle_X - \langle B^* \Pi x, B^* \Pi z \rangle_U + \langle Qx, z \rangle_Y = 0 \qquad (2.4)$$

for all x, z in $\mathcal{D}(A)$.

In [3, 5, 10, 11] it is assumed that Q and B are bounded linear operators. Rosen [10] assumes that A is strongly coercive, $\Pi BB^* \Pi$ is Hilbert-Schmidt whenever Π is Hilbert-Schmidt and that Q is Hilbert-Schmidt. On the other hand, De Santis, Germani and Jetto [3] make no additional assumptions on A, but require that Q be nuclear and B be bounded from U to X. Hence, the assumption on Q in [DGJ] is stronger than the corresponding condition in [10]. The two problems are not mutually exclusive and, as one might expect, there is no unified theory. In this paper we consider problems with B unbounded and Q not Hilbert-Schmidt. Unbounded B operators allow us to treat certain boundary control problems and problems with piezoelectric actuators/sensors. Also, the case where $Q = B = I_X$ (the identity on X) arises naturally in the solution of optimal sensor/actuator location problems (see [2, 6, 7]).

We consider here the case where A generates an analytic semigroup. The following theorem may be found in [8].

Theorem 2.1 *Assume that hypothesis (H) holds. If A generates an analytic semigroup, then there exists a self-adjoint, bounded, non-negative definite linear operator $\Pi = \Pi^*$ that solves the ARE (2.4). Moreover,*

 (a) For each $\epsilon > 0$, the operator $[A^*]^{1-\epsilon}\Pi$ belongs to $\mathcal{L}(X, X)$.

(b) If A is self-adjoint, normal or has a Riesz basis of eigenvectors, then
ε can be taken to be 0.

(c) The operator $B^* \Pi$ belongs to $\mathcal{L}(X, U)$.

Observe that if A has a compact resolvent, then Π is compact. In general it is not possible to conclude that Π has additional smoothing properties unless more is known about the operators A, B and Q. Consider the following example:

Example 2.1 *Let $X = Y = U = L_2(0,1)$ and set $A = Q = I$. If $B = \sqrt{3}I$, then $\Pi = I$ is the unique positive-definite solution to the ARE (2.4). Hypothesis (H) holds and A generates an analytic semigroup. However, Π is not compact.*

In certain specific cases it is possible to obtain additional information about the regularity of the Riccati operator Π. In the next section we consider a parabolic control problem similar to the one treated by Rosen (see [10, 11]) and use classical representation theory to show that Π is Hilbert-Schmidt. Representation of the feedback gain operator by integrals will follow.

3   A Parabolic Control Problem

We consider a one dimensional parabolic control problem. Let A be the operator defined on the state space $X = L_2(0,1)$ with domain

$$\mathcal{D}(A) = H_0^1[0,1] \cap H^2[0,1], \tag{3.1}$$

by

$$[A\phi](\xi) = \frac{d^2}{d\xi^2}\phi(\xi). \tag{3.2}$$

In order to simplify the proofs, we begin with the case where $Q = I_{L_2}$ and $B : U \to [\mathcal{D}(A^*)]'$ satisfies **(H)**-(iii). The extension to general Q operators and unbounded input operators B satisfying hypothesis **(H)** is straightforward. We note that if B is bounded into X, then **(H)** is satisfied.

The controlled heat equation is written as (see [10, 11])

$$\frac{\partial}{\partial t}w(t,\xi) = \frac{\partial^2}{\partial \xi^2}w(t,\xi) + [Bu(t)](\xi), \quad 0 < \xi < 1, \ 0 < t, \tag{3.3}$$

with boundary conditions

$$w(t,0) = 0, \quad w(t,1) = 0, \quad 0 < t \tag{3.4}$$

and cost function

$$J(u) = \frac{1}{2} \int_0^\infty \int_0^1 \left[|w(t,\xi)|^2 + |u(t,\xi)|^2 \right] d\xi dt. \qquad (3.5)$$

This problem has the form (2.1)-(2.2) and hypothesis **(H)** holds. The operator A is self-adjoint and generates an analytic semigroup on X.

Note that Q is not compact and hence the results in [3, 10] do not apply. However, we shall show below that the Riccati operator Π is Hilbert-Schmidt. We shall need the following representation theorem which goes back to Fullerton in 1946 (see Theorem 6 on page 277 in [4]).

Theorem 3.1 *Let T be a bounded linear operator mapping from $L_2[0,1]$ into $C^m[0,1]$. Then there exists a function $k(\xi,t)$ such that T has the representation*

$$[T\phi](\xi) = \int_0^1 k(\xi,t)\phi(t)dt, \qquad (3.6)$$

where the kernel $k(\xi,t)$ satisfies the following conditions;

(i) $k(\xi,t) \in L_2([0,1] \times [0,1])$.

(ii) For each $\xi \in (0,1)$, the mapping $\xi \to k(\xi,t)$ belongs to $C^m[0,1]$ for almost all t and for $\alpha \leq m$,

$$\frac{\partial^\alpha}{\partial \xi^\alpha} k(\xi,t) \in L_2([0,1] \times [0,1]).$$

(iii) For each $\phi \in L_2[0,1]$,

$$\frac{\partial^\alpha}{\partial \xi^\alpha} \int_0^1 k(\xi,t)\phi(t)dt = \int_0^1 \frac{\partial^\alpha}{\partial \xi^\alpha} k(\xi,t)\phi(t)dt.$$

The following result establishes the existence of an integral representation for the Riccati operator and provides information about the smoothness of the kernel.

Theorem 3.2 *Assume $Q = I_{L_2}$ and $B : U \to [\mathcal{D}(A^*)]'$ satisfies **(H)**-(iii). If $\Pi = \Pi^*$ is the unique non-negative definite solution to the ARE (2.4) defined by the system (3.1)-(3.5), then Π is Hilbert-Schmidt. Moreover, there exists a function $k(\xi,t)$ such that Π has the representation*

$$[\Pi\phi](\xi) = \int_0^1 k(\xi,t)\phi(t)dt, \qquad (3.7)$$

where the kernel $k(\xi,t)$ satisfies the following conditions,

(1) $k(\xi, t) = k(t, \xi) \in C^1([0, 1] \times [0, 1])$.

(2) For each $\phi \in L_2[0, 1]$, $\psi = \Pi\phi \in H^2 \cap H_0^1$ *and*

$$\frac{d}{d\xi}\psi(\xi) = \frac{\partial}{\partial\xi}\int_0^1 k(\xi, t)\phi(t)dt = \int_0^1 \frac{\partial}{\partial\xi}k(\xi, t)\phi(t)dt. \qquad (3.8)$$

Proof: First note that A is self-adjoint so that $A = A^*$. Let $W = \mathcal{D}(A)$ with graph norm, $\|\phi\|_W = \|\phi\|_{L_2} + \|A\phi\|_{L_2}$. It follows from Theorem 2.1 that Π and $A\Pi$ are bounded linear transformations on $L_2(0, 1)$. Therefore, there exist constants c_1 and c_2 such that for all $\phi \in L_2(0, 1)$, $\Pi\phi \in W = H_0^1 \cap H^2$, $A\Pi\phi \in L_2$ and

$$\|\Pi\phi\|_{L_2} \leq c_1\|\phi\|_{L_2} \quad \text{and} \quad \|A\Pi\phi\|_{L_2} \leq c_2\|\phi\|_{L_2}. \qquad (3.9)$$

Since $W = H_0^1 \cap H^2 \subseteq H^2$ and H^2 is continuously embedded in $C^1[0, 1]$, there is a constant c_3 such that for all $\psi \in W$

$$\|\psi\|_{C^1} = \{\|\psi\|_\infty + \|\psi'\|_\infty\} \leq c_3\|\psi\|_W = c_3\{\|\psi\|_{L_2} + \|A\psi\|_{L_2}\}.$$

Let $\phi \in L_2(0, 1)$ and $\psi = \Pi\phi \in W$. Then

$$\begin{aligned}\|\Pi\phi\|_{C^1} &\leq c_3\{\|\psi\|_{L_2} + \|A\psi\|_{L_2}\} \\ &\leq c_3\{\|\Pi\phi\|_{L_2} + \|A\Pi\phi\|_{L_2}\},\end{aligned}$$

and hence it follows from (3.9) that

$$\|\Pi\phi\|_{C^1} \leq c_3\{\|\Pi\phi\|_{L_2} + \|A\Pi\phi\|_{L_2}\} \leq c_3(c_1 + c_2)\|\phi\|_{L_2}.$$

Consequently, Π is a bounded linear transformation from $L_2(0, 1)$ into $C^1[0, 1]$. Applying Theorem 3.1 yields the representation (3.7).

Since $\Pi = \Pi^*$ the kernel $k(\xi, t)$ satisfies $k(\xi, t) = k(t, \xi)$. It follows from part (ii) of Theorem 3.1 that $k(\xi, t) = k(t, \xi) \in C^1([0, 1] \times [0, 1])$ and (3.8) is a consequence of part (iii) of Theorem 3.1. Finally (see page 210 in [12]), the operator $\Pi : L_2[0, 1] \to L_2[0, 1]$ is Hilbert-Schmidt since it has the representation (3.7) with $k(\xi, t) \in L_2([0, 1] \times [0, 1])$.

Corollary 3.3 *Assume A is defined by (3.1)-(3.2), $Q = I_{L_2}$ and $B : U \to L_2[0, 1]$ is bounded. If $\Pi = \Pi^*$ is the unique non-negative definite solution to the ARE (2.4) defined by the system (3.1)- (3.5), then the gain operator $K = -B^*\Pi$ is Hilbert-Schmidt.*

The proof given above goes through without change for any bounded Q and B satisfying (**H**). In particular, there is no need to assume that Q is Hilbert-Schmidt (see [3, 5, 10]) and B can be unbounded.

Note that since $k(t, \xi) \in C^1([0, 1] \times [0, 1])$ and (3.8) holds, the above results can clearly be extended to certain unbounded operators. For example, consider the operator $B : H_L^1 \to L_2$ defined by $[B\phi(\cdot)] = \phi'(\cdot)$ where $H_L^1 = \{\phi \in H^1 : \phi(0) = 0\}$. Note that the extension of B to $[\mathcal{D}(A^*)]'$ satisfies (H)-(iii). Also B^* is defined by $\mathcal{D}(B^*) = H_R^1 = \{\phi \in H^1 : \phi(1) = 0\}$ with $[B^*\phi(\cdot)] = -\phi'(\cdot)$.

It is clear from (3.8) that $B^*\Pi$ has the representation

$$[B^*\Pi\phi](\xi) = \int_0^1 -\frac{\partial}{\partial\xi}k(\xi, t)\phi(t)dt. \tag{3.10}$$

Therefore, the feedback operator has an integral representation of the form

$$[K\phi](\xi) = \int_0^1 h(\xi, t)\phi(t)dt \tag{3.11}$$

with $h(\xi, t) \in C([0, 1] \times [0, 1])$. The functional gain $h(\xi, t) = -\frac{\partial}{\partial\xi}k(\xi, t)$ is continuous and K is Hilbert-Schmidt.

We conjecture that if $B = [-A]^\beta$ with $\beta < 1$, then (3.11) holds with $h(\cdot, \cdot) \in L_2([0, 1] \times [0, 1])$. The numerical results below support this conjecture.

4 Numerical Results and Concluding Remarks

In this section we present some numerical results to demonstrate the role that the operator B plays in the smoothness of the operator Π. We conducted several experiments for the operators $B = [-A]^\beta$ where $\beta = -1/2, -1/4, 0, 1/4, 1/2, 3/4$, and 1. Note that if $\beta < 1$, then hypothesis (H) is satisfied. When $\beta = 0$, $B = I_{L_s}$ and $B = [-A]^{-1/2}$ is compact. We selected this collection of B operators because as $\beta \to 1$ the operator $A^{-1}B = A^{-1}A^\beta = A^{\beta-1}$ in condition (H)-(iii) becomes "less smooth". Observe that if $\beta = 1$, then (H)-(iii) is not satisfied for any $\gamma < 1$.

We use standard linear finite elements to compute $k^N(\xi, t) = k(\xi, t)$, the "N^{th} order approximation" of the kernel $k(\xi, t)$. Figures 4.1, 4.2 and 4.3 (which appear at the end of this section) show the $N = 8, 16, 32$ and 64 finite element approximations of $k(\xi, t)$ for the cases $B = [-A]^{-1/2}$, $B = [-A]^{-1/4}$, and $B = I_{L_2}$, respectively. Observe the fast convergence $k^N(\xi, t) \xrightarrow{N \to \infty} k(\xi, t)$ and, as implied by Theorem 3.2, the kernel $k(\xi, t)$ is smooth.

Figures 4.4, 4.5 and 4.6 contain analogous plots for $B = [-A]^\beta$ where $\beta = 1/4, 1/2$ and 3/4, respectively. It is interesting to note that as $\beta \to 1$ the kernels $k(\xi, t)$ become less smooth. Moreover, when $\beta = 1$ Theorem 3.2 breaks down and $k^N(\xi, t)$ appears to be converging to a "singular measure" concentrated on the line $\xi = t$ as shown in Figure 4.7. Similar results were observed for weakly damped hyperbolic problems in [2] and for beam equations in [7].

As noted in Corollary 3.3, when A is defined by (3.1)- (3.2), $C = Q = I_{L_2}$ and $B : U \to L_2[0, 1]$ is bounded, then the gain operator $K = -B^* \Pi$ is Hilbert-Schmidt. In some cases (e.g. $U = L_2[\Omega']$ where $\Omega' \subseteq [0, 1]$) it is possible to obtain explicit integral representations of $K = -B^* \Pi$. The case where B is unbounded requires additional analysis.

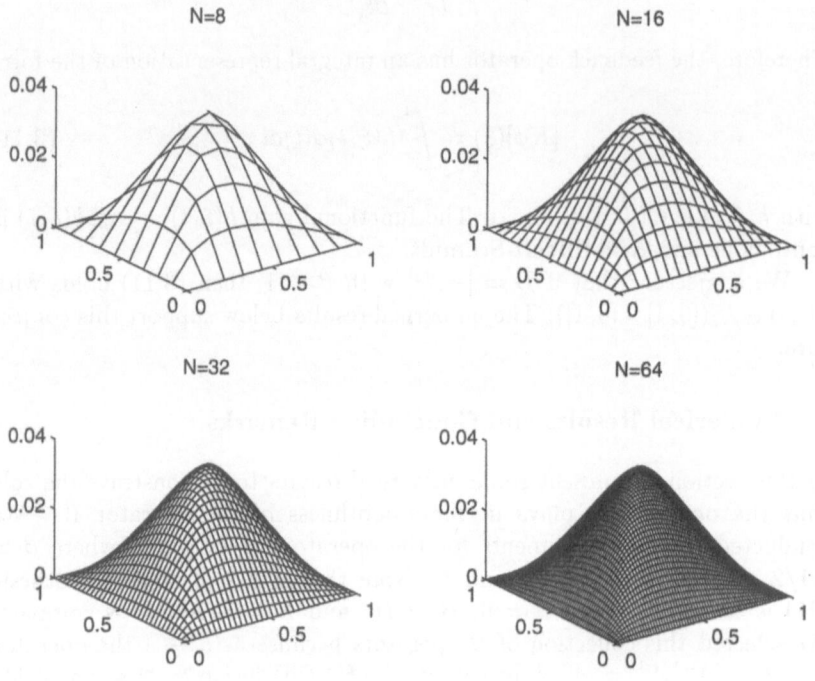

Figure 4.1: The kernels $k^N(\xi, t)$ for $B = [-A]^{-(1/2)}$

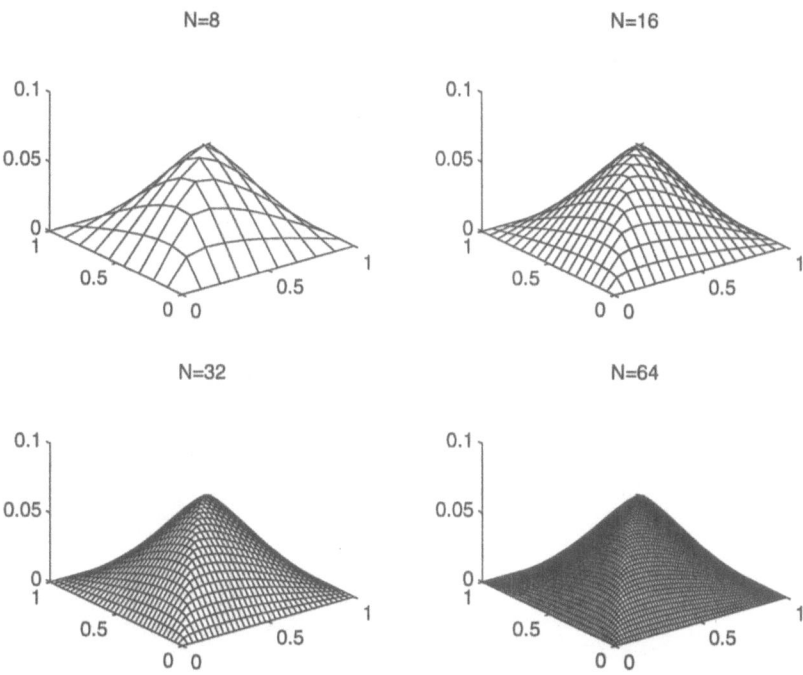

Figure 4.2: The kernels $k^N(\xi, t)$ for $B = [-A]^{-(1/4)}$

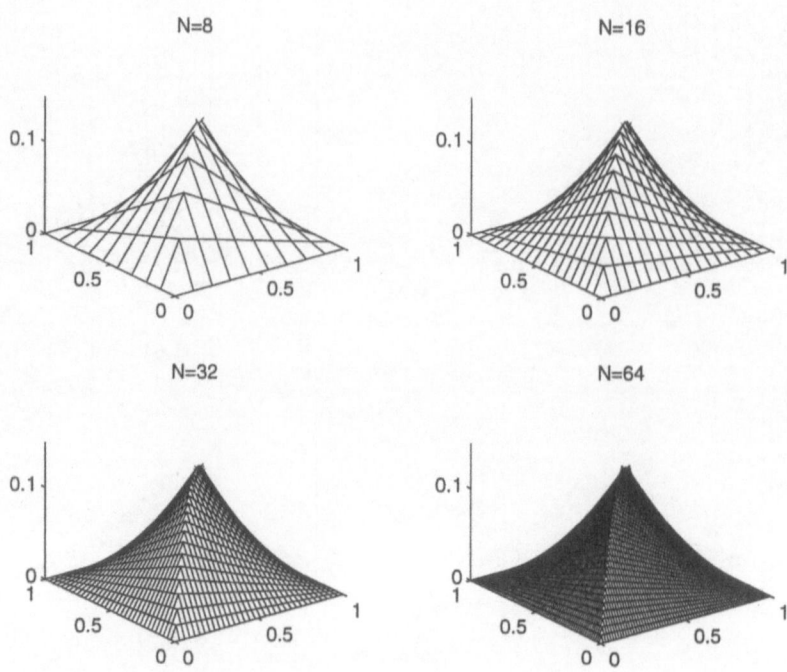

Figure 4.3: The kernels $k^N(\xi, t)$ for $B = I_{L_2}$

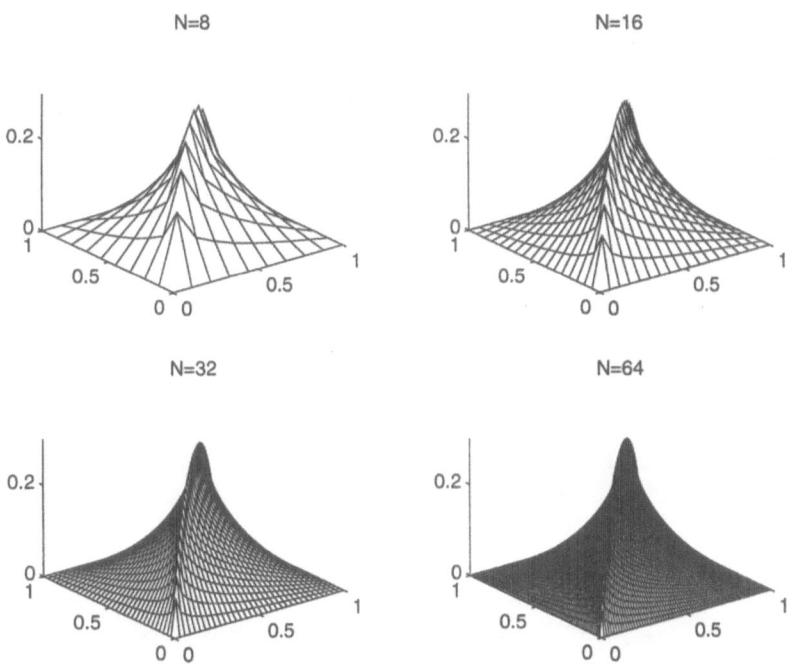

Figure 4.4: The kernels $k^N(\xi, t)$ for $B = [-A]^{1/1}$

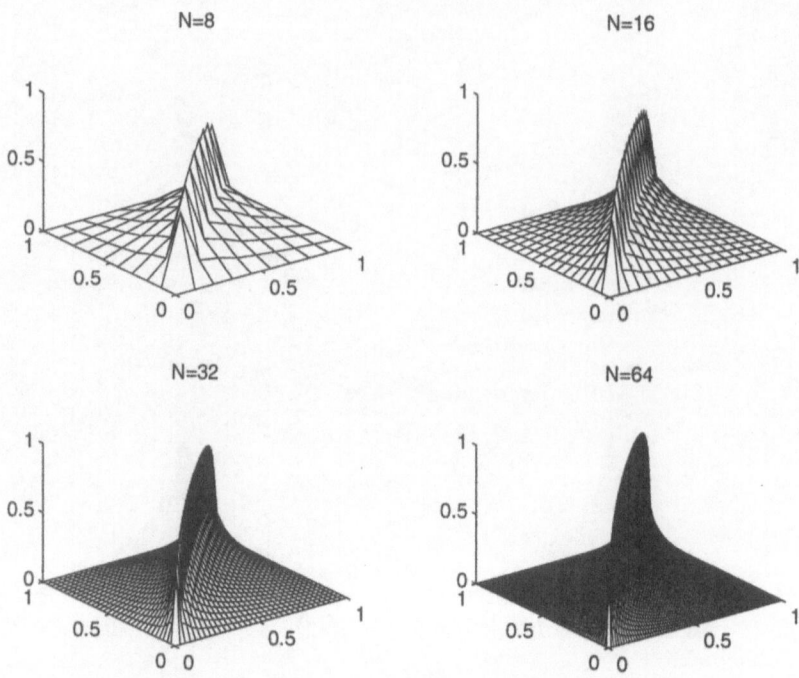

Figure 4.5: The kernels $k^N(\xi, t)$ for $B = [-A]^{1/2}$

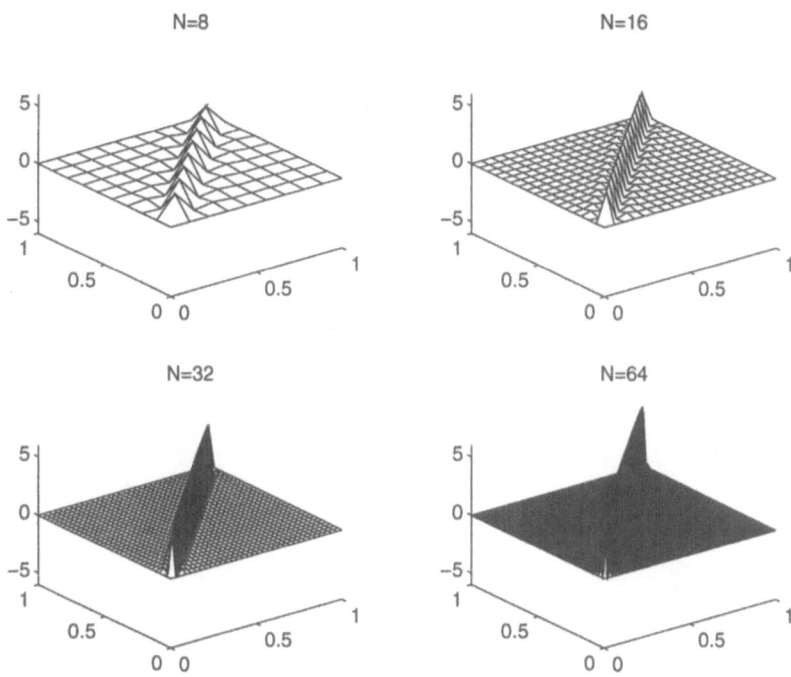

Figure 4.6: The kernels $k^N(\xi, t)$ for $B = [-A]^{3/4}$

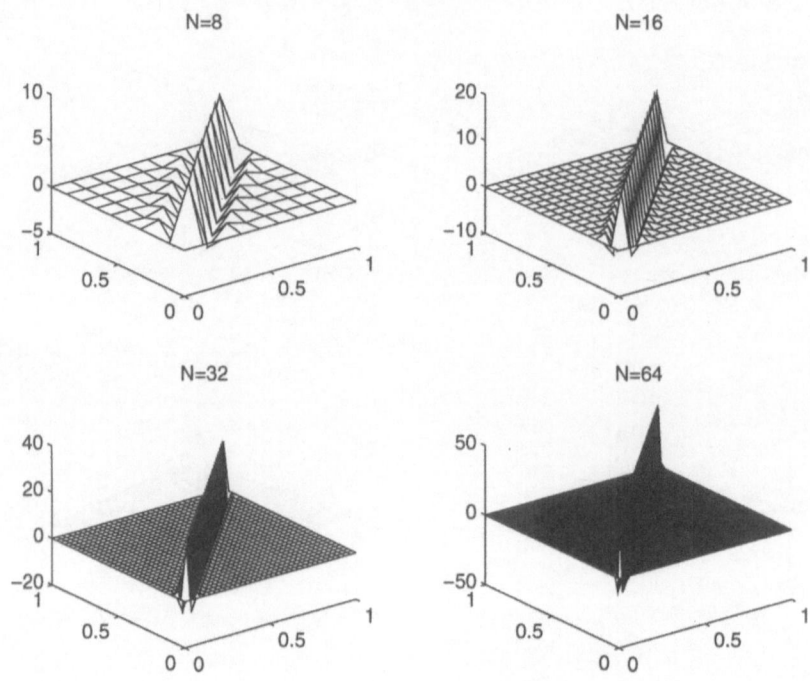

Figure 4.7: The kernels $k^N(\xi, t)$ for $B = [-A]^1$

Acknowledgement

Finally, the authors wish to thank J.S. Gibson for Example 2.1 and several helpful comments.

References

[1] R. A. ADAMS, *Sobolev Spaces*, Academic Press, New York, 1975.

[2] J. A. BURNS and B. B. KING, "Optimal Sensor Location for Robust Control of Distributed Parameter Systems," *Proc. of the 33rd IEEE Control and Decision Conference*, Dec. 1994, pp. 3967–3972.

[3] A. DE SANTIS, A. GERMANI and L. JETTO, "Approximation of the Algebraic Riccati Equation in the Hilbert Space of Hilbert-Schmidt Operators," *SIAM J. Control Optim.*, v. 31, 1993, pp. 847–874.

[4] R.E. FULLERTON, "Linear Operators with Range in a Space of Differentiable Functions," *Duke Math. Journal*, v. 13, 1946, pp. 269–280.

[5] A. GERMANI, L. JETTO and M. PICCIONI, "Galerkin Approximation for Optimal Filtering of Infinite Dimensional Linear Systems," *SIAM J. Control Optim.*, v. 26, 1988, pp. 1287–1305.

[6] C.S. KUBRUSLY and H. MALEBRANCHE, "Sensors and Controllers Location in Distributed Systems: A Survey, " *Automatica*, v. 21, 1985, p. 117–128.

[7] V.D. LUPI, H.M. CHUN, J.D. TURNER, "Distributed Control without Mode Shapes or Frequencies," *Adv. in the Astro. Sci.*, v. 76, 1991, pp. 447–470.

[8] I. LASIECKA and R. TRIGGIANI, *Differential and Algebraic Riccati Equations with Application to Boundary/Point Control Problems: Continuous Theory and Approximation Theory*, Lecture Notes in Control and Information Sciences, v. 164, Springer-Verlag, Berlin, Heidelberg, 1991.

[9] D.W. MILLER and M.C. VAN SCHOOR, "Formulation of Full State Feedback for Infinite Order Structural Systems," *Proc. 1st US/Japan Conf. on Adapt. Struct.*, Maui, Hawaii, 1990, pp. 304–331.

[10] I. G. ROSEN, "On Hilbert-Schmidt Norm Convergence of Galerkin Approximation for Operator Riccati Equations," *Int. Series of Numer. Math. 91*, Birkhäuser, Basel, 1989, pp. 335–349.

[11] I. G. ROSEN, "Convergence of Galerkin Approximations for Operator Riccati Equations - A Nonlinear Evolution Equation Approach," *J. Math. Anal. Appl.*, v. 155, 1991, pp. 226–248.

[12] M. REED and B. SIMON, *Functional Analysis I*, Academic Press, New York, 1980.

BINOCULAR OBSERVABILITY

Pamela Lockwood* and Clyde Martin*

Department of Mathematics
Texas Tech University
Lubbock, TX 79409-1042

1   Introduction

A rigid body is in motion at constant linear and angular velocities. From a distance an individual with normal binocular vision is observing this body. Under what conditions is the individual able to distinguish its position? To answer this question we would like to formulate a mathematical realization for the process of human binocular vision. A model of this type would be useful in the medical field to diagnose problems in a individual's eyesight.

An object is seen by an individual with normal binocular vision as long as one eye is able to distinguish the object's position. Both eyes distinguishing the body is not necessary. So when is at least one eye able to observe the position of the moving body?

This is the question we wish to answer with the formulation of a mathematical binocular vision problem. To formulate this problem and gain some insight into its solution we will begin by examining the case of monocular vision. If a body in constant motion is observed by a single visual device, can the position of the body be obtained and under what conditions is this possible? This problem was formulated and solved in [1]. The derivation of the monocular vision problem and its solution will be discussed. Next, the binocular vision problem will be derived by the addition of a second visual device. The question will then become, when can the position of the object be distinguished by at least one of these devices? A theorem will be presented which provides sufficient conditions for observability of the body using two visual devices.

The dynamics of human binocular vision is a complicated process which involves the use of parallel rays and the lenses of the eye to project an inverse image of the object in view onto the retina, located at the back of the eye. In order to devise a mathematical realization for human vision, this process will be simplified. Instead of two parralel rays, consider that each eye will view a single point on the body along a line formed by the point of position of the eye and the observed point on the body. In human vision the point on the body would then be projected by the lenses of the eye onto the retina. Although the retina is a concave surface, we will represent the retina as a plane. In the case of binocular vision the same plane will represent both

---
*Supported in part by NASA Grants NAG2-902 and NAG2-899.

retini. The projection of the observed point onto the plane is the method which will be used in the derivation of both the monocular and binocular vision problems.

2   Background on Monocular Vision

Consider a rigid body moving in space at constant linear and angular velocities. Suppose a single visual device, such as a human eye, is observing the motion of one particular point on this object. Define an n-dimensional coordinate system so that the visual device is positioned at the origin, and label the point on the body as x_t. The visual device will then observe this point along the line formed by x_1 and the origin. This has an equation $z = \alpha x_t$, where α is a real number (Figure 2.1).

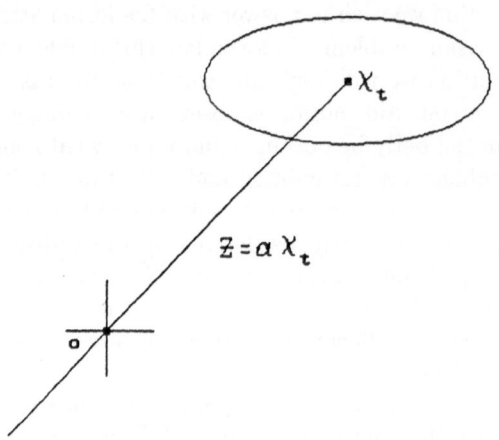

Figure 2.1: Monocular Vision Representation

Let $c_1 z = c_1 b$ be the equation of a plane oriented in the coordinate system so that the line, $z = \alpha x_t$, intersects this plane. Then the point of intersection of the plane and this line is the projection of x_t onto the plane $c_1 z = c_1 b$ (Figure 2.2).

If the linear and angular velocities of the object in motion are known, then we can describe this motion by a dynamical system, $\dot{x} = Ax$, where $x(o) = x_0$, $x_0 \neq 0$, is the initial position of the point on the body. We specity $X_0 \neq 0$ so that the motion of the body does not originate at the same point were the visual device is positioned. The projection of the point x_t onto the plane defines an observation function of the form $y(z) = c_2 z$. Specifically, at the point of intersection of the plane $c_1 z = c_1 b$ and the line $z = \alpha x_t$, we have

$$C_1 \alpha x_t = c_1 b.$$

This implies

$$\alpha = \frac{c_1 b}{c_1 x_t},$$

and the point of intersection is then $c_1 b \frac{c_2 x_t}{c_1 x_t}$. If we let $x = x_t$, the observation function can then be specifically defined as

$$y(x) = c_1 b \frac{c_2 x}{c_1 x}.$$

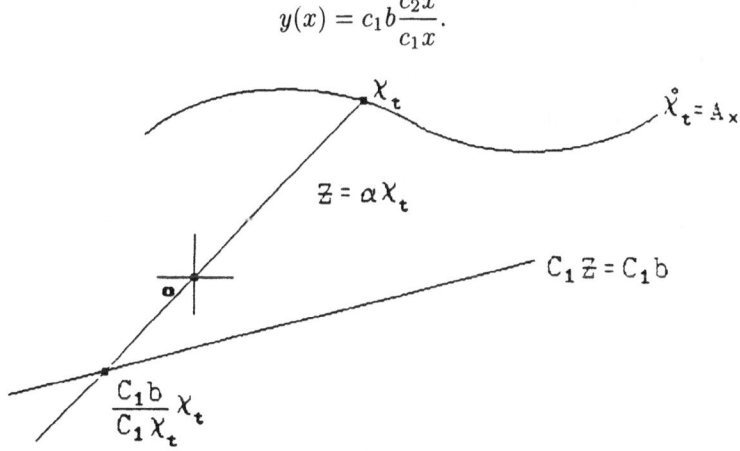

Figure 2.2: Monocular Vision Projection

For generality let us assume $c_1 b = 1$. The observed system is then

$$\dot{x} = Ax, \quad x(0) = x_0, \quad x_0 \neq 0 \tag{2.1}$$

$$y_1(x) = \frac{c_2 x}{c_1 x}. \tag{2.2}$$

The dynamical system $\dot{x} = Ax$, $x(0) = x_0$ has the solution $x(t) = e^{At} x_0$. Our question then becomes, if the observation function $y_1(x)$ is known for all $t \geq 0$, can we recover the initial position of the body in motion?

This is a question of system observability. Given a dynamical system with an observation function, where the observation function is known for all $t \geq 0$, when is the system observable? In other words, when can the initial data, $x(0) = x_0$ be recovered?

Consider a system

$$\dot{x}_i = Ax_i,$$

with an observation function

$$y = h(x_i),$$

and initial data

$$x_1(0) = x_0, \quad x_2(0) = x_1.$$

If we define the mappings O_o and O_1 such that

$$O_o :\rightarrow y_0(t) \quad \text{and} \quad O_1 :\rightarrow y_1(t),$$

then

$$y_0(t) = h(e^{At}x_0) \quad \text{and} \quad y_1(t) = h(e^{At}x_1).$$

By definition this system is observable if and only if

$$y_0(t) = y_1(t) \Rightarrow x_0 \equiv x_1.$$

Applying this definition, the system

$$\dot{x} = Ax, \quad x(0) = x_0, \quad x_0 \neq 0$$

$$y_1(x) = \frac{c_2 x}{c_1 x},$$

is not observable. This can be easily shown by considering initial data $x_0 = a$ and $x_1 = \alpha a$, where α is a real number.

From the initial data the observation functions y_0 and y_1 become

$$y_0(t) = \frac{c_1 e^{At} a}{c_2 e^{At} a} \quad \text{and} \quad y_1(t) = \frac{c_1 e^{At} \alpha a}{c_2 e^{At} \alpha a}.$$

Since α in the equation for $y_1(t)$ is a scalar quantity and cancels, we have $y_0(t) = y_1(t)$ for all values of t. This system is therefore not observable since for all values of α. Notice from our choice of initial data that the observation function $y_1(x)$ does not distinguish points on the ray $z = \alpha a$. This leads us to another question. Is the system observable up to rays? In other words, can we distinguish the ray through the origin on which the initial data lies? This is the concept of perspective observability. A system is perspectively observable provided it is observable up to rays.

A generalization of the perspective observability of the system (2.1), (2.2) was considered and solved in [1]. Let F be a field of real or complex numbers, and let A and C be $n \times n$ and $m \times n$ matrices, respectively with $2 \leq m \leq n$. Consider the dynamical system

$$\dot{x} = Ax, \quad x(0) = x_0, \qquad (2.3)$$

together with the observation function

$$Z : F^n - B \to FP^{m-1}, \qquad (2.4)$$

where Z is given by

$$x \to [Cx].$$

Here $B = \{x \in F^n : Cx = 0\}$, and $[Cx]$ represents the homogeneous coordinates of Cx as an element of FP^{m-1}, the $m-1$ dimensional projective space of all homogeneous lines in F^m. If $Cx(t)]$ is known for all $t \geq 0$, we have the following theorem.

Theorem 2.1 *The dynamical system (2.3), (2.4) is perspectively observable over the field of complex numbers if and only if*

$$Rank \begin{bmatrix} (A - sI)(A - \omega I) \\ c_1 \\ c_2 \end{bmatrix} = n.$$

for every pair of eigenvalues s, ω of A. If the set of all real numbers is the field then these conditions are sufficient for perspective observability.

If we return to the observed system

$$\dot{x} = Ax, \quad x(0) = x_0, \quad x_0 \neq 0$$

$$y_1(x) = \frac{c_2 x}{c_1 x},$$

the above theorem tells us that this system is perspectively observable provided

$$Rank \begin{bmatrix} (A - sI)(A - \omega I) \\ c_1 \\ c_2 \end{bmatrix} = n.$$

If we examine the observation function

$$y_1(x) = \frac{c_2 x}{c_1 x},$$

a question which may occur is what if $c_1 x = 0$ and $c_2 x = 0$? Utilizing our initial data this implies $c_2 e^{At} x_0 = c_1 e^{At} x_0 = 0$ for all values of t. If this occurs then by taking n derivatives and letting $t = 0$ these equations become $c_2 A^n x_0 = c_1 A^n x_0 = 0$. Let $M = \begin{bmatrix} (A - sI)(a_\omega I) \\ c_1 \\ c_2 \end{bmatrix}$

and consider the set $\{x_0, Ax_0 A^2 x_0, \ldots\}$. There exists an eigenvector $y \in$ span$\{x_0, Ax_0, A^2 x_0 \ldots\}$, such that $y = \sum_{i=0}^{n} \alpha_i A^i x_0$, and an eigenvalue ω_0 such that $Ay = \omega_0 y$. Since $c_1 x = 0$ and $c_2 x = 0$, we have $My = 0$. This implies the rank condition fails and by theorem 1 the system is not perspectively observable. Geometrically, the intersection of the null space of $c_1 x = 0$ and $c_2 x = 0$ is a subspace of R^n or C^n with dimension 2. The trajectory of the moving body may pass through this 2-dimensional subspace but it will not remain. It is does remain in this subspace then the rank condition will fail, and the system will not be perspectively observable.

3 Binocular Vision

In section 2 a rigid body in motion at known constant velocities was observed by a single visual device positioned at the origin of a defined n-dimensional coordinate system. The observed point on the body, x_t, was then projected along a line through the origin onto a plane. This projection was used to define a single observation function. The motion of the body was described by the dynamical system $\dot{x} = Ax, x(0) = x_0, x_0 \neq 0$. Theorem 2.1 then gave conditions for perspective observability of the initial position of the body given the observation function was known at all points in time.

Suppose we add a second visual device positioned at a point a in the coordinate system. The observed point x_t is now observed by two visual devices (Figure 3.1).

The visual device positioned at the origin will observe the point x_t along the line $z = \alpha x_t$, where α is a real number. The second visual device will observe x_t along the line joining x_t and a, which gives this line the equation $z = \alpha x_t + (1 - \alpha)a$.

As in the monocular vision problem position the plane $c_1 z = c_1 b$ in the coordinate system so that the lines $z = \alpha x_t$ and $z = \alpha x_t + (1 - \alpha)a \cdot$ intersect the plane (Figure 3.2).

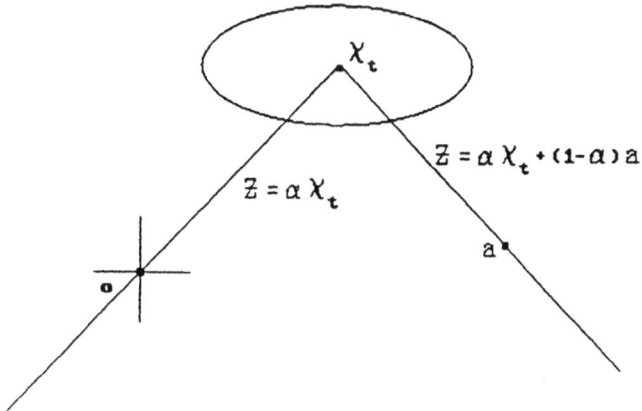

Figure 3.1: Binocular Vision Representation

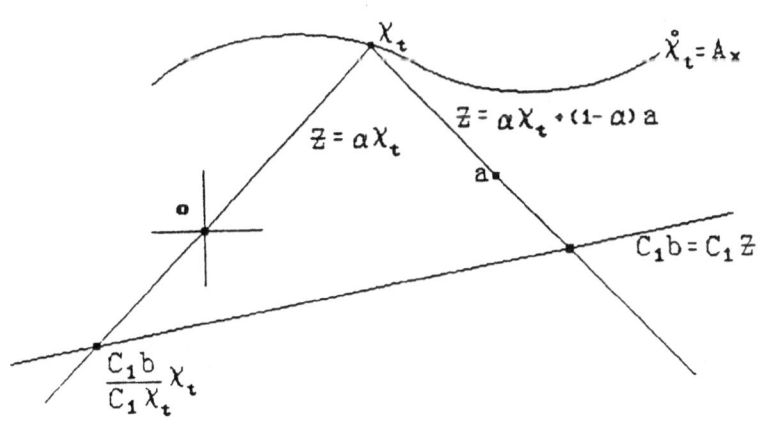

Figure 3.2

The point x_t is now projected onto the plane as two points rather than one in the case of monocular vision. These two projections will define two observation functions of the form

$$y(z) = c_2(z).$$

From the derivation of the monocular vision problem the projection of the point x_t through the origin will define the same observation function as in the monocular vision problem. This observation function is

$$y_1(x) = \frac{c_2 x}{c_1 x}.$$

Recall the system together $\dot{x} = Ax, x(0) = x_0, x_0 \neq 0$ with the observation function $y_1(x)$ is not observable, but is perspectively observable provided the rank condition stated in theorem 2.1 holds.

When the line through the point a intersects the plane we have

$$c_1 \alpha(x_t - a) + c_1 a = c_1 b,$$

which implies

$$\alpha = \frac{c_1(b - a)}{c_1(x_t - a)}.$$

This intersection point is then found to be

$$c_1(b - a) \left[\frac{x_t - a}{c_1(x_t - a)} \right] + a$$

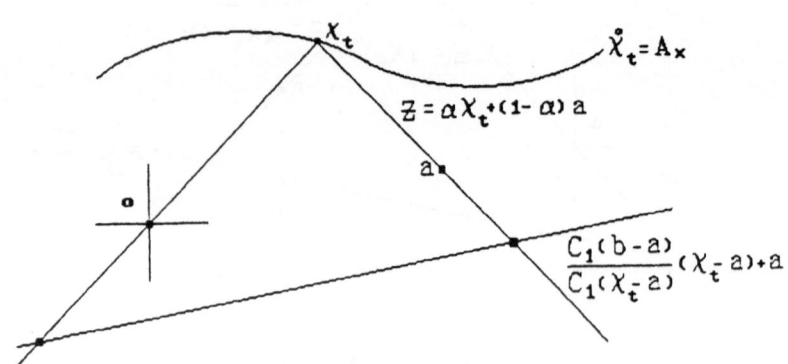

Figure 3.3

A second observation function can then be defined, if we let $x = x_t$, as

$$y - 2(x) = c_1(b - a) \left[\frac{c_2(x - a)}{c_1(x - a)} \right] + c_2 a.$$

The problem we would like to solve is given by the dynamical system

$$\dot{x} = Ax, x(0) = x_0,$$

together with the observation function

$$Y(x) = (y_1(x), y_2(x)),$$

under what conditions is this system observable? We know the dynamical system with the observation function $y_1(x)$ is not observable, but can be perspectively observable. What can be said about the observability of this system if we use the observation function $Y(x)$?

Theorem 3.1 *Consider the dynamical system*

$$\dot{x} = Ax,$$

with initial data

$$x(o) = x_0, x_0 \neq 0$$

together with the observation function

$$Y(x) = (y_1(x), y_2(x)),$$

$$y_1(x) = \frac{c_2 x}{c_1 x} \quad and \quad y_2(x) = c_1(b - a)\left[\frac{c_2(x - 1)}{c_1(x - a)}\right] + c_2 a,$$

where $x \in F^n$, the field of real or complex numbers. Let the

$$rank \begin{bmatrix} (A - sI)(A - \omega I) \\ c_1 \\ c_2 \end{bmatrix} = n \text{ for every pair } s, \omega \text{ of eigenvalues of } A. \text{ The}$$

system is not observable if and only if there exists an eigenvector, x_0, of A such that

$$\frac{c_1 x_0}{c_2 x_0} = \frac{c_1 a}{c_2 a}.$$

But what do the conditions for observability stated in theomem 3.1 mean from a geometrical standpoint? We can interpret these conditions in terms of the positions of the visual devices and the trajectory of the moving body. If $\frac{c_1 x_0}{c_2 x_0} = \frac{c_1 a}{c_2 a}$, then the initial position of the point on the body, x_0, lies on the ray through the origin and the point a, the positions of the visual devices. The origin, a, x_0 are colinear. If x_0 is an eigenvector of the matrix A, then the trajectory of the point on the moving body is along the line approaching the visual devices (Figure 3.3). So the system is not observable if the trajectory of the moving body is a ray which passes through both visual devices.

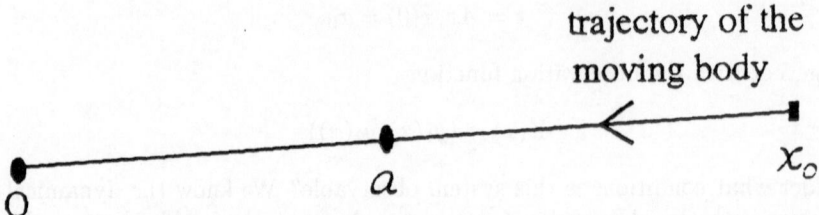

Figure 3.4: Geometric Conditions of Theorem 2

4 Conclusion

We have been considering two visual devices observing from a distance a rigid body in constant motion. We would like to determine conditions which ensure that the body will be observed by at least one of these devices. In section 2 the monocular vision problem was discussed. If a single optical device observes the motion of the moving body, when can its initial position be recovered? It was found in [1] that the initial data is perspectively observable, or the initial data is distinguishable only up to the ray on which it lies. The actual initial position cannot be recovered. A second visual device was then added in section 3 to create a binocular visual effect. We found that if we assume the first visual device can destinguish the initial position of the body up to rays, then the system will be observable provided the position of the two visual devices and the initial position of the point on thebody do not all lie on a common ray, with the trajectory of the body moving along this ray toward the visual devices. We therefore have a sufficient condition for observability of the body, but these conditions are not necessary. If the initial position of the body is not perspectively observable by the first visual device, then both visual devices together may or may not be able to distinguish the initial postion.

In order to produce a more precise mathematical realization for the process of human binocular vision, necessary and sufficient conditions need to be found which ensure that the body is observable by at least one visual device. Furter research is needed to find the complete solution for this binocular vision problem.

References

[1] W. Dayawansa, B. Ghosh, C. Martin and X. Wang, A Necessary and Sufficient Condition for the Condition for the Perspective Observability Problem. To appear, *Systems and Control Letters*

DYNAMICS OF OCULAR MOTION

Siyuan Lu* and Clyde Martin
Mathematics Department
Texas Tech University
Lubbock, Texas

1   Introduction

A primary goal of this research is to develop tracking models (observers) for moving objects based on the model of human binocular vision. First, a simplified physical model of a single eye was built, and it was used to derive the mathematic model of the eye. Then computer simulations were performed to find a suitable controller for this model. Second, the model was extended to binocular vision and the controller found was improved for the binocular model.

2   Basic Knowledge About the Eye

The eye is very nearly spherical in shape, and about 1 inch in diameter. From Gray's Anatomy [4], we know that the eye is controlled by a set of seven muscles as is indicated in Figure 2.1 [1]. They are *Levator palpebra superioris*, *Rectus superior*, *Rectus inferior*, *Rectus internus*, *Rectus externus*, *Obliquus oculi superior*, and *Obliquus oculi inferior*.

Each of the muscles has different action. The *Levator palpebra* raises the upper eyelid, and is the direct antagonist of the *Orbicularis palpebrarum*. The four *Recti* muscles are attached in such a manner to the globe of the eye that, acting singly, they will turn it either upward, downward, inward, or outward, as expressed by their names.

The movement produced by the *superior* or *inferior rectus* is not quite a simple one, for, inasmuch as they pass obliquely outward and forward to the eyeball, the elevation or depression of the cornea must be accompanied by a certain deviation inward, with a slight amount of rotation, which, however, is corrected by the oblique muscles, the *inferior oblique* correcting the deviation inward of the *superior rectus*, and the *superior oblique* that of the *inferior rectus*.

The contraction of the *external* and *internal recti*, on the other hand, produces a purely horizontal movement. If any two contiguous recti of one eye act together, they carry the globe of the eye in the diagonal of these directions–viz. upward and inward, upward and outward, downward and inward, or downward and outward. The movement of circumduction, as in

---

*Supported by the second author.
[1] This graph is copied from Gray's Anatomy [4].

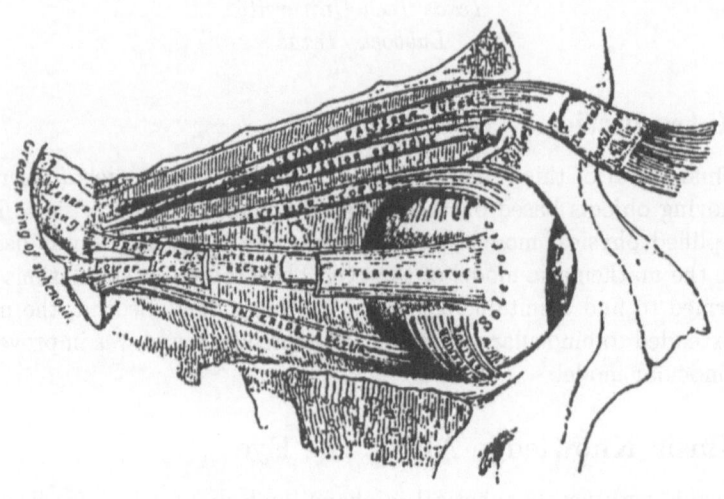

Figure 2.1: Muscles of the right orbit

looking round a room, is performed by the alternate action of the four recti. The oblique muscles rotate the eyeball on its *antero-posterior axis*, this kind of movement being required for the correct viewing of an object when the head is moved laterally, as from shoulder to shoulder, in order that the picture may fall in all respects on the same part of the retina of each eye.

3   Mathematic Model of Human Monocular Vision

We model the eye as a rigid rotating sphere of fixed radius attached at the center. The eye is controlled by a set of muscles as is indicated in Figure 2.1. The muscles are in opposing pairs and act in coordination with the muscles of the other eye. The eye is rotated by the contraction of the appropriate muscle. The muscle pair *obliquus oculi superior* and *inferior* serve to rotate the eye and to add stability [4]. In this paper we do not consider their

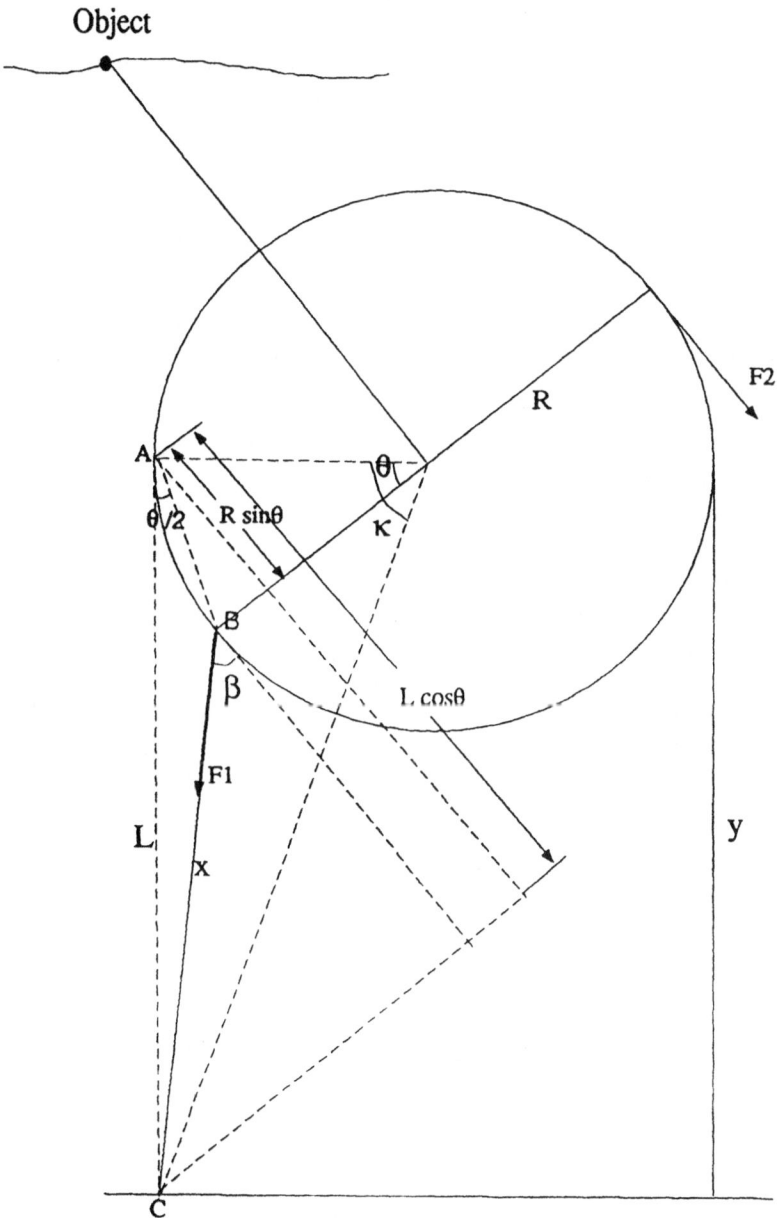

Figure 3.1: Schematic for Planar Eye Motion

action. So the eye is assumed to be controlled by four muscles attached in
opposing points on a diameter by pairs. This is shown in Figure 3.1.

We choose to model the muscles that control the eye movement as
damped springs having second order linear dynamics of the form

$$\ddot{x} = -k(l-x) - g\frac{d}{dx}(l-x) + v(t), \tag{3.1}$$

where l is the length of the unstretched spring, x is the length of the
stretched spring, k and g are the frequency of the spring and the damping
parameter respectively, and $v(t)$ is the controller added to the spring. This
might not be the best model for the muscle, the actual muscle is much
more complicated than the spring. But this model suffices to simplify our
problem, and works satisfactorially for the purpose of mathematical calculations.

In general the object being tracked is not a point source and such processes as edge detection and averaging are required to determine distance
to an object. These are difficulties that arise in determining the distance
to an object if its size and configuration are unknown. In this thesis we
will assume that the object being tracked is of known size and contains a
single distinguishing point. So it is reasonable to assume that the object is
a point source.

We start by deriving an equation that can describe the movement of a
single eye. The angle θ is chosen to be the system state variable. From
Figure 3.1, it can be seen that the eye is driven by two springs attached on
the diameter. So the movement of the eye is decided by the difference of
two tangential forces $F_1 \cos\beta$ and F_2. Thus,

$$(F_2 - F_1 \cos\beta)R = I\ddot{\theta}, \tag{3.2}$$

where I, the moment of inertia of the disk is

$$I = \frac{2}{5}mR^2. \tag{3.3}$$

The left spring is contracted. It can be described by the model of the
spring,

$$\begin{aligned}
F_1 &= \ddot{x} \\
&= -k(l-x) - g\cdot\frac{d}{dx}(l-x) + v(t) \\
&= -k(l-x) + g\dot{x} + v(t). \tag{3.4}
\end{aligned}$$

where l, x, k, g and $v(t)$ are as defined in Equation 3.1.

While the right spring is stretched, it is described by

$$F_2 = \ddot{y}$$
$$= -k(y-l) - g\dot{y}, \tag{3.5}$$

where y is the length of the stretched spring, and we assume that the controller is only added to the stretched spring.

By Equations 3.2-3.5

$$\ddot{\theta} = \frac{(F_2 - F_1 \cos\beta)R}{I}$$

$$= \frac{1}{\frac{2}{5}mR}\{-k(y-l) - g\dot{y} - [-k(l-x) + g\dot{x} + v(t)]\cos\beta\}. \tag{3.6}$$

We want $\ddot{\theta}$ to be expressed as a function about θ. So x, y, $\dot{x}$, $\dot{y}$, and $\cos\beta$ in Equation 3.6 all need to be expressed in terms of θ. According to Figure 3.1, the relations between x, y, l, R, and θ can be found. First, x can be solved by using Law of Cosines in triangle ABC.

$$x^2 = l^2 + (2R\sin\frac{\theta}{2})^2 - 2l(2R\sin\frac{\theta}{2})\cos\frac{\theta}{2}$$

$$- 4R^2\sin^2\frac{\theta}{2} + l^2 - 2lR\sin\theta, \tag{3.7}$$

$$\dot{x} = \frac{R(R\sin\theta - l\cos\theta)\dot{\theta}}{x}$$

$$= \frac{R\sqrt{R^2 + l^2}\sin(\theta - \kappa)\dot{\theta}}{x}. \tag{3.8}$$

The variable y is simply the length of the part of spring that wraps on the circular portion of the disk plus the length of the unstretched spring.

$$y = R\theta + l, \tag{3.9}$$
$$\dot{y} = R\dot{\theta}. \tag{3.10}$$

From Figure 3.1 we can find

$$\cos\beta = \frac{l\cos\theta - R\sin\theta}{x} \tag{3.11}$$

by using some basic geometry.

From Equations 3.6, 3.9 and 3.10,

$$\ddot{\theta} = \frac{1}{\frac{2}{5}mR}[-R(k\theta + g\dot{\theta}) - \cos\beta \cdot (-k(l - x) + g\dot{x} + v(t))]$$

$$= -\frac{k\theta + g\dot{\theta}}{\frac{2}{5}m} - \frac{1}{\frac{2}{5}mR}\cos\beta \cdot [-k(l - x) + g\dot{x} + v(t)]. \qquad (3.12)$$

Now, define

$$h(\theta) = \frac{x\cos\beta}{\frac{2}{5}mR}, \qquad (3.13)$$

and

$$g(\theta) = -\frac{h(\theta)}{x}, \qquad (3.14)$$

where $\cos\beta$ is given by Equation 3.11.

Thus from Equations 3.12-3.14,

$$\ddot{\theta} = -\frac{k\theta + g\dot{\theta}}{\frac{2}{5}m} - \frac{h(\theta)}{x}[-k(l - x) + g\dot{x} + v(t)]$$

$$= -\frac{k\theta + g\dot{\theta}}{\frac{2}{5}m} + \frac{h(\theta)}{x}kl - k \cdot h(\theta) - h(\theta) \cdot g \cdot \frac{\dot{x}}{x} - \frac{h(\theta)}{x} \cdot v(t)$$

$$= -\frac{k\theta + g\dot{\theta}}{\frac{2}{5}m} - kl \cdot g(\theta) - k \cdot h(\theta) - h(\theta) \cdot g \cdot \frac{gR\sqrt{R^2 + l^2}\sin(\theta - \kappa)\dot{\theta}}{l^2 + 4R^2\sin^2\frac{\theta}{2} - 2lR\sin\theta}$$

$$+ v(t)g(\theta)$$

$$= f(\theta, \dot{\theta}) + v(t)g(\theta). \qquad (3.15)$$

Summarizing,

$$\ddot{\theta} = f(\theta, \dot{\theta}) + v(t)g(\theta)$$

where,

$$f(\theta, \dot{\theta}) = -\frac{k\theta + g\dot{\theta}}{\frac{2}{5}m} - klg(\theta) - h(\theta)[k + \frac{gR\sqrt{R^2 + l^2}\sin(\theta - \kappa)\dot{\theta}}{l^2 + 4R^2\sin^2\frac{\theta}{2} - 2lR\sin\theta}], \quad (3.16)$$

$$g(\theta) = -h(\theta)[\frac{1}{\sqrt{l^2 + 4R^2\sin^2\frac{\theta}{2} - 2lR\sin\theta}}], \qquad (3.17)$$

and

$$h(\theta) = \frac{1}{\frac{2}{5}mR}(l\cos\theta - R\sin\theta). \qquad (3.18)$$

Thus the equations of motion are cast as an affine control system in the plane. Affine systems have been studied extensively in the literature, see for example W. Dayawansa, G. Knowles and C. Martin,1990 [1]. This is advantageous because it allows the use of techniques that have been used in control theory in the last decade.

4 Simulation of One Eye Tracking a Point Source in Two Dimensions

The parameters of the eye model need to be determined first. There are four parameters; m, the mass of the disk; l, the length of the unstretched spring; κ, the angle of rotation at which the spring has minimum length; and R, the radius of the disk.

Gray's Anatomy [4] gives that the diameter of the eye is 1 inch, so the radius of the disk is

$$R = \frac{1}{2}\text{inch} \approx 1.27\text{cm}. \tag{4.19}$$

The volume of the eye is calculated in Equation 4.20.

$$
\begin{aligned}
V &= \frac{4}{3}\pi R^3, \\
&= \frac{4}{3}\pi \cdot 1.25^3 \approx 8\text{cc}. \tag{4.20}
\end{aligned}
$$

The indices of refraction of the eye is approximately 1.336. So the mass of the eye is

$$m = 1.336 \cdot V \approx 10.7\text{gram}. \tag{4.21}$$

From [4], the length of the unstretched spring, l, is found to be

$$l \approx 3\text{cm}. \tag{4.22}$$

Refering to Figure 3.1, κ, the angle of rotation at which the spring has minimum length is

$$
\begin{aligned}
\kappa &= \arctan \frac{l}{R} \\
&= \arctan \frac{3}{1.25} \\
&= 67^o \\
&= 1.17\text{rad}. \tag{4.23}
\end{aligned}
$$

The next task is to choose the frequency of the spring, k; and the damping parameter, g. These two parameters should satisfy the following relation in order to bring the system to stabilize.

$$\lambda^2 + g\lambda + k = 0, \tag{4.24}$$

$$\lambda = \frac{-g \pm \sqrt{g^2 - 4k}}{2}, \tag{4.25}$$

where

$$g^2 - 4k < 0. \tag{4.26}$$

because the damping ratio $\xi = \frac{g}{2\sqrt{k}}$ is desired to be $\xi \leq 0.707$, so the two roots(poles) can locate in the left plane. Therefore the response of the system will have a little overshoot and settle very quickly.

The following sets of k and g are calculated according to this relation. Let $g = 2$, then $\sqrt{g^2 - 4k} = 1$, which implies $k = \frac{5}{4}$. Let $g = 8$, then $\sqrt{g^2 - 4k} = 4$, which implies $k = 20$.

We do not want to take g be larger than 8, since it will make the spring too stiff, and that is not realistic for human muscles.

Next, we need to find the controller to stabilize this nonlinear system. There are two goals, the first of which is to construct a control law that mimics the human process. We make extensive use of what is known about visual perception and use this to construct a control law. The second goal is to construct a control law that stabilizes the system in an exponential manner. Any stabilizing law will provide asymptotic tracking. But for a practical problem it is essential that the system "lock on" in as short a time as possible. It is important for purposes of robotics that the process be fast and robust.

We begin with a fixed object. The initial angle the ocular axis of the eye makes with the object is θ_0. First, a linear control law that uses only positional error is applied

$$v(t) = a \cdot (\theta - \theta_0),$$

where a is a constant to be chosen.

We repeat the experiment when the coefficients k and g are changing. These simulations show that the linear controller using only position error does not work very well. This is not a surprise, because our system is highly nonlinear.

Then we use another simple feedback $v = v_0$. v_0 is chosen to make the final position error be zero when the system is stable. In the Equation 3.15, let $\theta = \theta_0$, $\dot{\theta} = 0$ so that $\ddot{\theta} = 0$, we get

$$v(0) = -\frac{f(\theta_0, 0)}{g(\theta_0)}. \tag{4.27}$$

This time the system becomes stable in about 12 units of time, the overshoot is 0.21 rad (11°).

Based on the controller $v = v_0$, we try to improve it by using

$$v(t) = (1 - e^{-t/0.1})v_0$$

Since this type of controller does not work satisfactorily, the controller

$$v = v_0[1 - 5(\theta - \theta_0)]$$

is applied to the system. The system takes about 1 unit of time to response, and need about 12 units of time to be settle. The system takes less time to respond, almost the same amount of time to settle, but overshoots more and oscillates more.

We will improve this control by adding a term which makes the controller output large at the beginning of the tracking and becomes close to 0 when θ is close to θ_0. The controller is

$$v = v_0[1 - 5(\theta - \theta_0) - 2(\theta - \theta_0)e^{|\theta - \theta_0|}]$$

When the system is stable, by choosing v_0 appropriately we can minimize the final position error. The system now takes about 1.5 units of time to rise, 10 units of time to settle and its overshoot is about 0.09 rad less.

We have only used control laws involving position feedback so far, and the results are not satisfactory, the graphs of trace of θ show that there are large oscillatories, large overshoots, and settling times are quite long. So we improve the control law by adding a differential control to the controller which can reduce the overshoot and settling time.

$$v = v_0[1 - 5(\theta - \theta_0) - 2e^{|\theta - \theta_0|}(\theta - \theta_0) - 5\dot{\theta}]$$

Figure 6.1 is the graph of the error signal and controller feedback. Figure 6.2 gives the trace of θ. The system takes less than 1.6 units of time to rise, it overshoots a little bit and quickly becomes stable. This result is what we expected to get.

All the simulations have been done so far are for a fixed initial angle, $\theta_0 = \frac{\pi}{6}$. If the initial angle is smaller or larger, the controller with the same coefficients does not work as good as before.

When the angle is smaller, i.e., $\theta = \frac{\pi}{20}$, the coefficients of the controller need to be adjusted.The trace of θ is shown in Figure 6.3. On the other hand, if the angle is larger, the coefficients also need to be adjusted to get good results. The trace of θ is shown in Figure 6.4.

To construct a control law that will track both small angles and large angles, the coefficients ought to be changed into functions of $\theta - \theta_0$. We use the control law

$$v = v_0[1 - a_1(\theta - \theta_0) - a_2 e^{|\theta - \theta_0|}(\theta - \theta_0) - a_3\dot{\theta}], \qquad (4.28)$$

where the coefficients are chosen to be

$$a_1 = \frac{400}{1 + 500\sqrt[3]{|\theta - \theta_0|}}, \qquad (4.29)$$

$$a_2 = 5, \qquad (4.30)$$

$$a_3 = -a_1. \qquad (4.31)$$

If the angle is bigger than $\frac{\pi}{3} = 60°$, the result is close to a singularity. That is because it is close to κ, the maximum angle the disk can rotate. And for the angle too small, less than $\frac{\pi}{60} = 3°$, the control cannot work very well. The reason is that the eye is able to recognize the object within the range of $\theta \approx 2°$ without moving the eyeball, and we do not consider this case at this stage.

All the simulations have been done so far are about an object which does not move. Now the simulations for an moving object can be performed. The same control law as Equation 4.28 are used in the following experiments.

First simulate the model tracking a moving object whose locus is a straight line. Figure 6.5 is the graph of error signal and controller feedback, and Figure 6.6 is the trace of θ. Figure 6.6 shows that the system is able to response, but always with a delay.

Then simulate a moving object whose locus is a sine wave. Figure 6.7 is the graph of the error signal and controller feedback. The system has a better control when the initial angle is not too small. It is able to response in a short time, but the delay always exists. That is due to the nonlinearity of the actual model.

5   Simulation of Two Eyes Tracking A Point Source in Two Dimensions

The model for binocular vision in the plane is given by

$$\ddot{\theta} = f(\theta, \dot{\theta}) + (v(t) + u_1(t))g(\theta), \qquad (5.32)$$

$$\ddot{\beta} = f(\beta, \dot{\beta}) + (v(t) + u_2(t))g(\beta). \qquad (5.33)$$

The controller $v(t)$ is the controller for the gross movement.

The controllers $u_1(t)$ and $u_2(t)$ are added for the purpose of bringing the eyes out of parallel to obtain focus.

$$u_1(t) = k_1 \times (\theta - \theta_0), \qquad (5.34)$$

$$u_2(t) = k_2 \times (\beta - \beta_0). \qquad (5.35)$$

θ the angle of rotation of the left eye
β the angle of rotation of the right eye
θ_0 the initial angle of the left eye
β_0 the initial angle of the right eye

The controller which worked quiet well in monocular vision for angles in a wide range is applied to this binocular vision system.

$$v = v_0[1 - a_1(\theta - \theta_0) - a_2 e^{|\theta - \theta_0|}(\theta - \theta_0) - a_3\dot{\theta}] \qquad (5.36)$$

where

$$a_1 = \frac{400}{1 + 500\sqrt[3]{|\theta - \theta_0|}}, \qquad (5.37)$$

$$a_2 = 5, \qquad (5.38)$$

$$a_3 = -a_1. \qquad (5.39)$$

Figures 6.8-6.10 show the result when the object makes $\frac{\pi}{3}$, $\frac{\pi}{6}$ or $\frac{\pi}{60}$ with the ocular axis of left eye, respectively. Then do simulations for a moving object whose locus is a sine wave. The system can response in about $\frac{1}{2}$ units of time, and delay about $\frac{1}{2}$ units of time.

6   Conclusion

This model is sufficiently complete to model the movement of the eye and the tracking problem. The weakest part of the model is the modeling of the muscle as a second order actuator. The true vision problem is much more complex when the source is considered to not to be a point source. This research can be extended to

- Formulate the problem in three dimensions.

- Determine realistic values of the parameters modeling the muscles controlling the eye.

- Develop simulations of binocular eye movement for testing of various control algorithms.

These problems have potential applications in the automated screening of large numbers of patients for eye muscle disorders.

References

[1] W. Dayawansa, G. Knowles and C. Martin, On the asymptotic stabilization of smooth two dimensional systems in the plane, *SIAM Journal on Control and Optimization*, 28, pp 1321-1349, 1990.

[2] Foley, J. M. Primary distance perception. In: R. Held, H. Leibowitz
    and H. L. Teuber (eds.), *Handbook of Sensory Physiology*, Vol. VIII;
    Perception, Springer, Berlin, pp 181-213, 1978.

[3] Foley, J. M. Binocular distance perception. *Psychol. Rev.* 87, pp 411-
    433, 1980.

[4] Henry Gray, *Anatomy, Descriptive and Surgical*, pp 303-306, 1977.

[5] F. W. Sears and M. W. Zemansky, *University Physics*, pp 783-786,
    1955.

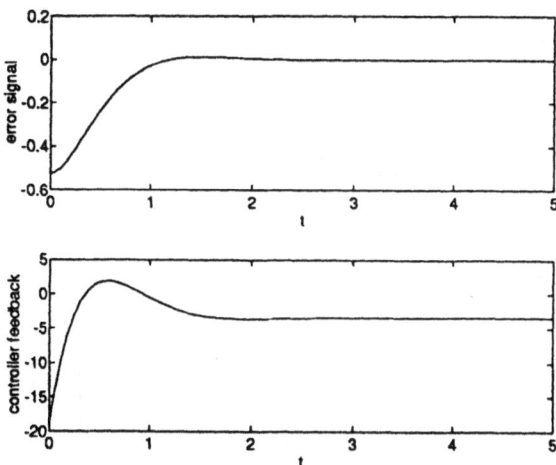

Figure 6.1: $v = v_0[1 - 5(\theta - \theta_0) - 2e^{|\theta - \theta_0|}(\theta - \theta_0) - 5\dot{\theta}]$

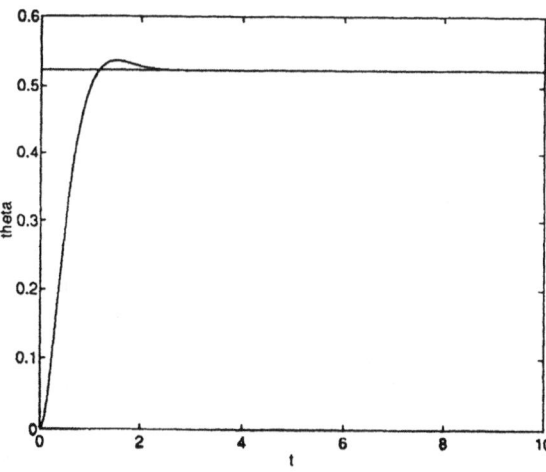

Figure 6.2: $v = v_0[1 - 5(\theta - \theta_0) - 2e^{|\theta - \theta_0|}(\theta - \theta_0) - 5\dot{\theta}]$, Trace of θ

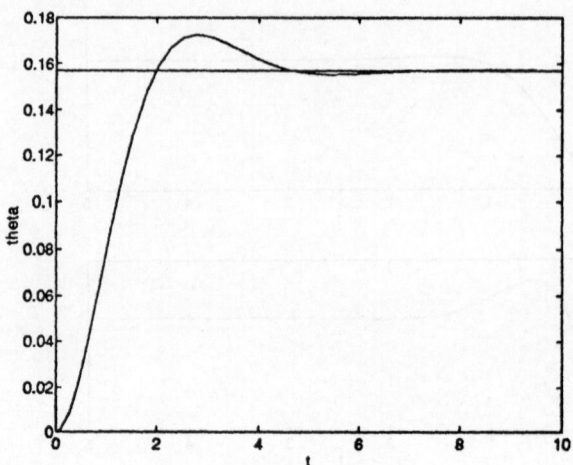

Figure 6.3: $v = v_0[1 - 80(\theta - \theta_0) - 5e^{|\theta - \theta_0|}(\theta - \theta_0) - 80\dot{\theta}]$, $\theta_0 = \frac{\pi}{20}$

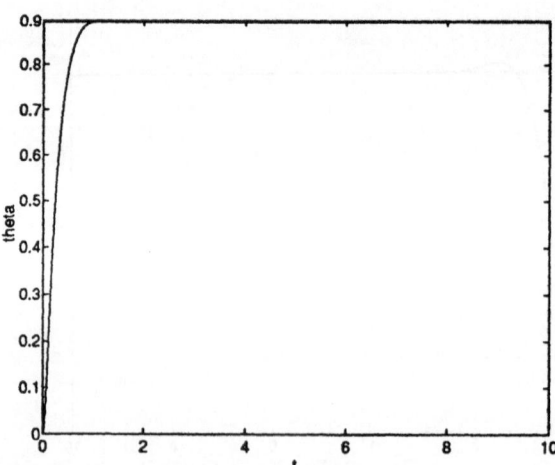

Figure 6.4: $v = v_0[1 - 2(\theta - \theta_0) - 2e^{|\theta - \theta_0|}(\theta - \theta_0) - 2\dot{\theta}]$, $\theta_0 = \frac{\pi}{3.5}$

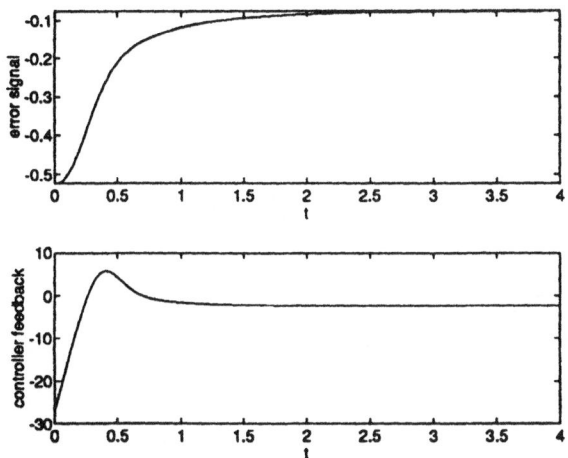

Figure 6.5: Graph of error signal and controller feedback when the locus of the object is $\frac{\pi}{6} + 0.1t$

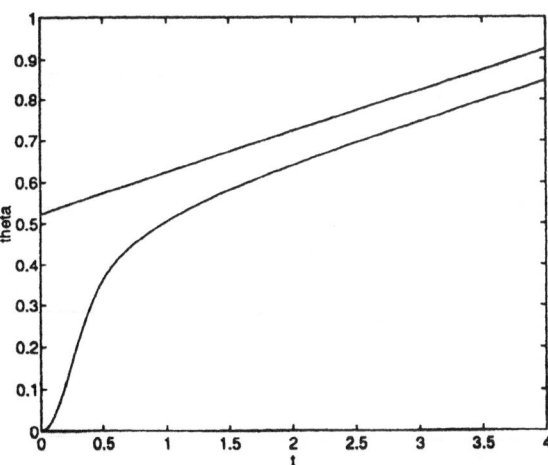

Figure 6.6: Trace of θ when the locus of the object is $\frac{\pi}{6} + 0.1t$

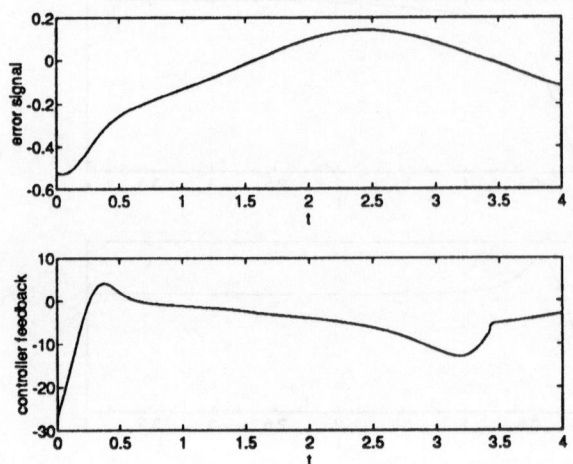

Figure 6.7: Graph of error signal and controller feedback when the locus of the object is $\frac{\pi}{6} + 0.2 \sin \frac{\pi}{2} t$

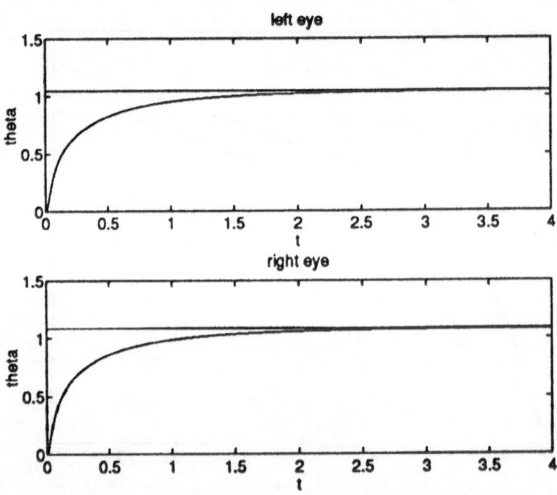

Figure 6.8: Traces of θ and β, $\theta = \frac{\pi}{3}$

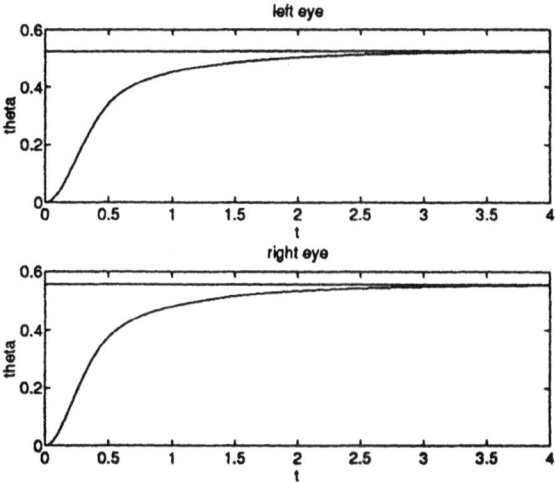

Figure 6.9: Traces of θ and β, $\theta = \frac{\pi}{6}$

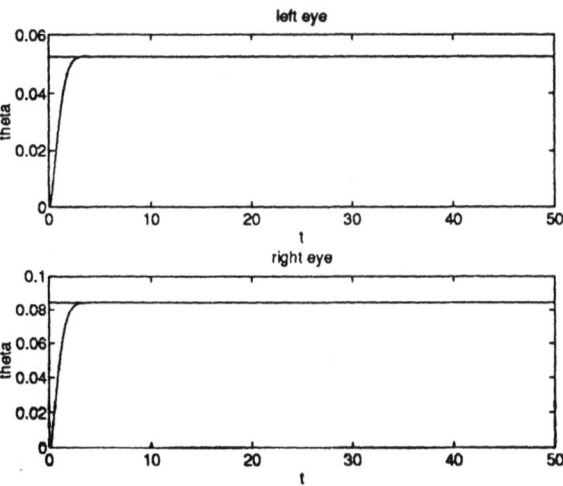

Figure 6.10: Trace of θ and β, $\theta = \frac{\pi}{60}$

THE SINC-GALERKIN SCHWARZ ALTERNATING METHOD FOR POISSON'S EQUATION

Nancy J. Lybeck
Center for Research in Scientific Computation
North Carolina State University
Raleigh, North Carolina 27695-8205

Kenneth L. Bowers
Department of Mathematical Sciences
Montana State University
Bozeman, Montana 59717

1 Introduction

Sinc-Galerkin and sinc-collocation methods provide a powerful and diverse set of tools for the numerical solution of differential equations. Sinc methods are particularly appealing because they can be used to solve problems with boundary singularities, while maintaining their characteristic exponential convergence rate. Since the introduction of the Sinc-Galerkin method in [12], sinc methods have been used on a variety of differential equations, including the two-point boundary-value problem, Poisson's equation, the wave equation, the heat equation, the advection-diffusion equation, and Burgers' equation. In addition, sinc methods have been successfully used in conjunction with more complex procedures such as domain decomposition (see [5], [6], [7], and [8]). A thorough convergence analysis for sinc domain decomposition methods for ordinary differential equations is in [9] and [10].

Domain decomposition methods have become increasingly popular with the advancement of parallel computing technology. The two traditional methods of domain decomposition, patching and overlapping, have been well studied in conjunction with a variety of numerical methods for elliptic partial differential equations. The Sinc-Galerkin patching and overlapping methods for Poisson's equation on a rectangle are developed in [6], where the discrete systems are solved directly, without iteration. Iterative procedures for solving the discrete system must be invoked to efficiently solve problems on parallel computers. A recent survey of iterative domain decomposition algorithms for elliptic partial differential equations can be found in [2]. In [10] convergence results are given for the sinc overlapping method used in conjunction with the Schwarz alternating method to solve a two-point boundary-value problem. The intent of this paper is to develop the Sinc-Galerkin Schwarz alternating method for Poisson's equation on a rectangle.

2   The Sinc-Galerkin Method

Let Ω be a rectangular region, $\Omega = \{(x,y) : a < x < b, c < y < d\}$. Let $\partial\Omega$ be the boundary of Ω. Poisson's equation on Ω with homogeneous Dirichlet boundary conditions is given by

$$-\nabla^2 u(x,y) \equiv -\Delta u(x,y) \;=\; f(x,y) \;,\; (x,y) \in \Omega$$

$$\tag{2.1}$$

$$u(x,y) \;=\; 0 \;,\; (x,y) \in \partial\Omega \;.$$

The basis used to solve (2.1) is a product of sinc basis functions in the x and y directions. The sinc basis functions are given by

$$
\begin{aligned}
S_j(x) &= S(j,h_x) \circ \phi(x) \\
&\equiv \mathrm{sinc}\left(\frac{\phi(x) - jh_x}{h_x}\right) \\
S_k(y) &= S(j,h_y) \circ \psi(y) \\
&\equiv \mathrm{sinc}\left(\frac{\psi(y) - kh_y}{h_y}\right)
\end{aligned}
$$

where

$$\mathrm{sinc}(z) = \begin{cases} \frac{\sin(\pi z)}{\pi z}, & z \neq 0 \\ 1, & z = 0 \end{cases} \;.$$

The conformal maps

$$
\begin{aligned}
\phi(x) &= \ln\left(\frac{x-a}{b-x}\right) \\
\psi(y) &= \ln\left(\frac{y-c}{d-y}\right)
\end{aligned}
$$

are used to define the basis functions on the finite intervals (a,b) and (c,d), respectively.

Assume a product approximate solution of the form

$$u_{m_x,m_y}(x,y) = \sum_{j=-M_x}^{N_x} \sum_{k=-M_y}^{N_y} u_{jk} S_j(x) S_k(y) \;. \tag{2.2}$$

Here $m_x = M_x + N_x + 1$, $m_y = M_y + N_y + 1$, and $h_x > 0, h_y > 0$. The sinc nodes x_j and y_k are chosen so that $x_j = \phi^{-1}(jh_x)$ and $y_k = \psi^{-1}(kh_y)$.

Orthogonalization of the residual against each basis function

$$(-\Delta u_{m_x,m_y} - f, S_p S_q) = 0 \;,\; -M_x \leq p \leq N_x \;,\; -M_y \leq q \leq N_y$$

uses the weighted inner product

$$(v, w) = \int_a^b \int_c^d v(x, y)w(x, y)(\phi'(x))^{-1/2}(\psi'(y))^{-1/2} dy dx \ .$$

Integration by parts is used to remove all partial derivatives from u_{m_x, m_y}, and applying the sinc quadrature rule (found in [3], [13], or [14]) yields the discrete Sinc-Galerkin system. The complete development can be found in [4].

The following notation will be necessary for writing down the system. For a function $\gamma(x)$ defined on (a, b), denote by $\vec{\gamma}$ the $m_x \times 1$ vector

$$\vec{\gamma} = [\gamma(x_{-M}) \ \cdots \ \gamma(x_N)]^T \ ,$$

and let $\mathcal{D}(\gamma)$ be the $m_x \times m_x$ diagonal matrix

$$\mathcal{D}(\gamma) = \begin{bmatrix} \gamma(x_{-M_x}) & & \\ & \ddots & \\ & & \gamma(x_{N_x}) \end{bmatrix} \ .$$

Also define

$$\delta_{jk}^{(2)} \equiv h_x^2 \frac{d^2}{d\phi^2} [S(j, h_x) \circ \phi(x)] \Big|_{x = x_k} = \begin{cases} \dfrac{-\pi^2}{3}, & k = j \\ \dfrac{-2(-1)^{k-j}}{(k-j)^2}, & k \neq j \end{cases} \ ,$$

and define the $m_x \times m_x$ matrix $I^{(2)}$ by

$$I^{(2)} = [\delta_{jk}^{(2)}] \ , \ j, k \ = \ -M_x, \ldots, N_x \ .$$

Similar definitions are used for the variable y on (c, d). The fundamental Sinc-Galerkin matrices are the $m_x \times m_x$ matrix

$$\Gamma_x = \left\{ \frac{-1}{h_x^2} I^{(2)} + \mathcal{D}\left(\frac{1}{4}\right) \right\} \mathcal{D}\left(\sqrt{\phi'}\right) \tag{2.3}$$

and the $m_y \times m_y$ matrix

$$\Gamma_y = \left\{ \frac{-1}{h_y^2} I^{(2)} + \mathcal{D}\left(\frac{1}{4}\right) \right\} \mathcal{D}\left(\sqrt{\psi'}\right) \ . \tag{2.4}$$

The discrete Sinc-Galerkin system for (2.1) is the Sylvester equation

$$\mathcal{D}\left((\phi')^{3/2}\right) \Gamma_x U + U \Gamma_y^T \mathcal{D}\left((\psi')^{3/2}\right) = F \ , \tag{2.5}$$

where the coefficient matrix $U = [u_{jk}]$ and the right–hand side $F = [f(x_j, y_k)]$ are matrices of size $m_x \times m_y$.

3   The Schwarz Alternating Method

The Schwarz alternating method was first proposed by H. A. Schwarz in [11]. Consider the rectangle $\Omega = (a, b) \times (c, d)$. For simplicity, split the domain into just two subdomains. For $a < \xi_1 < \xi_2 < b$, let $\Omega^1 = (a, \xi_2) \times (c, d)$ and $\Omega^2 = (\xi_1, b) \times (c, d)$, as shown in Figure 3.1. Let $\Gamma^1 = \{\xi_1\} \times (c, d)$ and $\Gamma^2 = \{\xi_2\} \times (c, d)$ be the two introduced boundaries.

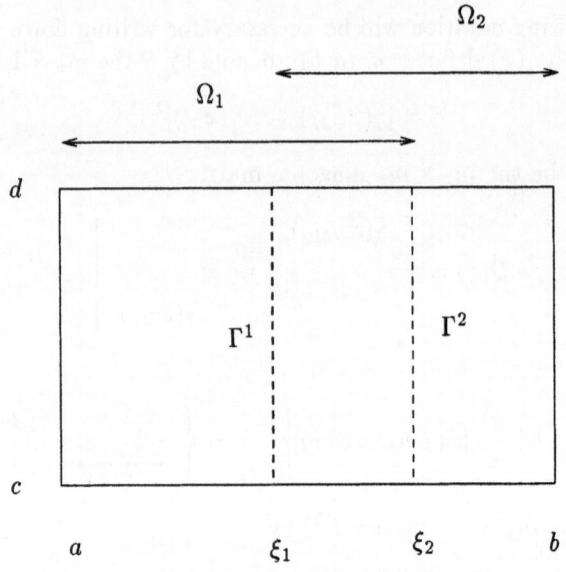

Figure 3.1: The domain Ω

The Schwarz method for (2.1) is then given as follows. Choose an arbitrary initial function u^0 defined on Ω^1 so that $u^0|_{\partial\Omega^1 \setminus \Gamma^2} = 0$. Then (from [1]) define the two sequences $\{u^{2n-1}\}$ and $\{u^{2n}\}$, for $n \geq 1$, by

$$
\begin{cases}
-\Delta u^{2n-1}(x, y) = f(x, y) & (x, y) \in \Omega^2 \\[2mm]
u^{2n-1}(x, y) = 0 & (x, y) \in \partial\Omega^2 \setminus \Gamma^1 \\[2mm]
u^{2n-1}(x, y) = u^{2n-2}(x, y) & (x, y) \in \Gamma^1
\end{cases} \qquad (3.1)
$$

$$\begin{cases} -\Delta u^{2n}(x,y) = f(x,y) & (x,y) \in \Omega^1 \\[2mm] u^{2n}(x,y) = 0 & (x,y) \in \partial\Omega^1 \setminus \Gamma^2 \\[2mm] u^{2n}(x,y) = u^{2n-1}(x,y) & (x,y) \in \Gamma^2 \end{cases} \qquad (3.2)$$

As n tends to infinity, these sequences converge geometrically to the solution of (2.1) in Ω^2 and Ω^1, respectively. For problems with larger overlap, $\gamma \equiv \xi_2 - \xi_1$, the convergence is more rapid (see [1]).

4   The Sinc-Galerkin Schwarz Alternating Method

The sinc approximation in (2.2) satisfies homogeneous Dirichlet boundary conditions. The subproblems from the Schwarz alternating method, (3.1) and (3.2), have nonhomogeneous Dirichlet boundary conditions at the introduced boundaries, Γ^1 and Γ^2. Hence boundary basis functions must be added to the approximate solution. Let

$$\omega_1(x) = (x-a)^3 \left(\frac{-3}{(\xi_2-a)^4} x + \frac{4\xi_2 - a}{(\xi_2-a)^4} \right) \qquad (4.1)$$

and

$$\omega_2(x) = (x-b)^3 \left(\frac{-3}{(\xi_1-b)^4} x + \frac{4\xi_1 - b}{(\xi_1-b)^4} \right) . \qquad (4.2)$$

Notice that the choice of the boundary basis function is somewhat arbitrary here. These basis functions were picked for their smoothness over the intervals Ω^1 and Ω^2, respectively, and their behavior at the endpoints of these intervals. The basis function $\omega_1(x)$ is chosen to satisfy $\omega_1(a) = \omega_1'(a) = \omega_1''(a) = 0$, $\omega_1(\xi_2) = 1$, and $\omega_1'(\xi_2) = 0$. Similarly, $\omega_2(x)$ satisfies $\omega_2(b) = \omega_2'(b) = \omega_2''(b) = 0$, $\omega_2(\xi_1) = 1$, and $\omega_2'(\xi_1) = 0$.

The sinc approximations for u^{2n-1} and u^{2n}, $n \geq 1$, are then

$$u_{m_x^2, m_y^2}^{2n-1}(x,y) = \sum_{k=-M_y^2}^{N_y^2} S_k^2(y) \left(b_k^{2n-1} \omega_2(x) + \sum_{j=-M_x^2}^{N_x^2} u_{jk}^{2n-1} S_j^2(x) \right) \qquad (4.3)$$

and

$$u_{m_x^1, m_y^1}^{2n}(x,y) = \sum_{k=-M_y^1}^{N_y^1} S_k^1(y) \left(b_k^{2n} \omega_1(x) + \sum_{j=-M_x^1}^{N_x^1} u_{jk}^{2n} S_j^1(x) \right) , \qquad (4.4)$$

where, for $j = 1,2$, $m_x^j = M_x^j + N_x^j + 1$, and $m_y^j = M_y^j + N_y^j + 1$.

The distinction between the basis functions $S_j^1(x)$ and $S_j^2(x)$ is mandatory because the functions have different domains. Thus there are two conformal maps, given by

$$\phi^1(x) \;=\; \ln\left(\frac{x-a}{\xi_2 - x}\right)$$

$$\phi^2(x) \;=\; \ln\left(\frac{x-\xi_1}{b-x}\right).$$

The reason for the distinction between $S_k^1(y)$ and $S_k^2(y)$ is not so obvious. The two subdomains, Ω^1 and Ω^2, both have $y \in (c,d)$. Thus there are not two distinct conformal maps. In certain problems, however, it is both convenient and beneficial to use more y nodes in one of the two subdomains. Even the grid spacings, h_y^1 and h_y^2, need not be the same, leading to different nodes in the two subdomains. For these reasons, a distinction is made between the basis functions. This is an option that is not available when the discrete system is developed as in [6].

The additional basis functions complicate the discrete Sinc-Galerkin system. The development of a discrete system for a two-point boundary-value problem with non-homogeneous Dirichlet boundary conditions is given in [3]. Similar techniques can be used to derive the discrete system for Poisson's equation. The notation used here has been changed from that found in [3] in order to better accommodate the Schwarz alternating method. It is helpful to notice that $u_{m_x^2, m_y^2}^{2n-1}(\xi_1, y_k^2) = b_k^{2n-1}$ and similarly $u_{m_x^1, m_y^1}^{2n}(\xi_2, y_k^1) = b_k^{2n}$. The Sinc-Galerkin Schwarz alternating method is outlined below.

1. Choose b_k^0, $k = -M_y^2, \ldots, N_y^2$.

2. $b_k^{2n-1} = u_{m_x^1, m_y^1}^{2n-2}(\xi_1, y_k^2)$, $k = -M_y^2, \ldots, N_y^2$, $n > 1$.

3. Solve

$$U^{2n-1}\Gamma_y^2 \mathcal{D}\left((\psi_2')^{-3/2}\right) + \mathcal{D}\left((\phi_2')^{-3/2}\right)\Gamma_x^2 U^{2n-1} =$$

$$F^2 - \vec{\omega}_2(\vec{b}^{2n-1})^T(\Gamma_y^2)^T \mathcal{D}\left((\psi_2')^{-3/2}\right) + \mathcal{D}\left((\phi_2')^{-3/2}\right)\vec{\omega}_2''(\vec{b}^{2n-1})^T .$$

4. $b_k^{2n} = u_{m_x^2, m_y^2}^{2n-1}(\xi_2, y_k^1)$, $k = -M_y^1, \ldots, N_y^1$.

5. Solve

$$U^{2n}\Gamma_y^1 \mathcal{D}\left((\psi_1')^{-3/2}\right) + \mathcal{D}\left((\phi_1')^{-3/2}\right)\Gamma_x^1 U^{2n} =$$

$$F^1 - \vec{\omega}_1(\vec{b}^{2n})^T(\Gamma_y^1)^T \mathcal{D}\left((\psi_1')^{-3/2}\right) + \mathcal{D}\left((\phi_1')^{-3/2}\right)\vec{\omega}_1''(\vec{b}^{2n})^T .$$

Repeat steps 2-5 until the convergence criterion is met.

5   Numerical Examples

For the following examples, the domain Ω is chosen to be the rectangle $\Omega = (-1, 4) \times (0, 1)$. The domain is split into two subdomains, $\Omega^1 = (-1, 1) \times (0, 1)$ and $\Omega^2 = (1 - \gamma, 4) \times (0, 1)$, where γ is the amount of overlap. The Schwarz alternating method is expected to converge more rapidly for larger values of γ, and this is phenomenon is seen in the numerical examples. The initial vector was chosen to be $\vec{b}^0 = \vec{0}$.

The matrix systems in steps 3 and 5 are solved by concatenating each side of the equations. The following definitions are necessary for this step.

Definition 5.1 *For a matrix* $B = (b_{ij})$, $1 \leq i \leq m$, $1 \leq j \leq n$, *the concatenation of* B *is the* $mn \times 1$ *vector*

$$co(B) \equiv \begin{bmatrix} \vec{b}_{i1} \\ \vec{b}_{i2} \\ \vdots \\ \vec{b}_{in} \end{bmatrix}$$

where $\vec{b}_{ik}$ *is the kth column of* B.

Definition 5.2 *Let* A *be an* $m \times n$ *matrix and* B *be a* $p \times q$ *matrix. The Kronecker or tensor product of* A *and* B *is the* $mp \times nq$ *matrix*

$$A \otimes B \equiv \begin{bmatrix} a_{11}B & a_{12}B & \cdots & a_{1n}B \\ a_{21}B & a_{22}B & \cdots & a_{2n}B \\ \vdots & & & \vdots \\ a_{m1}B & a_{m2}B & \cdots & a_{mn}B \end{bmatrix} .$$

A useful property of concatenation is given in Theorem 5.3.

Theorem 5.3 *Let* A *be* $m \times m$, X *be* $m \times n$, *and* B *be* $n \times n$. *Then*

$$co(AXB) = (B^T \otimes A)co(X) .$$

Proof of this theorem is given in [3].

Concatenating each side of the equation in step 3 yields

$$\left\{ \left(\mathcal{D}\left((\psi_2')^{-3/2} \right) \left(\Gamma_y^2 \right) \right)^T \otimes I + I \otimes \mathcal{D}\left((\phi_2')^{-3/2} \right) \Gamma_x^2 \right\} co(U^{2n-1}) =$$

$$co\left(F^2 - \vec{\omega}_2(\vec{b}^{2n-1})^T (\Gamma_y^2)^T \mathcal{D}\left((\psi_2')^{-3/2} \right) + \mathcal{D}\left((\phi_2')^{-3/2} \right) \vec{\omega}_2''(\vec{b}^{2n-1})^T \right)$$

The equation in step 5 is solved in the same manner. This procedure is used in [3] to solve similar systems for Poisson's equation. All numerical results are run using MATLAB version 4.0a on an IBM RS6000 model 220.

The convergence criterion in these examples was fairly simple. The iterations were continued until, at iterate $2NI$,

$$\max\left\{ \|\vec{b}^{2NI-1} - \vec{b}^{2NI-3}\|_\infty, \|\vec{b}^{2NI} - \vec{b}^{2NI-2}\|_\infty \right\} \le 10^{-5},$$

and

$$\max\left\{ \|co(U^{2NI-1} - U^{2NI-3})\|_\infty, \|co(U^{2NI} - U^{2NI-2})\|_\infty \right\} \le 10^{-5}.$$

More stringent stopping criterion did not noticeably improve the results.

Let the sinc error, assuming the last iterate is u^{2NI}, be defined by

$$\|E_S\| = \max\left\{ \max_{(x,y)\in S^2} |u(x,y) - u^{2NI-1}_{m_x^2, m_y^2}(x,y)|, \right.$$

$$\left. \max_{(x,y)\in S^1} |u(x,y) - u^{2NI}_{m_x^1, m_y^1}(x,y)| \right\},$$

where for $p = 1, 2$,

$$S^p = \{x_k^p : -M_x^p \le k \le N_x^p\} \times \{y_j^p : -M_y^p \le j \le N_y^p\}$$

is the set of points generated from the Sinc-Galerkin method in Ω^p. Similarly define a uniform error by

$$\|E_U\| = \max\left\{ \max_{(x,y)\in U\cap\Omega^2} |u(x,y) - u^{2NI-1}_{m_x^2, m_y^2}(x,y)|, \right.$$

$$\left. \max_{(x,y)\in U\cap\Omega^1} |u(x,y) - u^{2NI}_{m_x^1, m_y^1}(x,y)| \right\},$$

where

$$U = \{(-1 + 5j/100, k/100) : 0 \le j \le 100, 0 \le k \le 100\}$$

is a uniform grid over Ω.

Example 5.1 Consider the problem

$$-\Delta u(x, y) \;=\; f(x, y) \,, \; (x, y) \in \Omega = (-1, 4) \times (0, 1)$$

$$u(x, y) \;=\; 0 \,, \; (x, y) \in \partial\Omega \,,$$

where $f(x, y)$ is chosen so that the true solution is given by

$$u(x, y) = \frac{\sqrt{(x+1)y}(x-4)^2(1-y)^2}{5.4371} \,.$$

In this example $M \equiv M_x^1 = M_x^2 = M_y^1 = M_y^2$ and $N \equiv N_x^1 = N_x^2 = N_y^1 = N_y^2$. Let $h = h_x^1 = h_x^2 = h_y^1 = h_y^2 = \pi/\sqrt{2M}$. This problem has boundary singularities along the lines $x = -1$ and $y = 0$. The numerical results given in Table 5.1 demonstrate that as the overlap γ decreases, the number of iterations required to converge increases.

M	γ	$\|E_S\|$	$\|E_U\|$	NI
12	.20	$5.6768e-02$	$5.5571e-02$	11
	.15	$5.6768e-02$	$5.5570e-02$	14
	.10	$5.6768e-02$	$5.5569e-02$	19
	.05	$5.6768e-02$	$5.5569e-02$	34
14	.20	$4.2554e-02$	$4.1752e-02$	11
	.15	$4.2554e-02$	$4.1751e-02$	14
	.10	$4.2554e-02$	$4.1751e-02$	19
	.05	$4.2554e-02$	$4.1750e-02$	34

Table 5.1: Errors on both the sinc grid S and the uniform grid U with varying overlap γ and $M = 12, 14$ for the Sinc–Galerkin Schwarz alternating domain decomposition method applied to Example 5.1. Total number of iterations is NI.

Example 5.2 Consider the problem

$$-\Delta u(x,y) \;=\; f(x,y) \;, \; (x,y) \in \Omega = (-1,4) \times (0,1)$$

$$u(x,y) \;=\; 0 \;, \; (x,y) \in \partial\Omega \;,$$

where $f(x,y)$ is chosen so that the true solution is given by

$$u(x,y) = \sqrt{x+1}(x-4)(y-y^2) \;.$$

Because the singularity is only along the line $x = -1$, it is convenient to allow $M^1 \equiv M_x^1 = M_y^1 = N_x^1 = N_y^1$ to vary, and to fix $M^2 \equiv M_x^2 = M_y^2 = N_x^2 = N_y^2 = 6$. Thus choose $h_x^1 = h_y^1 = \pi/\sqrt{2M^1}$, and $h_x^2 = h_y^2 = \pi/\sqrt{2M^2}$. The numerical results given in Table 5.2 demonstrate the convergence of the Sinc-Galerkin method as M^1 increases for overlap $\gamma = .2$.

M^1	M^2	$\|E_S\|$	$\|E_U\|$	NI
2	6	$1.6456e-01$	$1.7626e-01$	11
4	6	$1.1580e-01$	$1.0504e-01$	12
6	6	$7.8422e-02$	$6.9377e-02$	11
8	6	$5.4563e-02$	$4.7363e-02$	11
10	6	$3.9089e-02$	$3.3589e-02$	11
12	6	$2.9663e-02$	$3.0560e-02$	11
14	6	$2.9904e-02$	$3.0751e-02$	11
16	6	$3.0041e-02$	$3.0855e-02$	11

Table 5.2: Errors on both the sinc grid S and the uniform grid U with fixed overlap $\gamma = .2$ for the Sinc–Galerkin Schwarz alternating domain decomposition method applied to Example 5.2. Total number of iterations is NI.

References

[1] C. CANUTO, M. Y. HUSSAINI, A. QUARTERONI, and T. A. ZANG, *Spectral Methods in Fluid Dynamics*, Springer-Verlag, New York, 1988.

[2] T. F. CHAN and T. P. MATHEW, " Domain decomposition algorithms," *Acta Numerica*, pages 61–143, 1994.

[3] J. LUND and K. L. BOWERS, *Sinc Methods for Quadrature and Differential Equations*, SIAM, Philadelphia, 1992.

[4] J. LUND, K. L. BOWERS, and K. M. MCARTHUR, "Symmetrization of the Sinc-Galerkin method with block techniques for elliptic equations," *IMA J. Numer. Anal.*, 9(1):29–46, 1989.

[5] N. J. LYBECK and K. L. BOWERS, "Sinc methods for domain decomposition," submitted to *Applied Mathematics and Computation*.

[6] N. J. LYBECK and K. L. BOWERS, "Domain decomposition in conjunction with sinc methods for Poisson's equation," submitted to *SIAM Journal on Scientific Computation*.

[7] N. J. LYBECK and K. L. BOWERS, "Domain decomposition via the Sinc-Galerkin method for second order differential equations," in D. Keyes, editor, *Proceedings of the 7th International Symposium on Domain Decomposition Methods for Partial Differential Equations*, A.M.S.,1995.

[8] N. J. LYBECK and K. L. BOWERS, "The Sinc-Galerkin patching method for Poisson's equation," in W. F. Ames, editor, *Proceedings of the 14th IMACS World Congress on Computation and Applied Mathematics*, volume 1, pages 325–328, Georgia Institute of Technology, Atlanta, 1994.

[9] A. C. MORLET, N. J. LYBECK, and K. L. BOWERS, "Sinc domain decomposition methods I: The direct approach," submitted to *Math Comp*.

[10] A. C. MORLET, N. J. LYBECK, and K. L. BOWERS, "Sinc domain decomposition methods II: The Schwarz alternating method," submitted to *Math Comp*.

[11] H. A. SCHWARZ,*Gesammelte Mathematische Abhandlungen*, volume 2, pages 133–143, Springer, 1890. First published in Vierteljahrsschrift der Naturforschenden Gesellschaft in Zürich, volume 15, 1870, pp. 272–286.

[12] F. STENGER, " A Sinc-Galerkin method of solution of boundary value problems," *Math. Comp.*, 33:85–109, 1979.

[13] F. STENGER, " Numerical methods based on Whittaker cardinal, or sinc functions," *SIAM Rev.*, 23(2):165–224, 1981.

[14] F. STENGER, *Numerical Methods Based on Sinc and Analytic Functions*, Springer-Verlag, New York, 1993.

A CROSS VALIDATION METHOD FOR FIRST KIND INTEGRAL EQUATIONS

B. A. Mair [*]
Department of Mathematics
University of Florida
Gainesville, Florida 32611

F. H. Ruymgaart[†]
Department of Mathematics
Texas Tech University
Lubbock, Texas 79409

1 Introduction

This paper analyzes a cross validation method of obtaining a stable, data–based, approximate solution to a general class of first kind Fredholm integral equations of the form

$$g(t) = \int_0^1 h(t, \tau) f(\tau) d\tau \qquad (1.1)$$

from finite, discrete, inaccurate data

$$Y_j = g(T_j) + \varepsilon_j \quad , \quad j = 1, 2, \ldots, n \qquad (1.2)$$

where, as usual, the $\{\varepsilon_j, \ j = 1, 2, \ldots, n\}$ is discrete white noise. That is, they are uncorrelated real valued random variables with mean zero and nonzero *unknown* variance σ^2.

Due to the ill-posedness of this problem, the regularization methods of Tikhonov and truncated singular value decomposition (TSVD) have been developed for its solution (cf. [2] and references therein). The ill–posed nature of the original problem is quantified in these formulations by the presence of a regularization parameter in Tikhonov regularization and a truncation level in TSVD. The major difficulty in applying these methods is to obtain a data–based method of computing these parameters. To date, the most successful has been that of generalized cross validation (GCV) developed by Wahba (cf. [10, 11]) for Tikhonov regularization and adapted by Vogel [9] for TSVD.

At the same time, statisticians have been concerned with the problem of density estimation. The method of cross validation has been developed for determining the bandwidth in kernel density estimators [8, 3].

This paper proposes a method of cross validation, analogous to that used for density estimation, for obtaining a data–based method for computing the truncation level in TSVD. We demonstrate that the truncation level, m_s, exhibits the same asymptotic behavior as that obtained by GCV [9] (see Theorem 4.1). Furthermore it is important to note, as discussed in [9],

[*]Supported in part by NSF Grant DMS-9311513
[†]Supported in part NSF Grant DMS-9204950

[2], that convergence of the approximation obtained by the GCV method is only guaranteed when the kernel h is a continuous function on $(0, 1) \times (0, 1)$ and explicit rates of convergence are obtained under additional smoothness constraints on h [7]. In this regard, note that the convergence rate obtained in [9], is a measure of the accuracy of the GCV approximation for a finite-dimensional approximation of the true solution f.

In this paper, we demonstrate that the approximate solution obtained by cross validation actually converges at the same rate as that obtained in [9] for the finite–dimensional case, without any continuity assumptions on the kernel h. Furthermore, this method is easily extended to cover very general regularizing methods (including Tikhonov) for general operator equations, and other statistical nonparametric estimation problems such as regression and density estimation [1, 5, 6].

Although it is customary to assume the sampling points T_j are deterministic (in fact equi–spaced) [9, 10, 11], this assumption limits the suitability of the inversion procedure to cases in which the unknown f is fairly "regular". This paper treats the case of sampling at random points having a known distribution. The idea here is that apriori knowledge of abrupt changes, or singularities could be incorporated in the sampling procedure, in much the same way that the sinc methodology [4] handles singularities with an appropriate choice of transformation. Therefore, we assume that

$\{Y_j\}$, $\{T_j\}$ are independent sequences of random variables

$$\text{and} \tag{1.3}$$

each sequence is independent, identically distributed.

It is convenient to rewrite (1.1) as an operator equation

$$g = Kf \tag{1.4}$$

where the operator

$$K : L^2(0,1) \to L^2(0,1) \ , \quad Kf(t) = \int_0^1 h(t,\tau)f(\tau)d\tau \tag{1.5}$$

is assumed to be injective and compact.

For non–injective operators K, one seeks an approximation to the Moore–Penrose inverse operator, denoted $K^\dagger$, applied to g. Now, $K^\dagger g$ is the unique solution of the equation $Kf = g$ in the orthogonal complement of the kernel of K. Thus, the assumption of injectivity can be obtained by restricting K to the orthogonal complement of its kernel.

In the next section, we formulate a regularized solution, denoted $\hat{f}_m$ of (1.1) based on the data in (1.2) and a truncation level m, of the SVD of K. This approximate solution is the exact analog of that proposed in [9] for deterministic sampling points. The accuracy of this approximation will be measured by the integrated mean square error (IMSE)

$$\mathbb{E}\|\hat{f}_m - f\|^2 \tag{1.6}$$

where $\mathbb{E}$ denotes mathematical expectation and $\|\cdot\|$ denotes the usual norm on $L^2(0,1)$.

In Section 3, we obtain an expression for the *exact* IMSE which only contains terms that can be computed from the data. This forms the basis for a data–based method for determining the level of truncation, which consists of minimizing a cross validation function $S_n(m)$ (see Definition 3.1). This cross validation function is quite different from that in [9]. Thus, the significance of this approach is not in the form of the approximate solution but rather in the cross-validation function used to determine the truncation level, and the analysis which obtains convergence rates without severe restrictions on the operator K.

Section 4 contains an analysis of the rates of convergence of the purely data–based approximation under assumptions on the rates of decrease of the singular values of K and the smoothness of the true solution f.

2 Regularized solution

Throughout this paper, $\overline{\cdot}$, denotes complex conjugation and $[\ ,\ ]$ denotes the usual inner product in $L^2(0,1)$.

Let $(\mu_k, \varphi_k, \psi_k)$, $k = 1, 2, \ldots$, be a singular system for K. More specifically, $\{\varphi_k\}$ is a complete orthonormal set in $L^2(0,1)$ and

$$K^* K \varphi_k = \lambda_k \varphi_k \quad , \quad \mu_k = \sqrt{\lambda_k} \quad , \quad \mu_k \psi_k = K \varphi_k \tag{2.1}$$

In addition we assume that

$$\sup_k(\|\varphi_k\|_\infty + \|\psi_k\|_\infty) < \infty \tag{2.2}$$

where $\|\cdot\|_\infty$ denotes the uniform norm in the space of continuous functions on the interval $[0, 1]$. Note that this condition is satisfied in cases where the kernel h is not continuous.

Let the joint probability density function of each two-dimensional random variable (Y_j, T_j) be p, and the marginal density of T_j be p_0. As stated in the introduction, p_0 is determined by choice, however p is unknown. The following relation plays an important role in the formulation of the regularized solution.

$$Kf(t) = \mathbb{E}(Y_j | T_j = t) = \int_{-\infty}^{\infty} y \frac{p(y,t)}{p_0(t)} dy \tag{2.3}$$

A regularized solution is obtained by seeking an unbiased estimator of

$$q = K^* K f. \tag{2.4}$$

Definition 2.1 *For each j=1,2,...,n define the random function*

$$\hat{q}_j = \frac{Y_j}{p_0(T_j)} \sum_{k=1}^{\infty} \mu_k \overline{\psi_k(T_j)} \varphi_k \tag{2.5}$$

and the mean of these estimators

$$\hat{q} = \frac{1}{n} \sum_{j=1}^{n} \hat{q}_j \tag{2.6}$$

Theorem 2.2 $\hat{q}$ *is an unbiased estimator of q.*

Proof: We show that each $\hat{q}_j$ is an unbiased estimator of q. By using (2.2), (2.3), and (2.1)

$$
\begin{aligned}
\mathbb{E}[\hat{q}_j] &= \int_0^1 \int_{-\infty}^{\infty} \sum_{k=1}^{\infty} \mu_k y \frac{\overline{\psi_k(t)}}{p_0(t)} p(y,t) \varphi_k \, dy \, dt \\
&= \sum_{k=1}^{\infty} \mu_k \int_0^1 \int_{-\infty}^{\infty} y \frac{\overline{\psi_k(t)}}{p_0(t)} p(y,t) \, dy \, dt \, \varphi_k \\
&= \sum_{k=1}^{\infty} \mu_k [Kf, \psi_k] \varphi_k \\
&= \sum_{k=1}^{\infty} \lambda_k [f, \varphi_k] \varphi_k
\end{aligned}
$$

Definition 2.3 *For each m define the regularized estimator*

$$\hat{f}_m = \sum_{k=1}^{m} \frac{1}{\lambda_k} [\hat{q}, \varphi_k] \varphi_k \tag{2.7}$$

Each $\hat{f}_m$ is a random function and by using (2.5) and (2.6) we obtain that

$$\hat{f}_m = \frac{1}{n} \sum_{k=1}^{m} \frac{1}{\mu_k} \sum_{j=1}^{n} Y_j \frac{\overline{\psi_k(T_j)}}{p_0(T_j)} \varphi_k \tag{2.8}$$

If the sampling points are uniformly distributed on $[0, 1]$, and the singular system of K is replaced by the singular system of the semidiscrete operator K_n defined in [9], then (2.8) reduces to (3.1) in [9]. We now compute the IMSE in approximating f by $\hat{f}_m$.

Theorem 2.4 *For each* $j = 1, 2, \ldots, n$

$$\mathbb{E}\|\hat{f}_m - f\|^2 \quad = \quad \mathbb{E}\frac{1}{n}\sum_{k=1}^m \frac{|[\hat{q}_j, \varphi_k]|^2}{\lambda_k^2} - \frac{n+1}{n}\sum_{k=1}^m \frac{|[q, \varphi_k]|^2}{\lambda_k^2} + \|f\|^2$$

Proof: By using (2.7) and Theorem 2.2 we obtain

$$\mathbb{E}[\hat{f}_m, f] = [\mathbb{E}\hat{f}_m, f] = \sum_{k=1}^m \frac{|[q, \varphi_k]|^2}{\lambda_k^2} \tag{2.9}$$

By using (2.6), (1.3), and Theorem 2.2, we obtain

$$\mathbb{E}|[\hat{q}, \varphi_k]|^2 \quad = \quad \frac{1}{n^2}\mathbb{E}|\sum_{j=1}^n [\hat{q}_j, \varphi_k]|^2$$

$$= \quad \frac{1}{n^2}\mathbb{E}\sum_{j=1}^n (\sum_{l \neq j}[\hat{q}_j, \varphi_k]\overline{[\hat{q}_l, \varphi_k]} + |[\hat{q}_j, \varphi_k]|^2)$$

$$= \quad \frac{1}{n}\mathbb{E}|[\hat{q}_j, \varphi_k]|^2 + \frac{n-1}{n}|[q, \varphi_k]|^2 \tag{2.10}$$

for all $j = 1, 2, \ldots, n$. The result follows by using (2.7), (2.10), (2.9) and the equality

$$\|\hat{f}_m - f\|^2 = \|\hat{f}_m\|^2 - [\hat{f}_m, f] - [f, \hat{f}_m] + \|f\|^2.$$

3   Cross–validation method

Observe that the expression for the IMSE in Theorem 2.4 contains the term $|[q, \varphi_k]|^2$, which depends on the true solution. The method of cross–validation used here, is based on replacing this term with the expectation of a data–based estimator. This is also valid for more general estimation problems [1]. By the statistical independence in (1.3), we find n random variables whose expectation is $|[q, \varphi_k]|^2$.

$$n(n-1)|[q, \varphi_k]|^2 \quad = \quad \mathbb{E}\sum_{j=1}^n \sum_{l \neq j}[\hat{q}_j, \varphi_k]\overline{[\hat{q}_l, \varphi_k]}$$

$$= \quad \mathbb{E}(|\sum_{j=1}^n [\hat{q}_j, \varphi_k]|^2 - n|[\hat{q}_l, \varphi_k]|^2) \tag{3.1}$$

for each $l = 1, 2, \ldots, n$.

By using (3.1) in Theorem 2.4, we see that for each $l = 1, 2, \ldots, n$

$$
\begin{aligned}
\mathbb{E}\|\hat{f}_m - f\|^2 &= \mathbb{E}\left(\frac{2}{n-1} \sum_{k=1}^{m} \frac{|[\hat{q}_l, \varphi_k]|^2}{\lambda_k^2} - \frac{n+1}{n^2(n-1)} \sum_{k=1}^{m} \frac{1}{\lambda_k^2} |\sum_{j=1}^{n} [\hat{q}_j, \varphi_k]|^2\right) \\
&+ \|f\|^2
\end{aligned}
\tag{3.2}
$$

Thus, by the strong law of large numbers, the mean of the random variables on the right side of (3.2) converges almost surely to $\mathbb{E}\|\hat{f}_m - f\|^2 - \|f\|^2$ as $n \to \infty$. Thus, it is reasonable to expect that the minimizer of this mean is asymptotically the same as the minimizer of the IMSE. The precise asymptotic relationship between these two minima is very delicate, and will not be investigated here. Although this heuristic approach is customary in the literature (cf. [9, 10, 11]), this is an interesting mathematical problem which needs to be addressed.

Definition 3.1 *Define the cross validation function*

$$
S_n(m) = \sum_{k=1}^{m} \frac{1}{\lambda_k} \left(\frac{2}{n} \sum_{j=1}^{n} Y_j^2 \frac{|\psi_k(T_j)|^2}{p_0(T_j)^2} - \frac{n+1}{n^2} \left| \sum_{j=1}^{n} Y_j \frac{\overline{\psi_k(T_j)}}{p_0(T_j)} \right|^2 \right)
$$

By taking the mean over l in (3.2) it follows that

$$
\mathbb{E}\|\hat{f}_m - f\|^2 = \frac{1}{n-1} \mathbb{E} S_n(m) + \|f\|^2
\tag{3.3}
$$

An immediate consequence of (3.3) is

Theorem 3.2

$$
\mathrm{argmin}_m \mathbb{E}\|\hat{f}_m - f\|^2 = \mathrm{argmin}_m \mathbb{E} S_n(m).
$$

Thus, the cross validation method proposed here, consists of finding a minimizer m_s of the score $S_n(m)$ in Definition 3.1, and then approximating f by $\hat{f}_{m_s}$ computed according to (2.7). The following section demonstrates the asymptotic behavior of the minimizer m_s and the IMSE in approximating f by $\hat{f}_{m_s}$.

For equispaced points $\{T_j\}$, the cross validation function $V(m; n)$ defined in [9] in terms of the operator $K_n f = (Kf(T_1), \ldots, Kf(T_n))$, is asymptotically

$$
V(m; n) \approx \frac{1}{(n-m)^2} \sum_{k=m+1}^{n} |\sum_{j=1}^{n} Y_j \psi_k(T_j)|^2
\tag{3.4}
$$

4   Convergence rates

To obtain rates of convergence, assume that the sampling points $\{T_j\}$ are uniformly distributed on $[0, 1]$ and that there exist constants $\alpha > 0$ and $\beta > 1$ such that

$$\lambda_k \approx k^{-\alpha} \quad \text{and} \quad |[f, \varphi_k]|^2 \approx k^{-\beta} \ , \quad \text{as} \ k \to \infty. \tag{4.1}$$

where the notation $\mu_k \approx \nu_k$ means that the sequence μ_k / ν_k is upper bouded and is lower bounded away from zero. Hence (4.1) is equivalent to the conditions in [9].

Since $p_0 \equiv 1$, (2.5), (1.3), (2.2), and (4.1) imply that, for each j

$$\begin{aligned}
\mathbb{E}|[\hat{q}_j, \varphi_k]|^2 &= \lambda_k \mathbb{E}(Y_j^2 |\psi_k(T_j)|^2) \\
&= \lambda_k \mathbb{E}(g(T_j)^2 + \varepsilon_j^2) |\psi_k(T_j)|^2 \\
&= \lambda_k (\|g\psi_k\|^2 + \sigma^2) \\
&\approx k^{-\alpha} \tag{4.2}
\end{aligned}$$

By using (4.2) and (4.1), we obtain sequences $\{a_k\}$, $\{b_k\}$ and constants A, B such that $0 < A < a_k < b_k < B$ for all k and

$$a_k k^\alpha \leq \frac{1}{\lambda_k^2} \mathbb{E}|[\hat{q}_j, \varphi_k]|^2 \leq b_k k^\alpha$$

$$\tag{4.3}$$

$$a_k k^{-\beta} \leq |[f, \varphi_k]|^2 \leq b_k \, k^{-\beta}$$

As in [9, 10], we obtain asymptotic rates for the minimizer of the expectation of the cross validation function.

Theorem 4.1 *The minimizer m_e, of $\mathbb{E}S_n(m)$ satisfies*

$$m_e \approx n^{1/(\alpha+\beta)}.$$

Proof: For each m define

$$D_n(m) = \mathbb{E}S_n(m+1) - \mathbb{E}S_n(m). \tag{4.4}$$

Then, as in [9], m_e satisfies the following property

$$D_n(m_e - 1) \leq 0 \leq D_n(m_e). \tag{4.5}$$

We now estimate $D_n(m)$ for any m.
By using (3.3) and Theorem 2.4 we obtain

$$\begin{aligned}
D_n(m) &= (n-1)\mathbb{E}(\|\hat{f}_{m+1} - f\|^2 - \|\hat{f}_m - f\|^2) \\
&= \frac{n-1}{n\lambda_m^2} \mathbb{E}|[\hat{q}_j, \varphi_{m+1}]|^2 - (n+1)|[f, \varphi_{m+1}]|^2 \tag{4.6}
\end{aligned}$$

By using (4.3) and (4.6) we obtain

$$D_n(m) \leq \frac{n-1}{n}(b_{m+1}(m+1)^\alpha - (n+1)a_{m+1}(m+1)^{-\beta})$$

$$D_n(m) \geq \frac{n-1}{n}(a_{m+1}(m+1)^\alpha - (n+1)b_{m+1}(m+1)^{-\beta})$$

The result follows from these inequalities and (4.5).

To estimate the rate of convergence of the estimator $\hat{f}_{m_e}$ we need the following easily obtained inequalities.

There exist constants $0 < U < V$ such that, for all m

$$U(1 - m^{1-\beta}) \leq \sum_{k=1}^{m} k^{-\beta} \leq V - Um^{1-\beta}$$

$$Um^{1+\alpha} \leq \sum_{k=1}^{m} k^\alpha \leq Vm^{1+\alpha} \tag{4.7}$$

$$Um^{1-\beta} \leq \sum_{k=m+1}^{\infty} k^{-\beta} \leq Vm^{1-\beta}$$

Theorem 4.2

$$\mathbb{E}\|\hat{f}_{m_e} - f\|^2 \approx n^{-(\beta-1)/(\alpha+\beta)}.$$

Proof: From Theorem 2.4 we obtain, for each j and m

$$\mathbb{E}\|\hat{f}_m - f\|^2 = \frac{1}{n}\sum_{k=1}^{m}(\frac{1}{\lambda_k^2}\mathbb{E}|[\hat{q}_j, \varphi_k]|^2 - |[f, \varphi_k]|^2) + \sum_{k=m+1}^{\infty} |[f, \varphi_k]|^2 \tag{4.8}$$

By (4.8), (4.3) and (4.7) we obtain

$$\mathbb{E}\|\hat{f}_m - f\|^2 \leq \frac{B}{n}\sum_{k=1}^{m} k^\alpha - \frac{A}{n}\sum_{k=1}^{m} k^{-\beta} + B\sum_{k=m+1}^{\infty} k^{-\beta}$$

$$\leq BV(\frac{m^{1+\alpha}}{n} + m^{1-\beta}) \tag{4.9}$$

and

$$\mathbb{E}\|\hat{f}_m - f\|^2 \geq \frac{A}{n}\sum_{k=1}^{m} k^\alpha - \frac{B}{n}\sum_{k=1}^{m} k^{-\beta} + B\sum_{k=m+1}^{\infty} k^{-\beta}$$

$$\geq AU\frac{m^{1+\alpha}}{n} + BU(m^{1-\beta} - \frac{1}{n}) \tag{4.10}$$

The result follows by using Theorem 4.1 in (4.9) and (4.10).

References

[1] A. K. Dey, B. A. Mair and F. H. Ruymgaart, "Cross-Validation for Parameter Selection in Inverse Estimation Problems," submitted.

[2] C. W. Groetsch, "The Theory of Tikhonov Regularization for Fredholm Equations of the First Kind," *Pitman*, 1984, Boston, MA.

[3] P. Hall, "Large Sample Optimality of Least Squares Cross-Validation in Density Estimation," *Ann. Statist.*, v. 11, 1983, pp. 1156–1174.

[4] J. Lund and K. L. Bowers, "Sinc Methods for Quadrature and Differential Equations," *SIAM*, 1992, Philadelphia, PA.

[5] B. A. Mair, "A cross–validation method for non–parametric estimation," *Control and Cybernetics*, v. 23, 1994, pp. 733–743.

[6] B. A. Mair and F. H. Ruymgaart, "Statistical Inverse Estimation in Hilbert Scales," submitted.

[7] M. Z. Nashed and G. Wahba, "Convergence Rates of Approximate Least Squares Solutions of Linear Integral and Operator Equations of the First Kind," *Math. Comp.*, v. 28, 1974, pp. 69–80.

[8] B. W. Silverman, "Density Estimation for Statistics and Data Analysis," *Chapman and Hall*, 1986, New York, NY.

[9] C. R. Vogel, "Optimal Choice of a Truncation Level for the Truncated SVD Solution of Linear First Kind Integral Equations when Data are Noisy," *SIAM J. Numer. Anal.*, v. 23, 1986, pp. 109–117.

[10] Grace Wahba, "Practical Approximate Solutions to Linear Operator Equations when the Data are Noisy," *SIAM J. Numer. Anal.*, v. 14, 1977, pp. 651–667.

[11] Grace Wahba, "Spline Models for Observational Data," *SIAM*, 1990, Philadelphia, PA.

LINEAR CONTROL THEORY, SPLINES AND INTERPOLATION

Clyde Martin, * Per Enqvist, John Tomlinson and Zhimin Zhang

Department of Mathematics
Texas Tech University
Lubbock, TX   79409

1   Introduction

Spline functions are well known and are widely used for practical approximation of functions or more commonly for fitting smooth curves through preassigned points. Spline techniques have the advantage over most approximation and interpolation techniques in that they are computationally feasible. Most of the published spline algorithms are for polynomial splines and the vast preponderance are for cubic splines. There is a small but excellent literature on the so called exponential splines and there is an even smaller literature on splines with more or less arbitrary nodal functions, [9, 3].

In this paper we will present a common framework for splines that includes polynomial splines of all orders and generalized exponential splines of all orders. This common framework is based on ideas from linear control theory. Using these ideas we also establish a spline interpolation methodology for two-dimensional problems. Let's recall some basic ideas from control theory. A linear control system is a differential equation

$$\frac{d}{dt}\vec{x}(t) = A\vec{x}(t) + B\vec{u}(t)$$

where $\vec{x} \in R^n$, $\vec{u} \in R^m$ and the matrices A and B are constant matrices of compatible dimension. The vector $\vec{x}$ is the state of the system and the vector $\vec{u}$ is the control. The idea is that we can use the control $\vec{u}$ to steer the state from point to point in the state space R^n. We can think of the first components of $\vec{x}$ as representing the position of the system and for appropriate A the second set of coordinates as the velocity, the third acceleration, etc. A common situation, for example in air traffic control, is to specify the position that the system must be in at a sequence of times. Another application might be the storage of traced curves such as signatures with a minimal number of points. So in fact what we have is a set of points through which the system must traverse at specified times. One could fit these points with a spline curve and then ask for the control

---

*Partially supported by NASA Grants NAG 2-902 and NAG 2-899 and a grant from the Texas Tech Leather Research Institute.

that would move the system along that trajectory. In fact this can be done but we will show that the control law can be developed from natural control theoretic principles that will move the system through the points at the desired times and the resulting curve will be piecewise analytic and will have $2n - 1$ continuous derivatives, i.e. a generalized spline. With this framework we can construct a wide variety of spline functions. If the matrix A is nilpotent then the resulting construction is just that for polynomial splines. If the matrix is 2×2 and one eigenvalue is zero and the other is a nonzero real number then the spline is the usual exponential spline. In general the nodal functions are the coordinate functions of the matrix function e^{At}.

In this paper we give a unified treatment of all of the common one dimensional spline functions using simple ideas from control theory. It is coming to be understood that there is a large overlap between linear control theory and elementary numerical analysis. Eigenvalue methods are know to be closely related to the theory of the matrix Riccati equation [2], there are close relations between observability and quadrature techniques [8], system identification and Prony's method are very similar [1] and now we see that the spline constructions and basic linear controllability are manifestations of the same phenomena.

In Section 2 we review basic material from the theory of linear control systems that is needed for the development and give a condition that characterizes the optimal control law that generates the spline functions. In Section 3 we give the details of the construction of spline functions using control theory and in Section 4 we classify the possible classes of spline functions that arise from the control theoretic construction. In Section 5 we examine in detail some of the particular classes from Section 4 and finally in Section 6 we present a series of numerical examples comparing the various classes.

2   Some Results from Control Theory

In this section we collect a series of results from linear control theory. Most can be found in any control theory textbook. See, for example, the book by Brockett, [5].

Consider the linear system:

$$\frac{d}{dt}\vec{x}(t) = A\vec{x}(t) + \vec{b}u(t), \quad t \in [0, T], \tag{2.1}$$

with

$$A = \begin{pmatrix} 0 & 1 & 0 & \cdots & 0 \\ 0 & 0 & 1 & \cdots & 0 \\ \vdots & \vdots & \vdots & \ddots & \vdots \\ 0 & 0 & 0 & \cdots & 1 \\ a_1 & a_2 & a_3 & \cdots & a_m \end{pmatrix}, \quad \vec{b} = \begin{pmatrix} 0 \\ 0 \\ \vdots \\ 0 \\ 1 \end{pmatrix}, \quad \vec{x}(t) = \begin{pmatrix} x_1(t) \\ x_2(t) \\ \vdots \\ x_{m-1}(t) \\ x_m(t) \end{pmatrix}, \tag{2.2}$$

and the observation function

$$y(t) = \vec{c}^T \vec{x}(t), \quad \vec{c}^T = (1, 0, \cdots, 0). \tag{2.3}$$

Let us divide $[0, T]$ into n subintervals as

$$0 = t_0 < t_1 < \cdots < t_{n-1} < t_n = T,$$

and define $h_k = t_k - t_{k-1}$, the length of the rth subinterval. Our goal is to find a control law $u \in C^{m-2}[0, T]$ that drives the system (2.1) from $\vec{x}(0) = \vec{x}^0$ to $\vec{x}(T) = \vec{x}^T$ such that the observation function $y(t)$ satisfies the interpolation conditions

$$y(t_k) = \alpha_k, \quad k = 1, \cdots, n - 1. \tag{2.4}$$

Furthermore, $u(t)$ minimizes the functional

$$\int_0^T u(s)^2 ds. \tag{2.5}$$

Such a control is called an optimal control.

Definition. The system (2.1) is called controllable if for any $\vec{x}^0$, $\vec{x}^\tau$, and $\tau > 0$, there is a $u(t)$ such that,

$$\vec{x}^\tau = \vec{x}(\tau) = e^{A\tau} \vec{x}^0 + \int_0^\tau e^{A(\tau - s)} \vec{b} u(s) ds.$$

Theorem 2.1 : *The system (2.1) is controllable if and only if*

$$rank \ (\vec{b}, A\vec{b}, \cdots, A^{m-1}\vec{b}) = m. \tag{2.6}$$

For the special matrix A as in (2.2), it is easy to verify that

$$(\vec{b}, A\vec{b}, \cdots, A^{m-1}\vec{b}) = \begin{pmatrix} 0 & 0 & \cdots & 0 & 1 \\ 0 & 0 & \cdots & 1 & * \\ \vdots & \vdots & \ddots & \vdots & \vdots \\ 0 & 1 & \cdots & * & * \\ 1 & * & \cdots & * & * \end{pmatrix}, \tag{2.7}$$

and hence the condition (2.6) is satisfied. From Theorem 2.1, the system (2.1) is controllable.

Theorem 2.2 : *The system (2.1) is controllable, if and only if the matrix* $\int_0^\tau e^{-As}\vec{b}\vec{b}^T e^{-A^T s}ds$ *is invertible.*

For the special matrix in (2.2), we then define

$$M(t) = (\int_0^t e^{-As}\vec{b}\vec{b}^T e^{-A^T s}ds)^{-1}. \tag{2.8}$$

Theorem 2.3 : *When the system (2.1) is controllable, a control that moves the system from* $\vec{x}(\underline{t}) = \vec{\rho}_L$ *to* $\vec{x}(\bar{t}) = \vec{\rho}_R$ *given by*

$$u(t) = \vec{b}^T e^{-A^T t}(\int_{\underline{t}}^{\bar{t}} e^{-As}\vec{b}\vec{b}^T e^{-A^T s}ds)^{-1}(e^{-A\bar{t}}\vec{\rho}_R - e^{-A\underline{t}}\vec{\rho}_L), \tag{2.9}$$

minimizes the functional $J(v) = \int_{\underline{t}}^{\bar{t}} v^2(s)ds$ *among all controls that move the system from* $\vec{x}(\underline{t}) = \vec{\rho}_L$ *to* $\vec{x}(\bar{t}) = \vec{\rho}_R$.

Theorem 2.4 : *When the system (2.1) is controllable, a control* $u \in C^{m-2}[0,T]$ *that moves the system from* $\vec{x}(0) = \vec{x}^0$, *passing through* $\vec{c}^T\vec{x}(t_k) = \alpha_k$, *to* $\vec{x}(T) = \vec{x}^T$ *is given by*

$$u(t) = \sum_{k=1}^{n-1} \beta_k f_k(t) + \sum_{i=1}^m \gamma_i g_i(t), \tag{2.10}$$

with

$$f_k(t) = \{ \begin{array}{ll} \vec{e}_1^T e^{A(t_k-t)}\vec{b} & t < t_k, \\ 0 & t \geq t_k, \end{array} \quad k = 1, \cdots, n-1,$$

$$g_i(t) = \vec{e}_i^T e^{A(t_n-t)}\vec{b}, \quad i = 1, \cdots, m,$$

where $\vec{e}_1^T = (1, 0, \cdots, 0), \cdots, \vec{e}_m^T = (0, \cdots, 0, 1)$, *and* β_k*'s,* γ_i*' are determined by* $n-1$ *interpolation conditions* $\vec{c}^T\vec{x}(t_k) = \alpha_k$ *and* m *boundary conditions* $\vec{x}(T) = \vec{x}^T$. *Moreover, the control (2.10) minimizes the functional* $J(v) = \int_0^T v^2(s)ds$ *among all functions* $v \in C^{m-2}[0,T]$ *that drives the system (2.1) from* $\vec{x}(0) = \vec{x}^0$, *passing through* $\vec{c}^T\vec{x}(t_k) = \alpha_k$, *to* $\vec{x}(T) = \vec{x}^T$.

3   Construction of Splines by Control Theory

Theorem 2.4 implies that an optimal control for the system (2.1) (with A given by (2.2)) is unique. But in general, β_k's and γ_i's in (2.10) are difficult to find, we then introduce a practical procedure to construct a control law that satisfies all the requirements. This control law actually leads us to a construction of spline functions.

By the existence of a control law, there exists a set of points $\vec{x}^1, \cdots, \vec{x}^{n-1}$ with $x_1^k = \alpha_k$, $k = 1, \cdots, n-1$ such that the solution of the system (2.1) satisfies $\vec{x}(t_k) = \vec{x}^k$, $k = 0, 1, \cdots, n-1, n$. By virtue of Theorem 2.3, a control law that satisfies all the requirement can be defined piecewise as

$$u(t)|_{[t_{k-1}, t_k]} = u_k(t), \quad k = 1, \cdots, n, \tag{3.1}$$

where $u_k(t)$ is given by (2.9) with $\underline{t} = t_{k-1}$, $\bar{t} = t_k$, $\vec{\rho}_L = \vec{x}^{k-1}$, and $\vec{\rho}_R = \vec{x}^k$. Then equations to find $(n-1)(m-1)$ unknowns in $\vec{x}^1, \cdots, \vec{x}^{n-1}$ (recall that $x_1^k = \alpha_k$ $k = 1, \cdots, n-1$, are known) come from $(n-1)(m-1)$ continuity conditions on $u(t)$, i.e.,

$$u_k^{(r)}(t_k) = u_{k+1}^{(r)}(t_k), \quad r = 0, \cdots, m-2, \quad k = 1, \cdots, n-1. \tag{3.2}$$

Setting $\vec{y}^k = e^{-At_k} \vec{x}^k - e^{-At_{k-1}}$, it follows from (2.9),

$$u_k(t_k) = (A^r \vec{b})^T e^{-A^T t_k} \left(\int_{t_{k-1}}^{t_k} e^{-As} \vec{bb}^T e^{-A^T s} ds \right)^{-1} \vec{y}^k; \tag{3.3}$$

$$u_{k+1}(t_k) = (A^r \vec{b})^T e^{-A^T t_k} \left(\int_{t_k}^{t_{k+1}} e^{-As} \vec{bb}^T e^{-A^T s} ds \right)^{-1} \vec{y}^{k+1}. \tag{3.4}$$

Next, we shall simplify (3.3) and (3.4). Toward this end, we introduce a change of variable $s = t_{k-1} + s'$ into

$$\left(\int_{t_{k-1}}^{t_k} e^{-As} \vec{bb}^T e^{-A^T s} ds \right)^{-1} = e^{A^T t_{k-1}} M(h_k) e^{At_{k-1}}, \tag{3.5}$$

where $M(h_k)$ is defined by (2.8). Substituting (3.5) into (3.3), we have

$$u_k^{(r)}(t_k) = (A^r \vec{b})^T e^{-A^T h_k} M(h_k) (e^{-Ah_k} \vec{x}^k - \vec{x}^{k-1}). \tag{3.6}$$

Similarly,

$$u_{k+1}^{(r)}(t_k) = (A^r \vec{b})^T M(h_{k+1}) (e^{-Ah_{k+1}} \vec{x}^{k+1} - \vec{x}^k). \tag{3.7}$$

Substituting (3.6) and (3.7) into (3.2) yields a linear system for $(n-1)(m-1)$ unknowns in $\vec{x}^1, \cdots, \vec{x}^{n-1}$:

$$(A^r \vec{b})^T \left[-e^{-A^T h_k} M(h_k) \vec{x}^{k-1} + [e^{-A^T h_k} M(h_k) e^{-Ah_k} + M(h_{k+1})] \vec{x}^k \right]$$
$$-(A^r \vec{b})^T M(h_{k+1}) e^{-Ah_{k+1}} \vec{x}^{k+1} = 0, \tag{3.8}$$

for $r = 0, \cdots, m - 2$, $k = 1, \cdots, n - 1$. By virtue of the existence and uniqueness of the optimal control, the linear system (3.8) has a unique solution and hence its coefficient matrix is invertible.

In order to solve (3.8), The following quantities need to be calculated, A^r, e^{-Ah} $(e^{-A^T h})$, and $M(h)$. Sometimes it is easier to use the Jordan matrix of A, denoted by Λ. There exists an invertible matrix Q such that $A = Q\Lambda Q^{-1}$, and hence

$$A^r = Q\Lambda^r Q^{-1}, \quad e^{-Ah} = Qe^{-\Lambda h}Q^{-1}, \quad e^{-A^T h} = Q^{-T}e^{-\Lambda^T h}Q^T. \quad (3.9)$$

$$M(h) = (\int_0^h Qe^{-\Lambda s}Q^{-1}\vec{b}\vec{b}^T Q^{-T}e^{-\Lambda^T s}Q^T ds)^{-1} = Q^{-T}\hat{M}(h)Q^{-1}, \quad (3.10)$$

where

$$\hat{M}(h) = (\int_0^h e^{-\Lambda s}Q^{-1}\vec{b}(Q^{-1}\vec{b})^T e^{-\Lambda^T s}ds)^{-1}. \quad (3.11)$$

Substituting (3.9) and (3.10) into (3.8), we then have

$$-(\Lambda^r Q^{-1}\vec{b})^T e^{-\Lambda^T h_k}\hat{M}(h_k)Q^{-1}\vec{x}^{k-1}$$
$$+(\Lambda^r Q^{-1}\vec{b})^T [e^{-\Lambda^T h_k}\hat{M}(h_k)e^{-\Lambda h_k} + \hat{M}(h_{k+1})]Q^{-1}\vec{x}^k$$
$$-(\Lambda^r Q^{-1}\vec{b})^T \hat{M}(h_{k+1})e^{-\Lambda h_{k+1}}Q^{-1}\vec{x}^{k+1} = 0, \quad (3.12)$$

for $r = 0, \cdots, m - 2$, $k = 1, \cdots, n - 1$. Solving (3.8) or (3.12) for $(n - 1)(m - 1)$ unknowns in $\vec{x}^1, \cdots, \vec{x}^{n-1}$, we then have the control $u(t)$ defined piecewise by (3.1). The solution of the system (2.1) is thus given by

$$\begin{aligned}
\vec{x}(t) &= e^{At}\vec{x}^0 + \int_0^t e^{A(t-s)}\vec{b}u(s)ds \\
&= Q(e^{\Lambda t}Q^{-1}\vec{x}^0 + \int_0^t e^{\Lambda(t-s)}Q^{-1}\vec{b}u(s)ds). \quad (3.13)
\end{aligned}$$

Note that $x_i'(t) = x_{i+1}(t)$, $i = 1, \cdots, m - 1$. So the continuity of $x_i'(t)$ is continuity of $x_{i+1}(t)$ for $i < m$. Further, continuity of $x_1^{(m+r)}(t)$ is continuity of $u^{(r)}(t)$, $r = 0, \cdots, m - 2$. Therefore, the observation function $y(t) = \vec{c}^T \vec{x}(t) = x_1(t)$ is a $2m - 2$ times continuously differentiable function that satisfies the boundary conditions

$$y^{(r)}(0) = x_{r+1}^0, \quad y^{(r)}(T) = x_{r+1}^T, \quad r = 0, \cdots, m - 1 \quad (3.14)$$

and the interpolation conditions

$$y(t_k) = x_1^k, \quad k = 1, \cdots, n - 1. \quad (3.15)$$

Hence $y(t)$ is a spline function. We see that from the control theory, we can derive quite general spline functions. Summing up, we have proved

Theorem 3.1 : *(1) There exists a unique function $y(t) \in C^{m-2}[0,T]$ that satisfies the boundary conditions (3.14) and the interpolation conditions (3.15); (2) $y(t)$ is the first component of the vector function $\vec{x}(t)$ given by (3.13) in which $u(s)$ is defined piecewise on each subinterval $[t_{k-1}, t_k]$, $k = 1, \cdots, n$, by*

$$
\begin{aligned}
u_k^{(r)}(t_k) &= \vec{b}^T e^{-A^T(t-t_{k-1})} M(h_k)(e^{-Ah_k} \vec{x}^k - \vec{x}^{k-1}) \qquad (3.16) \\
&= (Q^{-1}\vec{b})^T e^{-\Lambda^T(t-t_{k-1})} \hat{M}(h_k)(e^{-\Lambda h_k} Q^{-1}\vec{x}^k - Q^{-1}\vec{x}^{k-1}),
\end{aligned}
$$

where $\vec{x}^k$, $k = 1, \cdots, n-1$ are determined by solving the linear systems (3.8) or (3.12) (Note that $\vec{x}^0$, $\vec{x}^n$ and x_1^k, $k = 1, \cdots, n-1$ are given by the boundary conditions (3.14) and the interpolation conditions (3.15)).

In the next section, we will see that these splines can be piecewise polynomials, trigonometric functions, exponentials or any combination. As special cases, we are able to recover classical polynomial splines (odd order) and exponential splines by properly selecting parameters $a_1, \cdots, a_m$ in (2.2) for the matrix A.

4   Classification of Splines.

The type of a spline is determined by its nodal shape functions. From the control theory, we are able to construct the nodal shape functions of splines.

In order to see the kind of interpolation functions in $\vec{x}(t)$, we only need to consider one subinterval. Without loss of generality, we use the first interval $(t_0, t_1) = (0, h)$ where the solution of the system (2.1) is given by

$$
\vec{x}(t) = e^{At}\vec{x}^0 + \int_0^t e^{A(t-s)}\vec{b}u(s)ds. \qquad (4.1)
$$

From Theorem 2.3,

$$
u(t) = \vec{b}^T e^{-A^T t}\left(\int_0^h e^{-As}\vec{b}\vec{b}^T e^{-A^T s}ds\right)^{-1}(e^{-Ah}\vec{x}^1 - \vec{x}^0). \qquad (4.2)
$$

Substituting (4.2) into (4.1), we have

$$
\begin{aligned}
\vec{x}(t) &= e^{At}\vec{x}^0 + \int_0^t e^{A(t-s)}\vec{b}\vec{b}^T e^{-A^T s}ds M(h)(e^{-Ah}\vec{x}^1 - \vec{x}^0) \\
&= Qe^{\Lambda t}[Q^{-1}\vec{x}^0 + \hat{M}(t)^{-1}\hat{M}(h)(e^{-\Lambda h}Q^{-1}\vec{x}^1 - Q^{-1}\vec{x}^0)], \qquad (4.3)
\end{aligned}
$$

where $M(h)$ and $\hat{M}(h)$ are defined by (2.8) and (3.11), respectively.

Theorem 4.1 : *Let A be given by (2.2), let $(p_1(t), \cdots, p_m(t))$ be the first row of the matrix*

$$e^{At}[I - M(t)^{-1}M(h)] = Qe^{\Lambda t}[I - \hat{M}(t)^{-1}\hat{M}(h)]Q^{-1}, \qquad (4.4)$$

and let $(q_1(t), \cdots, q_m(t))$ be the first row of the matrix

$$e^{At}M(t)^{-1}M(h)e^{-Ah} = Qe^{\Lambda t}\hat{M}(t)^{-1}\hat{M}(h)e^{-\Lambda h}Q^{-1}. \qquad (4.5)$$

Then for $r = 0, \cdots, m - 1$,

$$p_i^{(r)}(0) = \delta_{i,r+1}, \quad p_i^{(r)}(h) = 0, \quad i = 1, \cdots, m, \qquad (4.6)$$

$$q_j^{(r)}(0) = 0, \quad q_j^{(r)}(h) = \delta_{j,r+1}, \quad j = 1, \cdots, m. \qquad (4.7)$$

We call p_i, q_i nodal shape functions by the characteristics (4.6) and (4.7). From (4.4) and (4.5), We see that the nodal shape functions are linear combinations of function entries of matrices $e^{\Lambda t}$ and $e^{\Lambda t}\hat{M}(t)^{-1}$. In order to see the type of functions in the spline, we only need to examine the entries of these two matrices.

In the following, we classify the spline functions derived from control theory. This classification is based on the spectrum of the coefficient matrix A of the system (2.1) under different circumstances. We shall concentrate on the case $m = 2$. The reasons are: (1) The general situation for large m is very complicated and is difficult to describe precisely. (2) The case $m = 2$ has almost all features for the general case. (3) From the practical point of view, the case $m = 2$ is the most useful and important case. Let

$$A = \begin{pmatrix} 0 & 1 \\ \beta & 2\gamma \end{pmatrix}, \quad \beta, \gamma \in R^1.$$

The eigenvalues of A are $\lambda_1 = \gamma + \sqrt{\gamma^2 + \beta}$, $\lambda_2 = \gamma - \sqrt{\gamma^2 + \beta}$.

1. $\gamma^2 + \beta > 0$. There are two distinct real eigenvalues. Then the Jordan matrix Λ of A, the transformation matrix Q and its inverse are given by

$$\Lambda = \begin{pmatrix} \lambda_1 & 0 \\ 0 & \lambda_2 \end{pmatrix}, \quad Q = \begin{pmatrix} 1 & 1 \\ \lambda_1 & \lambda_2 \end{pmatrix}, \quad Q^{-1} = \frac{1}{\lambda_2 - \lambda_1}\begin{pmatrix} \lambda_2 & -1 \\ -\lambda_1 & 1 \end{pmatrix}. \qquad (4.8)$$

Then

$$e^{\Lambda t} = \begin{pmatrix} e^{\lambda_1 t} & 0 \\ 0 & e^{\lambda_2 t} \end{pmatrix}. \qquad (4.9)$$

From (3.11) (by changing h to t), we have

$$\hat{M}(t)^{-1} = \frac{1}{(\lambda_2 - \lambda_1)^2}\int_0^t e^{-\Lambda s}\begin{pmatrix} -1 \\ 1 \end{pmatrix}(-1, 1)\, e^{-\Lambda s}ds$$

$$= \frac{1}{(\lambda_2 - \lambda_1)^2} \begin{pmatrix} \frac{(1-e^{-2\lambda_1 t})}{2\lambda_1} & \frac{(e^{-(\lambda_1+\lambda_2)t}-1)}{\lambda_1+\lambda_2} \\ \frac{(e^{-(\lambda_1+\lambda_2)t}-1)}{\lambda_1+\lambda_2} & \frac{(1-e^{-2\lambda_2 t})}{/2\lambda_2} \end{pmatrix}, \qquad (4.10)$$

and hence

$$e^{\Lambda t}\hat{M}(t)^{-1} = \frac{1}{(\lambda_2 - \lambda_1)^2} \begin{pmatrix} \frac{e^{\lambda_1 t}-e^{-\lambda_1 t}}{2\lambda_1} & \frac{e^{-\lambda_2 t}-e^{\lambda_1 t}}{\lambda_1+\lambda_2} \\ \frac{e^{-\lambda_1 t}-e^{\lambda_2 t}}{\lambda_1+\lambda_2} & \frac{e^{\lambda_2 t}-e^{-\lambda_2 t}}{2\lambda_2} \end{pmatrix} \qquad (4.11)$$

Then we can further consider the differences in the spectra caused by constraints on β and γ.

1.a. $\gamma \neq 0$, $\beta \neq 0$. In this case $\lambda_1, \lambda_2, -\lambda_1, -\lambda_2$ are all distinct. We then have the exponential spline with basis functions given by linear combinations of $e^{\lambda_1 t}, e^{-\lambda_1 t}, e^{\lambda_2 t}, e^{-\lambda_2 t}$.

1.b. $\gamma = 0$, $\beta > 0$. In this case $\lambda_1 = -\lambda_2 = \sqrt{\beta}$, the basis functions in 1.a. degenerate. However, by applying the following limits

$$\lim_{\lambda_2 \to -\lambda_1} \frac{e^{-\lambda_2 t} - e^{\lambda_1 t}}{\lambda_1 + \lambda_2} = \lim_{\lambda_1 + \lambda_2 \to 0} \frac{e^{\lambda_1 t}(e^{-(\lambda_1+\lambda_2)t} - 1)}{\lambda_1 + \lambda_2} = -te^{\lambda_1 t},$$

$$\lim_{\lambda_2 \to -\lambda_1} \frac{e^{-\lambda_1 t} - e^{\lambda_2 t}}{\lambda_1 + \lambda_2} = \lim_{\lambda_1 + \lambda_2 \to 0} \frac{e^{-\lambda_1 t}(1 - e^{(\lambda_1+\lambda_2)t})}{\lambda_1 + \lambda_2} = -te^{-\lambda_1 t},$$

(4.11) becomes

$$e^{\Lambda t}\hat{M}(t)^{-1} = \frac{1}{4\lambda_1^2} \begin{pmatrix} (e^{\lambda_1 t} - e^{-\lambda_1 t})/2\lambda_1 & -te^{\lambda_1 t} \\ -te^{-\lambda_1 t} & (e^{\lambda_1 t} - e^{-\lambda_1 t})/2\lambda_1 \end{pmatrix}. \qquad (4.12)$$

Hence we have the exponential spline with basis functions given by linear combinations of $e^{\sqrt{\beta}t}, e^{-\sqrt{\beta}t}, te^{\sqrt{\beta}t}, te^{-\sqrt{\beta}t}$.

1.c. $\beta = 0$, $\gamma \neq 0$. In this case, $\lambda_1 = 0$ (if $\gamma < 0$), or $\lambda_2 = 0$ (if $\gamma > 0$). Again the basis functions in 1.a. degenerate. Assume that $\lambda_1 = 0$, then $\lambda_2 = \lambda = 2\gamma$. From the limits

$$\lim_{\lambda_1 \to 0} \frac{(e^{\lambda_1 t} - e^{-\lambda_1 t})}{2\lambda_1} = t, \quad \lim_{\lambda_1 \to 0} e^{\lambda_1 t} = 1,$$

we have

$$e^{\Lambda t} = \begin{pmatrix} 1 & 0 \\ 0 & e^{\lambda t} \end{pmatrix}, \quad \hat{M}(t)^{-1} = \frac{1}{\lambda^3} \begin{pmatrix} \lambda t & e^{-\lambda t} - 1 \\ e^{-\lambda t} - 1 & (1 - e^{-2\lambda t})/2 \end{pmatrix}$$

$$e^{\Lambda t}\hat{M}(t)^{-1} = \frac{1}{\lambda^3} \begin{pmatrix} \lambda t & e^{-\lambda t} - 1 \\ 1 - e^{\lambda t} & (e^{\lambda t} - e^{-\lambda t})/2 \end{pmatrix} \qquad (4.13)$$

Therefore we end up with the exponential spline with basis functions given by linear combinations of $1, t, e^{2\gamma t}, e^{-2\gamma t}$. Later we shall further show that this is the classical exponential spline [9].

2. $\gamma^2 + \beta < 0$. There are two complex eigenvalues: $\lambda_1 = \gamma + i\omega$, $\lambda_2 = \gamma - i\omega$, where $\omega = \sqrt{-\gamma^2 - \beta}$. And we can investigate the effect of different constraints on β and γ.

2.a. $\gamma \neq 0$, $\beta < 0$. Evaluating (4.9) and (4.11), we have the exponential-trigonometric spline with basis functions given by linear combinations of $e^{\gamma t} \sin \omega t$, $e^{\gamma t} \cos \omega t$, $e^{-\gamma t} \sin \omega t$, $e^{-\gamma t} \cos \omega t$.

2.b. $\gamma = 0$, $\beta < 0$. Again this is a degenerated case where $\lambda_1 = \lambda_2 = i\omega = i\sqrt{-\beta}$. Therefore (4.9) is now

$$e^{\Lambda t} = \begin{pmatrix} \cos \omega t + i \sin \omega t & 0 \\ 0 & \cos \omega t + i \sin \omega t \end{pmatrix}. \qquad (4.14)$$

Taking the limit $\gamma \to 0$ in (4.11), we then have

$$e^{\Lambda t} \hat{M}(t)^{-1} = \frac{-1}{4\omega^2} \begin{pmatrix} \sin \omega t / \omega & -t(\cos \omega t + i \sin \omega t) \\ -t(\cos \omega t - i \sin \omega t) & \sin \omega t / \omega \end{pmatrix}.$$

Hence, we have the polynomial-trigonometric spline with basis functions given by linear combinations of $\sin \sqrt{-\beta} t, \cos \sqrt{-\beta} t, t \sin \sqrt{-\beta} t, t \cos \sqrt{-\beta} t$.

3. $\gamma^2 + \beta = 0$. In this case we have only one eigenvalue in the spectrum $\lambda_1 = \lambda_2 = \gamma$. We consider the effect of the eigenvalue being either zero or non-zero.

3.a. $\gamma \neq 0$. We have non-degenerated Jordan form in this case,

$$\Lambda = \begin{pmatrix} \gamma & 1 \\ 0 & \gamma \end{pmatrix}, \quad Q = \begin{pmatrix} 1 & -1/\gamma \\ \gamma & 0 \end{pmatrix}, \quad Q^{-1} = \begin{pmatrix} 0 & 1/\gamma \\ -\gamma & 1 \end{pmatrix}. \quad (4.15)$$

Therefore,

$$e^{\Lambda t} = \begin{pmatrix} e^{\gamma t} & te^{\gamma t} \\ 0 & e^{\gamma t} \end{pmatrix} = e^{\gamma t} \begin{pmatrix} 1 & t \\ 0 & 1 \end{pmatrix}, \qquad (4.16)$$

$$\begin{aligned} \hat{M}(t)^{-1} &= \int_0^t e^{-2\gamma s} \begin{pmatrix} 1 & -s \\ 0 & 1 \end{pmatrix} \begin{pmatrix} 1/\gamma \\ 1 \end{pmatrix} (1/\gamma, 1) \begin{pmatrix} 1 & 0 \\ -s & 1 \end{pmatrix} ds \\ &= \int_0^t e^{-2\gamma s} \begin{pmatrix} (1/\gamma - s)^2 & 1/\gamma - s \\ 1/\gamma - s & 1 \end{pmatrix} ds. \end{aligned} \qquad (4.17)$$

After some more detailed manipulation (see the next section) we can show that this is the exponential spline with basis functions given by linear combinations of $e^{\gamma t}, te^{\gamma t}, e^{-\gamma t}, te^{-\gamma t}$, similar to the case 1.b.

3.b. $\gamma = 0$. In this case, A itself is a Jordan matrix. We compute directly the following quantities:

$$A = \begin{pmatrix} 0 & 1 \\ 0 & 0 \end{pmatrix}, \quad e^{At} = \begin{pmatrix} 1 & t \\ 0 & 1 \end{pmatrix}, \tag{4.18}$$

$$\begin{aligned} M(t)^{-1} &= \int_0^t \begin{pmatrix} 1 & -s \\ 0 & 1 \end{pmatrix} \begin{pmatrix} 0 \\ 1 \end{pmatrix} (0,1) \begin{pmatrix} 1 & 0 \\ -s & 1 \end{pmatrix} ds \\ &= \begin{pmatrix} t^3/3 & -t^2/2 \\ -t^2/2 & t \end{pmatrix}. \end{aligned}$$

and

$$e^{At} M(t)^{-1} = \begin{pmatrix} -t^3/6 & t^2/2 \\ -t^2/2 & t \end{pmatrix}. \tag{4.19}$$

Then we have the polynomial spline with basis functions given by linear combinations of $1, t, t^2, t^3$. In the next section, we shall further show that this is the well-known cubic spline [4].

From the above discussion, we see that we may encounter all kinds of splines by varying parameters β and γ. Two general cases are 1.a. and 2.a. where we have full sized exponential or exponential-trigonometric splines. Degeneration occurs when zero or multiple eigenvalues appear. The extremal is the case 3.b. when both eigenvalues are zero. It is this extremal case that draws most of the attention. This is evidenced by extensive investigation regarding the cubic spline in the literature. Case 1.c. also has been investigated from a different point of view. But we can hardly find any work regarding the other cases (except 1.c. and 3.b.) listed above.

The situation for $m > 2$ is similar. Let $\lambda_1, \cdots, \lambda_m$ be eigenvalues of A. When $\lambda_1, \cdots, \lambda_m; -\lambda_1, \cdots, -\lambda_m$ are all distinct, we have the exponential spline with the basis functions given by linear combinations of

$$e^{\lambda_1 t}, \quad e^{-\lambda_1 t}, \quad \cdots, \quad e^{\lambda_m t}, \quad e^{-\lambda_m t}.$$

See case 1.a. When complex eigenvalues appear, we get the exponential-trigonometric splines with basis functions $e^{\lambda_k t} \sin \omega t$, $e^{-\lambda_k t} \cos \omega t$ (see case 2.a.). If we have multiple eigenvalues, the terms like

$$t e^{\lambda t}, \quad t \sin \omega t, \quad t e^{-\lambda t} \cos \omega t, \quad t^2 e^{\lambda t}, \quad \cdots$$

will appear in basis functions (see cases 1.b., 2.b. and 3.a.). Finally, zero eigenvalues will introduce polynomials into basis functions (see case 1.c.) and the extremal situation is that all eigenvalues are zero in which case we recover polynomial splines of order $2m - 1$ (see case 3.b.).

5 Two-Dimensional Splines

We consider the system with the state and dynamics given by:

$$\vec{x} = \begin{bmatrix} x \\ \dot{x} \\ y \\ \dot{y} \end{bmatrix}, \quad \begin{cases} \dot{\vec{x}} = A\vec{x} + B\vec{u} \\ \vec{y} = C\vec{x} \end{cases} \tag{5.1}$$

where

$$A = \begin{pmatrix} 0 & 1 & 0 & 0 \\ 0 & \lambda_1 & \alpha_1 & \alpha_2 \\ 0 & 0 & 0 & 1 \\ \beta_1 & \beta_2 & 0 & \lambda_2 \end{pmatrix}, B = \begin{pmatrix} 0 & 0 \\ 1 & 0 \\ 0 & 0 \\ 0 & 1 \end{pmatrix}, C = \begin{pmatrix} 1 & 0 & 0 & 0 \\ 0 & 0 & 1 & 0 \end{pmatrix} \tag{5.2}$$

This gives

$$\vec{y} = C\vec{x} = \begin{pmatrix} x \\ y \end{pmatrix} \tag{5.3}$$

thus

$$\begin{cases} \ddot{x} = \lambda_1 \dot{x} + \alpha_1 y + \alpha_2 \dot{y} + u_1 \\ \ddot{y} = \lambda_2 \dot{y} + \beta_1 x + \beta_2 \dot{x} + u_2 \end{cases} \tag{5.4}$$

Then given a set of points in a plane $(x_0, y_0), (x_1, y_1), \ldots, (x_n, y_n)$ and the corresponding time points $t_1, t_2, \ldots, t_n$ we would like to find the control functions $\vec{u}_0, \vec{u}_1, \ldots, \vec{u}_n$ that take the system through the points at the specified times.

Studying the control $\vec{u}_k : \begin{pmatrix} \vec{x}_k \\ t_k \end{pmatrix} \mapsto \begin{pmatrix} \vec{x}_{k+1} \\ t_{k+1} \end{pmatrix}$

As $t \in [t_k, t_{k+1}]$ the state of the system will be

$$\vec{x}(t) = e^{A(t-t_k)}\vec{x}_k + \int_{t_k}^{t} e^{A(t-s)} B\vec{u}_k(s)ds \tag{5.5}$$

and as we want the state of the system to be $\vec{x}_{k+1}$ at time t_{k+1} we get the following condition.

$$\vec{x}_{k+1} = e^{A(t_{k+1}-t_k)}\vec{x}_k + \int_{t_k}^{t_{k+1}} e^{A(t_{k+1}-s)} B\vec{u}_k(s)ds \tag{5.6}$$

The solution $\vec{u}_k$ to equation (5.6) that minimizes the norm of the control signal is then given by

$$\vec{u}_k(t) = B^T e^{-A^T t} \left(\int_{t_k}^{t_{k+1}} e^{-As} BB^T e^{-A^T s}ds \right)^{-1} \vec{y}_{k+1} \tag{5.7}$$

where

$$\vec{y}_{k+1} = e^{-At_{k+1}}\vec{x}_{k+1} - e^{-At_k}\vec{x}_k \ .$$

The control would be specified completely by equation (5.7) if we knew the whole state-vector at each interpolation point. We know all (x_k, y_k) but we do not know the $(\dot{x}_k, \dot{y}_k)$. To determine the $2(n+1)$ unknowns we have to apply some conditions on the solution, and our first choice will be to require that the control is continuous, thus we assume $\vec{u}_k(t_{k+1}) = \vec{u}_{k+1}(t_{k+1})$, $k = 0\ldots n-2$. This will leave us $2(n-1)$ conditions and will leave only four unknowns. We will apply the additional conditions on the boundary.

In many applications, the shape of the trajectory is of more interest than the velocity between the nodal points. Thus we may assume that the time between nodal points is constant and so let $t_{k+1} - t_k = h$.

Thus if we let $\{\tau = s - t_k\}$ and let

$$M \equiv \int_0^h e^{-A\tau} BB^T e^{-A\tau} d\tau$$

Then (5.7) can be rewritten as

$$\vec{u}_k(t) = B^T e^{-AT(t-t_k)} M^{-1}(e^{-Ah}\vec{x}_{k+1} - \vec{x}_k) \tag{5.8}$$

Using this expression we can now investigate the results of the continuity condition.

$$\vec{u}_k(t_{k+1}) = B^T e^{-A^T h} M^{-1}(e^{-Ah}\vec{x}_{k+1} - \vec{x}_k)$$

$$= B^T M^{-1}(e^{-Ah}\vec{x}_{k+2} - \vec{x}_{k+1}) = \vec{u}_{k+1}(t_{k+1})$$

Then defining

$$Z \equiv M^{-1}e^{-Ah}$$

$$W \equiv e^{-A^T h} M^{-1}e^{-Ah} + M^{-1}$$

gives

$$B^T(Z^T\vec{x}_k - W\vec{x}_{k+1} + Z\vec{x}_{k+2}) = 0, \qquad k = 0\ldots n-2 \tag{5.9}$$

In block diagonal form this becomes

$$B^T \begin{bmatrix} Z^T & -W & Z & \ldots & 0 & 0 & 0 \\ 0 & Z^T & -W & \ldots & 0 & 0 & 0 \\ \vdots & \vdots & \vdots & \ddots & \vdots & \vdots & \vdots \\ 0 & 0 & 0 & \ldots & -W & Z & 0 \\ 0 & 0 & 0 & \ldots & Z^T & -W & Z \end{bmatrix} \begin{bmatrix} \vec{x}_0 \\ \vec{x}_1 \\ \vec{x}_2 \\ \vdots \\ \vec{x}_{n-2} \\ \vec{x}_{n-1} \\ \vec{x}_n \end{bmatrix} = 0 \tag{5.10}$$

The vector in (5.10) is made up of subvectors

$$\vec{x}_k = \begin{bmatrix} x_k \\ \dot{x}_k \\ y_k \\ \dot{y}_k \end{bmatrix}$$

consisting of two knowns and two unknowns. By partitioning the submatrices W, Z as

$$W = \begin{bmatrix} \vdots & \vdots & \vdots & \vdots \\ w_{.1} & w_{.2} & w_{.3} & w_{.4} \\ \vdots & \vdots & \vdots & \vdots \end{bmatrix}, Z = \begin{bmatrix} \vdots & \vdots & \vdots & \vdots \\ z_{.1} & z_{.2} & z_{.3} & z_{.4} \\ \vdots & \vdots & \vdots & \vdots \end{bmatrix}$$

and using the notations given in the following definitions, we can keep the unknowns on the left side and move the position coordinates over to the right hand side.

Let

$$\vec{x}^{pos} = \begin{bmatrix} x \\ y \end{bmatrix}$$

$$\vec{x}^{vel} = \begin{bmatrix} \dot{x} \\ \dot{y} \end{bmatrix}$$

$$W_l^B = B^T \begin{bmatrix} w_{.2} & w_{.4} \end{bmatrix} = \begin{bmatrix} w_{22} & w_{24} \\ w_{42} & w_{44} \end{bmatrix}$$

$$W_r^B = B^T \begin{bmatrix} w_{.1} & w_{.3} \end{bmatrix} = \begin{bmatrix} w_{21} & w_{23} \\ w_{41} & w_{43} \end{bmatrix}$$

$$Z_{lu}^B = B^T \begin{bmatrix} z_{.2} & z_{.4} \end{bmatrix} = \begin{bmatrix} z_{22} & z_{24} \\ z_{42} & z_{44} \end{bmatrix}$$

$$Z_{ru}^B = B^T \begin{bmatrix} z_{.1} & z_{.3} \end{bmatrix} = \begin{bmatrix} z_{21} & z_{23} \\ z_{41} & z_{43} \end{bmatrix}$$

$$Z_{ll}^B = B^T \begin{bmatrix} z_{2.} \\ z_{4.} \end{bmatrix}^T = \begin{bmatrix} z_{22} & z_{42} \\ z_{24} & z_{44} \end{bmatrix}$$

$$Z_{rl}^B = B^T \begin{bmatrix} z_{1.} \\ z_{3.} \end{bmatrix}^T = \begin{bmatrix} z_{12} & z_{32} \\ z_{14} & z_{34} \end{bmatrix}$$

And we get the system

$$
\begin{bmatrix}
Z_{ll}^{B} & -W_{l}^{B} & Z_{lu}^{B} & \cdots & 0 & 0 & 0 \\
0 & Z_{ll}^{B} & -W_{l}^{B} & \cdots & 0 & 0 & 0 \\
\vdots & \vdots & \vdots & \ddots & \vdots & \vdots & \vdots \\
0 & 0 & 0 & \cdots & -W_{l}^{B} & Z_{lu}^{B} & 0 \\
0 & 0 & 0 & \cdots & Z_{ll}^{B} & -W_{l}^{B} & Z_{lu}^{B}
\end{bmatrix}
\begin{bmatrix}
\vec{x}_{0}^{vel} \\
\vec{x}_{1}^{vel} \\
\vec{x}_{2}^{vel} \\
\vdots \\
\vec{x}_{n-2}^{vel} \\
\vec{x}_{n-1}^{vel} \\
\vec{x}_{n}^{vel}
\end{bmatrix}
=
$$

$$
-
\begin{bmatrix}
Z_{rl}^{B} & -W_{l}^{B} & Z_{ru}^{B} & \cdots & 0 & 0 & 0 \\
0 & Z_{rl}^{B} & -W_{l}^{B} & \cdots & 0 & 0 & 0 \\
\vdots & \vdots & \vdots & \ddots & \vdots & \vdots & \vdots \\
0 & 0 & 0 & \cdots & -W_{l}^{B} & Z_{ru}^{B} & 0 \\
0 & 0 & 0 & \cdots & Z_{rl}^{B} & -W_{l}^{B} & Z_{ru}^{B}
\end{bmatrix}
\begin{bmatrix}
\vec{x}_{0}^{pos} \\
\vec{x}_{1}^{pos} \\
\vec{x}_{2}^{pos} \\
\vdots \\
\vec{x}_{n-2}^{pos} \\
\vec{x}_{n-1}^{pos} \\
\vec{x}_{n}^{pos}
\end{bmatrix}
$$

As we evaluate the right hand side, we get a constant vector. We have two unknowns still unaccounted for in this system and we can choose any boundary conditions to allow solution of this system. The above solution will give a solution for zero initial and final velocities. With slight modification the system will present a solution with arbitrary known initial and final velocities.

Rewriting the system, we get

$$
\begin{bmatrix}
Z_{ll}^{B} & -W_{l}^{B} & Z_{lu}^{B} & \cdots & 0 & 0 & 0 \\
0 & Z_{ll}^{B} & -W_{l}^{B} & \cdots & 0 & 0 & 0 \\
\vdots & \vdots & \vdots & \ddots & \vdots & \vdots & \vdots \\
0 & 0 & 0 & \cdots & -W_{l}^{B} & Z_{lu}^{B} & 0 \\
0 & 0 & 0 & \cdots & Z_{ll}^{B} & -W_{l}^{B} & Z_{lu}^{B}
\end{bmatrix}
\begin{bmatrix}
\vec{x}_{1}^{vel} \\
\vec{x}_{2}^{vel} \\
\vdots \\
\vec{x}_{n-2}^{vel} \\
\vec{x}_{n-1}^{vel}
\end{bmatrix}
$$

$$
= -
\begin{bmatrix}
\Omega_{1} \\
\Omega_{2} \\
\vdots \\
\Omega_{n-2} \\
\Omega_{n-1}
\end{bmatrix}
$$

Where

$$
\Omega_{1} = -Z_{rl}^{B}\vec{x}_{0}^{pos} + W_{r}^{B}\vec{x}_{1}^{pos} - Z_{ru}^{B}\vec{x}_{2}^{pos} - Z_{ll}^{B}\vec{x}_{0}^{vel}
$$

$$
\Omega_{k} = -Z_{rl}^{B}\vec{x}_{k-1}^{pos} + W_{r}^{B}\vec{x}_{k}^{pos} - Z_{ru}^{B}\vec{x}_{k+1}^{pos}
$$

$$
\Omega_{n-1} = -Z_{rl}^{B}\vec{x}_{n-2}^{pos} + W_{r}^{B}\vec{x}_{n-1}^{pos} - Z_{ru}^{B}\vec{x}_{n}^{pos} - Z_{lu}^{B}\vec{x}_{n}^{vel}
$$

Then, having solved this system, we now know the states of the system at each nodal point. We will now use equation (5.8) for the control, and insert it into the equation (5.5) to get the trajectory.

$$\vec{x}(t) = e^{A(t-t_k)} \left(\vec{x}_k + \int_0^{t-t_k} e^{-A\tau} BB^T e^{-A^T\tau} d\tau M^{-1} (e^{-Ah}\vec{x}_{k+1} - \vec{x}_k) \right)$$

for $k = 0 \ldots n - 1$ and $t \in [t_k, t_{k+1}]$. As we can see from this equation the fundamental matrix and the integral is the same for all k and only has to be evaluated for $t - t_k$ between 0 and h, once and for all.

From the discussion on one-dimensional splines, we can vary the characteristics of the resulting spline. With all the parameters in (5.2) set equal to zero, we get cubic parametric splines. This makes the calculation of the fundamental matrix easy. By choosing to decouple the parameters, i.e. $\alpha_1 = \alpha_2 = \beta_1 = \beta_2 = 0$, we were able to generate exponential parametric splines with the basis functions $1, \lambda_1 t, e^{\lambda_1 t}, e^{-\lambda_1 t}$ and $1, \lambda_2 t, e^{\lambda_2 t}, e^{-\lambda_2 t}$ for the x and y coordinates. Just as with one-dimensional splines, the result of taking large eigenvalues is almost linear interpolation, which can be good for certain applications.

We now have two choices for boundary conditions of the system. Zero velocity and a specified starting and ending velocity. We must, of course, make a good choice for the starting velocity. If we use a bad direction and high magnitude for the velocity on $y = x^2$, we get the graph of figure 1. Even though we use n=40 to reproduce the curve, the bad boundary conditions ruin the tracking.

On the other hand, by choosing boundary conditions to approximate the initial and final derivatives of the function, we can simulate the cycloid,

$$\begin{cases} x = \pi t - \sin \pi t \\ y = 1 - \cos \pi t \end{cases}$$

and prolate cycloid

$$\begin{cases} x = \pi t - 2 \sin \pi t \\ y = 1 - 2 \cos \pi t \end{cases}$$

as can be seen in figures 2 and 3.

It is evident that the cusp and crossover do not cause any problem, as could be expected since we are using a parameterized interpolant.

6   Conclusions

The two dimensional control law described herein does a good job simulating complex shapes. With appropriate choice of boundary conditions and control parameters, this spline can be used as an efficient means of storage

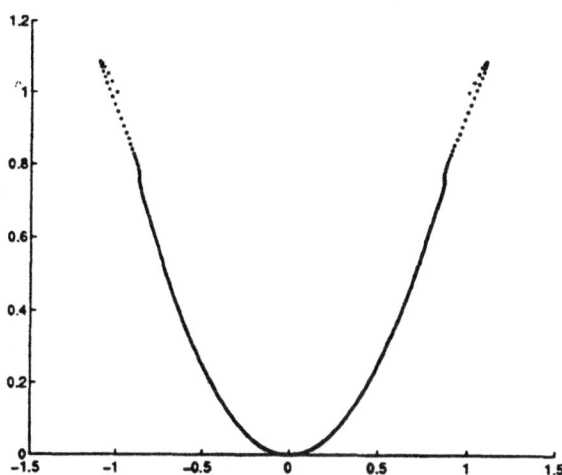

Figure 5.1: Trajectory of system tracking $y = x^2$, with badly specified starting conditions

for complex curves. The use of the control law allows effective experimentation with parameters to optimize the characteristics of the spline to the task at hand.

7 A Final Remark

For purposes of discussion, we constructed spline approximation in this paper by introducing the nodal shape functions. This is not necessary in practical computation. From the framework we have established, based on the control theory in Section 3, all we need to do is provide the matrix A, and the vector $\vec{b}$ to the linear system (3.8). We solve (3.8) numerically to obtain the $\vec{x}^k$'s, and hence the control law $u(t)$ (see (2.9)). Then with the control $u(t)$, the expected spline function is given by the first component of $\vec{x}(t)$ defined by (3.13). Based on our analysis, we are able to generate different splines by simply selecting appropriate entries of the matrix A.

The significance of this investigation is two fold: first, it exposes the relationship between two important fields - control theory and spline approximations. This enables us to discover new spline functions and to investigate, systematically, the properties of the spline approximations. Sec-

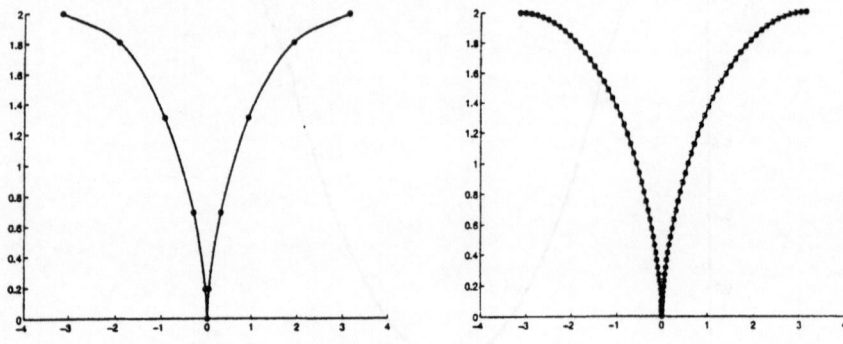

Figure 7.1: Figure 2: Graph of cycloid, reproduced with $n = 10$, $n = 100$ and $\lambda_1 = 10$, $\lambda_2 = 10$

ondly, it provides a practical way to construct different splines from a same simple framework. From our experience, we feel that this construction is more natural and easier than the traditional approach.

References

[1] Ammar, G., Dayawansa, W., Martin, C. (1991): Exponential interpolation: theory and numerical algorithms. Applied Mathematics and Computation **44**, 189-232

[2] Ammar, G., Martin, C. (1986): Geometry of matrix eigenvalue methods. Acta Applicandae Mathematicae **5**, 239-278

[3] Böckmann, C (1995): A modification of the trust-region Gauss-Newton method to solve separable nonlinear least squares problems. Journal of Math. Syst., Est. and Control **5**, 111-115

[4] de Boor, C. (1978): A Practical Guide to Splines. Springer-Verlag

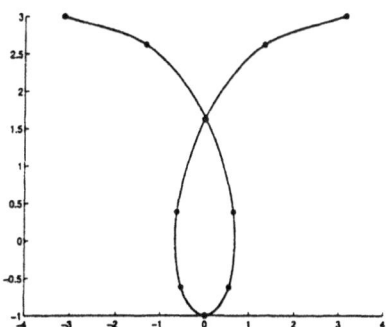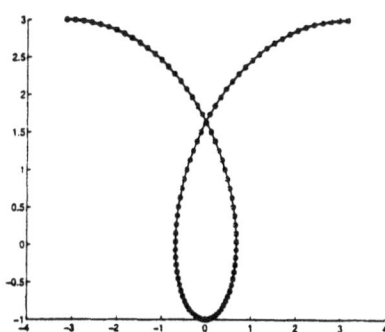

Figure 7.2: Figure 3: Graph of prolate cycloid, reproduced with $n = 10$, $n = 100$ and $\lambda_1 = 10$, $\lambda_2 = 10$

[5] Brockett, R.W. (1970): Finite Dimensional linear systems. John Wiley & Sons

[6] Doolin, B., and Martin, C. (1990): An introduction to differential geometry for control engineers. New York: Marcel Dekker, Inc.

[7] Luenberger, D.G. (1969): Optimization by Vector Space Methods. John Wiley & Sons

[8] Martin, C., Stamp, M., Wang, X. (1991): Discrete observability and numerical quadrature. IEEE Tran. Aut. Control **36**, 1337-1340

[9] McCartin, B.J. (1991): Theory of exponential splines. Journal of Approximation Theory **66**, 1-23

SINC APPROXIMATION OF SOLUTION OF HEAT EQUATION WITH DISCONTINUOUS INITIAL CONDITION

Anne C. Morlet [*]
Department of Mathematics
Ohio State University
Columbus, OH 43210-1174

Frank Stenger [†]
Department of Computer Science
University of Utah
Salt Lake City, UT 84112

1 Introduction

We solve the one dimensional heat equation with discontinuous initial condition, the discontinuity being at $x = 0$, using the heat kernel and the Sinc collocation method for convolution integrals. In [7], it is shown that the method is converging at an exponential rate. We compare the following formulations: the first one is based on the fact that for $t > 0$, the solution of the forced heat equation, the forcing term being a smooth function, is infinitely many time differentiable; the other one ignores the smoothing properties of the heat kernel. With the first approach, we apply the Sinc collocation algorithm presented in [7] to the double integral for the forcing term ignoring the discontinuity in the initial condition. With the other one, we break the double integral for the forcing term into two integrals, one over $(-\infty, 0) \times [0, T]$ and the other over $(0, \infty) \times [0, T]$, then we make the change of variable $\xi = -\eta$ in the integral over $(-\infty, 0) \times [0, T]$ to only compute integrals over $(0, \infty) \times [0, T]$. In the first case, the conformal map is the identity; in the other one it is $\log(\sinh(z))$. Our numerical results show that for the problem considered here, the first approach is more efficient and more accurate than the second one. The method presented here is easily generalized to the case of finitely many jumps in the initial condition.

In Section 2, we present a brief overview of the Sinc collocation method for convolution integrals and an algorithm for the Sinc convolution-quadrature method for integrals. In Section 3, we compute the solution of the heat equation with discontinuous initial condition, the discontinuity being at $x = 0$, with the algorithms presented in Section 2 and two different viscosities $\nu = 1.$ and $\nu = .2$. We also present tables of errors for different values of the spatial and temporal mesh sizes.

2 Description of Sinc Approximation

In this section, we give the main results on the collocation of convolution integrals with a Sinc method; for more details the reader is referred to [7]. We first remind the reader of a few definitions.

[*]Supported in part by NSF grant # CCR-9307602
[†]Supported in part by NSF grant # CCR-9307602

Let (a, b) be a real interval, $(a, b) \subset \mathbb{R}$. Let ϕ be the one to one mapping, mapping $\mathcal{D}$ onto $\mathcal{D}_d$, $(a, b) \subset \mathcal{D}$, $\mathcal{D}_d = \{z \in \mathbb{C} : |\Im(z)| < d\}$, $d \in (0, \pi)$. The function ϕ is also a one to one mapping of (a, b) onto $\mathbb{R}$. The reader is referred to [6] for more information on the most common mappings ϕ.

Let f be an analytic function in $\mathcal{D}$. The function f is said to belong to $\mathbf{L}_{\alpha,\beta}(\mathcal{D})$, α and β strictly positive numbers, if there exists a constant C such that

$$|f(z)| \leq C \frac{|e^{\phi(z)}|^\alpha}{[1 + |e^{\phi(z)}|]^{\alpha+\beta}}, \qquad z \in \mathcal{D}.$$

The function f is said to belong to $\mathbf{M}_{\alpha,\beta}(\mathcal{D})$, α and β strictly positive numbers, if the function g defined by

$$g(z) = f(z) - \frac{f(a) + e^{\phi(z)} f(b)}{1 + e^{\phi(z)}}, \qquad z \in \mathcal{D},$$

is in $\mathbf{L}_{\alpha,\beta}(\mathcal{D})$.

The Sinc approximation in $\mathbf{M}_{\alpha,\beta}(\mathcal{D})$ is defined as follows. Let n be a positive integer and let $\epsilon = \min(\alpha, \beta)$. If $\epsilon = \alpha$, then M and N are given by

$$M = n, \qquad N = [|\alpha n|],$$

with $[|u|]$, the greatest integer smaller or equal to u. On the other hand, if $\epsilon = \beta$, then

$$M = [|\beta n|], \qquad N = n.$$

Let

$$
\begin{aligned}
\text{sinc}(z) &= \frac{\sin(\pi z)}{\pi z}, \\
h &= \sqrt{\frac{\pi d}{\epsilon n}}, \\
z_j &= \phi^{-1}(jh), \qquad j = -M, \dots, N, \\
\gamma_j(z) &= \text{sinc}\left(\frac{\phi(z) - jh}{h}\right), \qquad j = -M, \dots, N, \qquad (2.1) \\
\omega_j(z) &= \gamma_j(z), \qquad j = -M+1, \dots, N-1, \\
\omega_{-M}(z) &= [1 + e^{-Mh}]\left[\frac{1}{1 + e^{\phi(z)}} - \sum_{j=-M+1}^{N} \frac{1}{1 + e^{jh}} \gamma_j(z)\right], \\
\omega_N(z) &= [1 + e^{-Nh}]\left[\frac{e^{\phi(z)}}{1 + e^{\phi(z)}} - \sum_{j=-M}^{N-1} \frac{e^{jh}}{1 + e^{jh}} \gamma_j(z)\right], \\
\epsilon_n &= n^{1/2} e^{-(\pi d \epsilon n)^{1/2}}
\end{aligned}
$$

With these notations, we now can give an estimate of the error committed if we substitute for the function f, $f \in M_{\alpha,\beta}(\mathcal{D})$, its Sinc approximation.

Theorem 2.1 *Let $f \in M_{\alpha,\beta}(\mathcal{D})$. Then there exists a constant C_1 independent of n such that*

$$\left| f(z) - \sum_{j=-M}^{N} f(z_j) \omega_j(z) \right|_{\infty} \leq C_1 \epsilon_n. \tag{2.2}$$

Here $|\cdot|_{\infty}$ denotes the maximum norm on (a, b); the constants and functions z_j, ω_j, ϵ_n have been defined in (2.1).

Let us now introduce some notations specific to Sinc indefinite integration and convolution. Let

$$\sigma_k = \int_0^k \text{sinc}(x) dx, \quad e_k = \frac{1}{2} + \sigma_k.$$

Let $m = M + N + 1$, N and M defined as before. Let $I^{(-1)}$ be the Toeplitz matrix of order m, $I^{(-1)} = [e_{i-j}]$, $1 \leq i \leq m$, $1 \leq j \leq m$ and e_{i-j} the (i, j)th element of the matrix $I^{(-1)}$. Let u be a function defined on (a, b), let $D(u)$ be the diagonal matrix whose ith diagonal entry is $u(z_i)$, V_m be the operator that converts u into the column vector $V_m u = [u(z_{-M}), \ldots, u(z_N)]^T$ and Π_m the operator that converts a column vector $\mathbf{c}$ into a function on (a, b), $\Pi_m \mathbf{c} = \sum_{j=-M}^{N} c_j \omega_j(z)$, ω_j defined as before. Let the matrices A_m and B_m be defined as

$$A_m = h I^{(-1)} D(1/\phi'), \quad B_m = h \left(I^{(-1)} \right)^T D(1/\phi'),$$

and the operators T, T', T_m, and T_m' as

$$(Tf)(x) = \int_a^x f(t) dt, \quad (T'f)(x) = \int_x^b f(t) dt,$$

$$T_m = \Pi_m A_m V_m, \quad T_m' = \Pi_m B_m V_m.$$

The operators T_m and T_m' are respectively discrete approximations of the continuous operators T and T' and converge to them. The result summarizes as

Theorem 2.2 *If $f/\phi' \in L_{\alpha,\beta}(\mathcal{D})$, then there exists a constant C_2 independent of n such that*

$$|Tf - T_m f|_{\infty} \leq C_2 \epsilon_n, \tag{2.3}$$

$$|T'f - T_m'f|_{\infty} \leq C_2 \epsilon_n.$$

As before $|\cdot|_{\infty}$ denotes the maximum norm on (a, b). The constant ϵ_n is defined in (2.1) and the operators T, T', T_m, and T_m' above.

Now let us define the convolution integrals $p(x)$ and $q(x)$

$$p(x) = \int_a^x f(x-t)g(t)dt, \quad q(x) = \int_x^b f(t-x)g(t)dt, \quad (2.4)$$

$x \in (a, b)$ and use the operators defined above to find an approximation to p and q with Sinc formulae. We first need to assume the existence of the "Laplace Transform" of f.

Assumption 2.1 *Assume that the "Laplace Transform"*

$$F(s) = \int_0^\infty f(t)e^{-t/s}dt,$$

exists for all $s \in \Omega_+$, $\Omega_+ = \{s \in \mathbb{C} : \Re(s) > 0\}$ and that $F(s) = O(s)$ as $s \to \infty$. Let $P(r, x)$ be defined as

$$P(r, x) = \int_a^x f(r+x-t)g(t)dt.$$

Furthermore, assume that

- $P(r, x) \in \mathbf{M}_{\alpha,\beta}(\mathcal{D}')$ *uniformly in x, $r \in [0, b-a]$;*

- $P(r, x)$ *is of bounded variation on $(0, b-a)$, uniformly in r, $x \in [a, b]$.*

Under these assumptions, we have the following convergence result

Theorem 2.3 *Let the "Laplace Transform" be defined as in Assumption 2.1 and satisfy the properties of Assumption 2.1, let the convolution integrals p and q be defined as in (2.4), and let the matrices A_m and B_m and the operators V_m and Π_m be defined as before, then there exists a constant C_3, independent of n, such that,*

$$\begin{aligned} \left|p - \Pi_m F(A_m)V_m g\right|_\infty &\leq C_3\epsilon_n, \quad (2.5)\\ \left|q - \Pi_m F(B_m)V_m g\right|_\infty &\leq C_3\epsilon_n. \end{aligned}$$

The constant ϵ_n is defined in (2.1).

Now we want to apply the result of Theorem 2.3 to a double integral of the type

$$\begin{aligned} r(x, t) &= \int_0^t \int_0^\infty K(x-\xi, t-\tau)g(\xi, \tau)d\xi d\tau,\\ &= \int_0^t \int_0^x K(x-\xi, t-\tau)g(\xi, \tau)d\xi +\\ &\quad \int_0^t \int_x^\infty K(x-\xi, t-\tau)g(\xi, \tau)d\xi, \end{aligned}$$

$t \in [0, T]$. Assume that $r(x, t)$ belongs to $\mathbf{M}_{\alpha_i, \beta_i}(\mathcal{D}_i)$, $i = x, t$, with respect to each variable, the other one being fixed in its interval of definition, that the mappings $\phi_i : \mathcal{D}'_i \to \mathcal{D}_d$, and, given n_i, positive integers N_i and M_i, have been determined as before, and that $h_i = \sqrt{\pi d_i / (\epsilon_i n_i)}$, $i = x, t$. Furthermore, assume that $[\|n_x d_x \epsilon_x\|] = [\|n_t d_t \epsilon_t\|]$, so that the error in the approximation is of the same size for each variable. Set $m_i = M_i + N_i + 1$, define the Sinc points as $z_l^{(i)} = \phi_i^{-1}(l h_i)$, $l = -M_i, \ldots, N_i$, $i = x, t$. Then determine the matrices A_x, B_x, and A_t as

$$A_x = h_x I_{m_x}^{(-1)} D(1/\phi'_x) = X_1 S_1 X_1^{-1},$$

$$B_x = h_x \left(I_{m_x}^{(-1)} \right)^T D(1/\phi'_x) = X_2 S_1 X_2^{-1},$$

$$A_t = h_t I_{m_t}^{(-1)} D(1/\phi'_t) = X_3 S_3 X_3^{-1},$$

$I_{m_i}^{(-1)}$ defined as before $i = x$, t, and S_j, $j = 1, 2$, diagonal matrices

$$S_j = \mathrm{diag}(s^j_{-M_j}, \ldots, s^j_{N_j}),$$

and $X_2 = D(1/\phi'_x) X_1$, since $B_x^T = D(1/\phi'_x) A_x (D(1/\phi'_x))^{-1}$. Then, define the "Laplace Transforms" as

$$F(s^{(x)}, t) = \int_0^\infty K(x, t) e^{-x/s^{(x)}} dx, \quad G(s^{(x)}, s^{(t)}) = \int_0^{c_t} F(s^{(x)}, t) e^{-t/s^{(t)}} dt,$$

with $c_t \in [2T, \infty]$, $j = 1$, 2. In the final algorithm, only the function $G(s^{(x)}, s^{(t)})$ is needed and the function $F(s^{(x)}, t)$ is only introduced to understand how separation of variables applies to the case considered.

We only describe in detail the procedure followed to compute the first integral in the expression of $r(x, t)$ since the expression for the second integral is easily deduced from the first one by replacing the matrix A_x by B_x.

Define the vectors $\mathbf{g}(\eta)$ and $\mathbf{p}(t)$ as

$$\mathbf{g}(\eta) = (g(z^{(x)}_{-M_x}, \eta), \ldots, g(z^{(x)}_{N_x}, \eta))^T, \quad \mathbf{p}(t) = \int_0^t F(A_x, t - \eta) \mathbf{g}(\eta) d\eta,$$

F and A_x defined above. Since A_x is a diagonalizable matrix, $\mathbf{p}(t)$ can be rewritten as

$$\mathbf{p}(t) = X_1 \int_0^t \mathrm{diag}(F(s^{(x)}_{-M_x}, t - \eta), \ldots, F(s^{(x)}_{N_x}, t - \eta)) X_1^{-1} \mathbf{g}(\eta) d\eta. \quad (2.6)$$

Then define the vectors $\mathbf{h}(\eta)$ and $\mathbf{q}(t)$ as

$$X_1 \mathbf{h}(\eta) = \mathbf{g}(\eta), \quad X_1 \mathbf{q}(t) = \mathbf{p}(t). \quad (2.7)$$

With these notation, (2.6) becomes a set of m_x decoupled scalar equations

$$q_i(t) = \int_0^t F(s_i^{(x)}, t - \eta)h_i(\eta)d\eta. \tag{2.8}$$

Now, we repeat the procedure followed in the x direction to get discrete equations in the t direction. Define the vectors $\mathbf{h}_i$ and $\mathbf{q}_i$ as

$$\mathbf{h}_i = (h_i(z_{-M_t}^{(t)}), \ldots, h_i(z_{N_t}^{(t)}))^T, \quad \mathbf{q}_i = G(s_i^{(x)}, A_t)\mathbf{h}_i,$$

G and A_t defined as before. Since A_t is diagonalizable, then the above system can be rewritten as

$$\mathbf{q}_i = X_3 \mathrm{diag}(G(s_i^{(x)}, s_{-M_t}^{(t)}), \ldots, G(s_i^{(x)}, s_{N_t}^{(t)}))X_3^{-1}\mathbf{h}_i. \tag{2.9}$$

Then define the vectors $\mathbf{k}_i$ and $\mathbf{r}_i$ as

$$X_3\mathbf{k}_i = \mathbf{h}_i, \quad X_3\mathbf{r}_i = \mathbf{q}_i. \tag{2.10}$$

We notice that with these notations (2.9) reduces to a set of decoupled scalar equations

$$r_{i,j} = G(s_i^{(x)}, s_j^{(t)})k_{i,j}, \tag{2.11}$$

with $i = -M_x, \ldots, N_x$, $j = -M_t, \ldots, N_t$. At this point, the $k_{i,j}$ are known since they are obtained from the function g by solving 2 linear system of equations, (2.7) and (2.10). We then gets the value of the $r_{i,j}$ from (2.11), and the value of $\mathbf{p}$ at the Sinc points from (2.10) and (2.7). The computation are carried in the order

$$g_{i,j} \to h_{i,j} \to k_{i,j} \to r_{i,j} \to q_{i,j} \to p_{i,j}.$$

So we implement the following algorithm to find the approximate value of $r(x, t)$ at the Sinc points

Algorithm 2.1

- *Find the location of the Sinc points $z_i^{(j)}$ and the value of $d\phi_{(j)}/dx$ at $x = z_i^{(j)}$, $j = x, t$, $i = -M_j, \ldots, N_j$. Set up the array of values of g at the Sinc points, $g_{i,j} = g(z_i^{(x)}, z_j^{(t)})$.*

- *Set up the matrices A_x, B_x, A_t, X_1, X_2, and X_3.*

- *Solve the systems $X_1\mathbf{h}_{\cdot,j}^1 = \mathbf{g}_{\cdot,j}$ and $X_2\mathbf{h}_{\cdot,j}^2 = \mathbf{g}_{\cdot,j}$.*

- *Solve the systems $X_3\mathbf{k}_{i,\cdot}^l = \mathbf{h}_{i,\cdot}^l$, $l = 1, 2$.*

- *Compute $r^l_{i,j} = G_l(s^{(x)}_i, s^{(t)}_j)k^l_{i,j}$, $l = 1$, 2.*

- *Compute the matrix vector multiplications $q^l_{i,\cdot} = X_3 r^l_{i,\cdot}$, $l = 1$, 2.*

- *Compute the matrix vector multiplications $p^1_{\cdot,j} = X_1 q^1_{\cdot,j}$ and $p^2_{\cdot,j} = X_2 q^2_{\cdot,j}$.*

Now that the coefficients $p^l_{i,j}$, $l = 1$, 2 are known, we can get an approximate value of $r(x,t)$ anywhere on $[0,\infty) \times [0,T]$ using the sinc interpolant

$$\sum_{i=-M_x}^{N_x} \sum_{j=-M_t}^{N_t} (p^1_{i,j} + p^2_{i,j})\omega^{(x)}_i(x)\omega^{(t)}_j(t), \qquad (2.12)$$

with $\omega^{(x)}_i$ and $\omega^{(t)}_j$ defined as in (2.1).

In the above algorithm, the vector $\mathbf{a}_{\cdot,j}$ is the vector of components $a_{1,j}, \ldots, a_{M_x+N_x+1,j}$; similarly, $\mathbf{b}_{i,\cdot}$ the vector of components $b_{i,1}, \ldots, b_{i,M_t+N_t+1}$.

Assuming that M_x, N_x, M_t, and N_t have been chosen such that the error for each variable is of the same size and $\mathcal{O}(n_x^{1/2}e^{-\sqrt{\pi d_x \beta_x n_x}})$, then the error in the approximation (2.12) is $\mathcal{O}([\log(n_x)]n_x^{1/2}e^{-\sqrt{\pi d_x \beta_x n_x}})$.

We also want to approximate a double integral of the type

$$s(x,t) = \int_0^t \int_0^\infty K(x+\xi, t-\tau)g(\xi,\tau)d\xi d\tau,$$

$t \in [0,T]$. Assume that $s(x,t)$ belongs to $\mathbf{M}_{\alpha_t,\beta_t}(\mathcal{D}_t)$, $x \in [0,\infty)$, that the mapping $\phi_t : \mathcal{D}_t \to \mathcal{D}_d$, and, given n_t, positive integers N_t and M_t have been determined. Assume that $K(\xi,t)g(\xi,t)/\phi'_x$ belongs to $\mathbf{L}_{\alpha_x,\beta_x}(\mathcal{D}_x)$ for $t \in [0,T]$, that the mapping $\phi_x : \mathcal{D}_x \to \mathcal{D}_d$, and, given n_x, positive integers N_x and M_x have been determined. Define A_t as before and define the "Laplace Transform" as

$$H(x, s^{(t)}) = \int_0^{c_t} K(x,t)e^{-t/s^{(t)}}dt,$$

with $c_t \in [2T, \infty]$. Define the vector $\bar{\mathbf{g}}(x)$ as

$$\bar{\mathbf{g}}(x) = (g(x, z^{(t)}_{-M_t}), \ldots, g(x, z^{(t)}_{N_t}))^T,$$

and the vector $\bar{\mathbf{q}}(x)$ as

$$\bar{\mathbf{q}}(x) = H(x, A_t)\bar{\mathbf{g}}(x),$$

H and A_t defined as before. Since A_t is diagonalizable, then the above system can be rewritten as

$$\bar{\mathbf{q}}(x) = X_3 \mathrm{diag}(H(x, s^{(t)}_{-M_t}), \ldots, H(x, s^{(t)}_{N_t}))X_3^{-1}\bar{\mathbf{g}}(x). \qquad (2.13)$$

Then define the vectors $\bar{\mathbf{k}}(x)$ and $\bar{\mathbf{r}}(x)$ as

$$X_3\bar{\mathbf{k}}(x) = \bar{\mathbf{g}}(x), \qquad X_3\bar{\mathbf{r}}(x) = \bar{\mathbf{q}}(x). \tag{2.14}$$

With these notations, (2.13) reduces to a set of decoupled linear equations. To compute the integral with respect to the spatial variable, we use the quadrature rule

$$\int_0^\infty F(z)dz \approx h_x \sum_{m=-M_x}^{N_x} \frac{F(z_m^{(x)})}{\phi_x'(z_m^{(x)})}, \tag{2.15}$$

with h_x, M_x, N_x, and $z_m^{(x)}$ defined as before. Combining the spatial quadrature rule with the temporal Sinc convolution approximation, we get

$$s(z_i^{(x)}, z_j^{(t)}) \approx h_x \sum_{m=-M_x}^{N_x} \frac{\bar{k}_j(z_i^{(x)} + z_m^{(x)})}{\phi_x'(z_m^{(x)})}. \tag{2.16}$$

So we implement the following algorithm to find the approximate value of $s(x,t)$ at the Sinc points

Algorithm 2.2

- *Find the location of the Sinc points $z_i^{(j)}$ and the value of $d\phi_{(j)}/dx$ at $x = z_i^{(j)}$, $j = x, t$, $i = -M_j, \ldots, N_j$. Set up the array of values of g at the Sinc points, $g_{m,j} = g(z_m^{(x)}, z_j^{(t)})$.*

- *Set up the matrices A_t and X_3.*

- *Solve the systems $X_3\bar{\mathbf{k}}(z_i^{(x)}) = \bar{\mathbf{g}}(z_i^{(x)})$.*

- *Compute $\bar{r}_j(\cdot) = H(\cdot, s_j^{(t)})\bar{k}_j(z_i^{(x)})$.*

- *Compute the matrix vector multiplications $\bar{\mathbf{q}}(\cdot) = X_3\bar{\mathbf{r}}(\cdot)$.*

- *Use the Sinc quadrature rule to compute the spatial integral*

$$\bar{s}_j(z_i^{(x)}) = h_x \sum_{m=-M_x}^{N_x} \frac{\bar{q}_j(z_i^{(x)} + z_m^{(x)})}{\phi_x'(z_m^{(x)})}.$$

Now that the coefficients $\bar{s}_j(z_i^{(x)})$ are known, we can get an approximate value of $s(x,t)$ anywhere on $[0,\infty) \times [0,T]$ using the Sinc interpolant

$$\sum_{i=-M_x}^{N_x} \sum_{j=-M_t}^{N_t} \bar{s}_j(z_i^{(x)})\omega_i^{(x)}(x)\omega_j^{(t)}(t), \tag{2.17}$$

with $\omega_i^{(x)}$ and $\omega_j^{(t)}$ defined as in (2.1).

Assuming that M_x, N_x, M_t, and N_t have been chosen such that the error for each variable is of the same size and $\mathcal{O}(n_x^{1/2} e^{-\sqrt{\pi d_x \beta_x n_x}})$, then the error in the approximation (2.17) is $\mathcal{O}([\log(n_x)] n_x^{1/2} e^{-\sqrt{\pi d_x \beta_x n_x}})$.

3 Numerical Results

Here we want to solve the one-dimensional heat equation

$$
\begin{aligned}
\frac{\partial}{\partial t} u(x,t) &= \nu \frac{\partial^2}{\partial x^2} u(x,t) + f(x,t), \\
u(x,0) &= g(x), \\
\lim_{x \to -\infty} u(x,t) &= 0, \quad \lim_{x \to \infty} u(x,t) = 0.
\end{aligned}
\tag{3.18}
$$

Let $\xi = x/\nu$, $\tau = t/\nu$, $v(\xi,\tau) = u(x,t)$, $\tilde{f}(\xi,\tau) = \nu f(\nu\xi, \nu\tau)$, and $\tilde{g}(\xi) = g(\nu\xi)$. Then, in the new variables, (3.18) becomes

$$
\begin{aligned}
\frac{\partial}{\partial \tau} v(\xi,\tau) &= \frac{\partial^2}{\partial \xi^2} v(\xi,\tau) + \tilde{f}(\xi,\tau), \\
v(\xi,0) &= \tilde{g}(\xi), \\
\lim_{\xi \to -\infty} v(\xi,\tau) &= 0, \quad \lim_{\xi \to \infty} v(\xi,\tau) = 0.
\end{aligned}
\tag{3.19}
$$

Using the one-dimensional heat kernel, we find that

$$
v(\xi,\tau) = \frac{1}{2\sqrt{\pi\tau}} \int_{-\infty}^{\infty} e^{-\frac{(\xi-\eta)^2}{4\tau}} \tilde{g}(\eta) d\eta
\tag{3.20}
$$

$$
+ \int_0^\tau \frac{1}{2\sqrt{\pi(\tau-\mu)}} \int_{-\infty}^{\infty} e^{-\frac{(\xi-\eta)^2}{4(\tau-\mu)}} \tilde{f}(\eta,\mu) d\eta d\mu.
$$

¿From the theory of parabolic equations [1], we know that if $\tilde{f}$ is a C^∞ function on $\mathbb{R}$, if the initial condition and the forcing term satisfy the boundedness conditions

$$
\begin{aligned}
|\tilde{f}(x,t)| &\leq K_0 \exp(h_0 x^2), \\
|\tilde{g}(x)| &\leq K_1 \exp(h_1 x^2),
\end{aligned}
$$

K_0, K_1, h_0, and h_1 positive constants, and if the initial condition has finitely many jumps, then the solution of (3.19) is a C^∞ function for $t > 0$. We use this property to compare the numerical results obtained by approximating the double integral in (3.20) with the Sinc convolution Algorithm 2.1 to the ones obtained by first breaking up the double integral in (3.20) into 2 double integrals, then by using Sinc convolution or Sinc convolution-quadrature described in Algorithm 2.2. In the case of a jump at $x = 0$, the

2nd approach gives us for (3.20)

$$v(\xi, \tau) = \frac{1}{2\sqrt{\pi\tau}} \left(\int_0^\infty e^{-\frac{(\xi-\eta)^2}{4\tau}} \tilde{g}(\eta) d\eta + \int_0^\infty e^{-\frac{(\xi+\eta)^2}{4\tau}} g(-\eta) d\eta \right)$$

$$+ \int_0^\tau \frac{1}{2\sqrt{\pi(\tau-\mu)}} \left(\int_0^\infty e^{-\frac{(\xi-\eta)^2}{4(\tau-\mu)}} \tilde{f}(\eta, \mu) d\eta \right. \tag{3.21}$$

$$\left. + \int_0^\infty e^{-\frac{(\xi+\eta)^2}{4(\tau-\mu)}} \tilde{f}(-\eta, \mu) d\eta \right) d\mu.$$

We get (3.21) from (3.20) by breaking up the spatial integral into two integrals, one on $\mathbb{R}^-$, the other on $\mathbb{R}^+$. Then, in the integral on $\mathbb{R}^-$, we make the change of variable $\mu = -\mu$. Making the change of variable $\xi = -\lambda$ when $\xi \leq 0$, we find that (3.21) can be rewritten as

$$v(-\lambda, \tau) = \frac{1}{2\sqrt{\pi\tau}} \left(\int_0^\infty e^{-\frac{(\lambda+\eta)^2}{4\tau}} \tilde{g}(\eta) d\eta + \int_0^\infty e^{-\frac{(\lambda-\eta)^2}{4\tau}} g(-\eta) d\eta \right)$$

$$+ \int_0^\tau \frac{1}{2\sqrt{\pi(\tau-\mu)}} \left(\int_0^\infty e^{-\frac{(\lambda+\eta)^2}{4(\tau-\mu)}} \tilde{f}(\eta, \mu) d\eta \right. \tag{3.22}$$

$$\left. + \int_0^\infty e^{-\frac{(\lambda-\eta)^2}{4(\tau-\mu)}} \tilde{f}(-\eta, \mu) d\eta \right) d\mu.$$

The expression for $v(\xi, \tau)$, $\xi \geq 0$ being given by (3.21).

If the initial condition and the forcing term are

$$\tilde{g}(x) = \begin{cases} e^{\nu x} & \text{if } x < 0 \\ -e^{-\nu x} & \text{if } x > 0 \end{cases},$$

$$\tilde{f}(x, t) = \nu e^{-\nu^2 x^2},$$

then we can get get a close form solution of the heat equation,

$$v(\xi, \tau) = -\frac{1}{2} e^{-\nu\xi} e^{\nu^2 \tau} \left(1 - \text{erf}\left(-\frac{\xi - 2\nu\tau}{2\sqrt{\tau}} \right) \right) + \frac{1}{2} e^{\nu\xi} e^{\nu^2 \tau}$$

$$\left(1 - \text{erf}\left(\frac{\xi + 2\nu\tau}{2\sqrt{\tau}} \right) \right) - \frac{1}{2\nu} e^{-\nu^2 \xi^2} \tag{3.23}$$

$$+ \frac{\sqrt{1 + 4\nu^2\tau}}{2\nu} e^{-\frac{\nu^2 \xi^2}{1 + 4\nu^2 \tau}} - \frac{\sqrt{\pi}}{2} \xi \text{erf}(\nu\xi) + \frac{\sqrt{\pi}}{2} \xi \text{erf}\left(\frac{\nu\xi}{\sqrt{1 + 4\nu^2\tau}} \right).$$

The above expression is valid for $\xi \in \mathbb{R}$, $\tau > 0$. To get the above expression, we only describe the procedure followed to compute the first integral in

M_x	N_x	M_t	N_t	h	τ	$\|E\|_\infty$	Asymptotic Error
4	4	4	4	.785	1.571	$3.070e-2$	$1.198e-1$
12	12	12	12	.453	.907	$1.295e-2$	$3.730e-2$
20	20	20	20	.351	.702	$7.910e-3$	$1.192e-2$
28	28	28	28	.297	.594	$5.452e-3$	$4.330e-3$
36	36	36	36	.262	.524	$3.980e-3$	$1.735e-3$
44	44	44	44	.237	.474	$3.033e-3$	$7.492e-4$

Table 3.1: Table of errors on the Sinc mesh in the approximation of the double convolution integral defined on $\mathbb{R} \times [0, T]$ for different values of M_x, N_x, M_t, and N_t and for a viscosity $\nu = 1$.

(3.20) since the expression for the 2nd one is obtained from the first one with only minor changes. We rewrite $(x-\xi)^2/(4t)+\nu\xi$ as $(\xi-[x-2\nu t])^2/(4t)+ \nu x - \nu^2 t$, make the change of variable $z = (\xi - [x - 2\nu t])/(2\sqrt{t})$, then use the definition of the error function. To compute the 2nd integral in (3.20), depending on the forcing term, we rewrite $(x - \xi)^2/(4(t - \tau)) + \nu^2\xi^2$ as $(1+4\nu^2(t-\tau))(\xi-x/(1+4\nu^2(t-\tau)))^2/(4(t-\tau))+\nu^2 x^2/(1+4\nu^2(t-\tau))$, make the change of variable $z = \sqrt{1 + 4\nu^2(t - \tau)}/(2\sqrt{t - \tau})(\xi-x/(1+4\nu^2(t-\tau)))$, use the value of the error function at ∞, make the change of variable $u = \nu x/\sqrt{1 + 4\nu^2(t - \tau)}$, integrate by parts once the integral obtained after the last change of variable, then use the definition of the error function.

In Table 3.1, we present, for a viscosity $\nu = 1$, the maximum norm error in the expression of the double integral of (3.20), approximated with the Sinc convolution Algorithm 2.1. We evaluate the error at the Sinc points only, since it has been previously shown empirically that the maximum of the error is attained at one of the Sinc points [2], [3], [4], and [5]. Proceeding as in [6], we can formally prove that the empirical rule holds. We did not run our code for grids larger than $M_x = N_x = M_t = N_t = 44$ since the estimated condition number of the matrices of the systems solved are of the order 10^{10} and warrants us from getting more accurate numerical results unless we carry the arithmetic in quadruple precision. We integrated the equation up to time $T = 40$ and that may explain the relatively large errors. Nevertheless, the asymptotic and computed errors agree relatively well.

In Table 3.2, we present, for a viscosity $\nu = 1$, the maximum norm error in the expression of the sum of the 2 double integrals of (3.21), the first one being discretized with the Sinc convolution Algorithm 2.1 and

M_x	N_x	M_t	N_t	h	τ	$\|E\|_\infty$	Asymptotic Error
4	4	4	4	.785	1.571	$1.898e-1$	$1.198e-1$
8	8	8	8	.555	1.111	$6.302e-2$	$6.918e-2$
12	12	12	12	.453	.907	$2.465e-2$	$3.730e-2$
16	16	16	16	.393	.785	$1.102e-2$	$2.071e-2$
20	20	20	20	.351	.702	$5.387e-3$	$1.192e-2$
24	24	24	24	.321	.641	$4.034e-3$	$7.083e-3$
28	28	28	28	.297	.594	$3.918e-3$	$4.330e-3$

Table 3.2: Table of errors on the Sinc mesh in the approximation of the sum of the double integrals defined on $\mathbb{R}^+ \times [0, T]$ for different values of M_x, N_x, M_t, and N_t and for a viscosity $\nu = 1$.

the second one with the Sinc convolution-quadrature Algorithm 2.2. We integrate the equation up to time $T = 40$. For the reason given before, we did not run the code for grids larger than $M_x = N_x = M_t = N_t = 28$. If we compare the errors in the approximation of the double integral (3.20), we see that the method using the analyticity properties of the solution gives us smaller or of the same size errors than the method that breaks up the real line $\mathbb{R}$ into two subdomains, $\mathbb{R}^+$ and $\mathbb{R}^-$. The poor performance of the method that breaks up the real line $\mathbb{R}$ into two subdomains, $\mathbb{R}^-$ and $\mathbb{R}^+$ may be explained by the fact that near $x = 0$, the area of analyticity of the mapping $\phi(x) = \log(\sinh(x))$ is much smaller than the area of analyticity for the identity operator, the parameter d being the same in both cases. The operation count for the approach that breaks up the real line $\mathbb{R}$ into two subdomains $\mathbb{R}^-$ and $\mathbb{R}^+$ is more than twice as much as the operation count for the straight Sinc convolution method. The computed errors are in relatively good agreement with the asymptotic errors.

In Table 3.3, we present, for a viscosity $\nu = .2$, the maximum norm error in the expression of the double integral of (3.20), approximated with the Sinc convolution Algorithm 2.1. We did not run our code for grids larger than $M_x = N_x = M_t = N_t = 52$ since the estimated condition number of the matrices of the systems solved are of the order 10^{12} and warrants us from getting more accurate numerical results unless we carry the arithmetic in quadruple precision. We integrated the equation up to time $T = 200$ since in the original time variable, that T is equivalent to 40. The large errors can explain by the fact that we do not yet have enough

M_x	N_x	M_t	N_t	h	τ	$\|E\|_\infty$	Asymptotic Error
4	4	4	4	.785	1.571	2.557	$1.198e-1$
12	12	12	12	.453	.907	1.014	$3.730e-2$
20	20	20	20	.351	.702	$3.740e-1$	$1.192e-2$
28	28	28	28	.297	.594	$1.492e-1$	$4.330e-3$
36	36	36	36	.262	.524	$6.401e-2$	$1.735e-3$
44	44	44	44	.237	.474	$3.015e-2$	$7.492e-4$
52	52	52	52	.218	.436	$1.758e-2$	$3.431e-4$

Table 3.3: Table of errors on the Sinc mesh in the approximation of the double convolution integral defined on $\mathbb{R} \times [0, T]$ for different values of M_x, N_x, M_t, and N_t and for a viscosity $\nu = .2$.

Sinc points to catch the main features of the forcing term and that we have to increase the length of the time interval by a factor $1/\nu$ to get a numerical solution up to time 40 in the original temporal variable.

In Table 3.4, we present, for a viscosity $\nu = .2$, the maximum norm error in the expression of the sum of the 2 double integrals of (3.21), the first one being discretized with the Sinc convolution Algorithm 2.1 and the second one with the Sinc convolution quadrature Algorithm 2.2. We integrate the equation up to time $T = 200$. For the reason given before, we did not run the code for grids larger than $M_x = N_x = M_t = N_t = 32$. We see, as when we discretized the double integral of (3.20), that we need quite a few Sinc points to catch the features of the forcing term. The remark made about the time interval is still valid in this case since the integrals with respect to the temporal variable are approximated the same way in both cases.

M_x	N_x	M_t	N_t	h	τ	$\|E\|_\infty$	Asymptotic Error
4	4	4	4	.785	1.571	1.834	$1.198e - 1$
8	8	8	8	.555	1.111	1.168	$6.918e - 2$
12	12	12	12	.453	.907	$6.567e - 1$	$3.730e - 2$
16	16	16	16	.393	.785	$3.787e - 1$	$2.071e - 2$
20	20	20	20	.351	.702	$2.244e - 1$	$1.192e - 2$
24	24	24	24	.321	.641	$1.361e - 1$	$7.083e - 3$
28	28	28	28	.297	.594	$8.488e - 2$	$4.330e - 3$
32	32	32	32	.278	.555	$6.093e - 2$	$2.712e - 3$

Table 3.4: Table of errors on the Sinc mesh in the approximation of the sum of the double integrals defined on $\mathbb{R}^+ \times [0, T]$ for different values of M_x, N_x, M_t, and N_t and for a viscosity $\nu = .2$.

References

[1] A. FRIEDMAN, *Partial Differential Equations of Parabolic Type*, Robert E. Krieger, second edition, 1983.

[2] J. LUND and K. L. BOWERS, *Sinc Methods for Quadrature and Differential Equations*, SIAM, 1992.

[3] A. C. MORLET, "Convergence of the Sinc Method for Fourth Order ODE with Application," to appear in *SIAM J. Numer. Anal.*

[4] R. C. SMITH, G. A. BOGAR, K. L. BOWERS, and J. LUND, "The Sinc-Galerkin Method for Fourth-Order Differential Equations," *SIAM J. Numer. Anal.*, v. 28, 1991, pp. 760–780.

[5] R. C. SMITH, K. L. BOWERS, and J. LUND, "A Fully Sinc–Galerkin Method for Euler–Bernoulli Beam Models," *Numer. Meth. in Part. Diff. Eq.*, v. 8, 1992, pp. 171–202.

[6] F. STENGER, *Numerical Methods Based on Sinc and Analytical Functions*, Springer-Verlag, 1993.

[7] F. STENGER, "Collocating Convolutions," to appear in *Math. Comp.*

[8] F. STENGER, B. BARKEY, and R. VALIKI, "Sinc Convolution Approximate Solution of Burgers' Equation," *Proc. of Computational and Control III*, 1993, pp. 341–354.

EXACT BOUNDARY CONTROLLABILITY OF A BEAM AND MASS SYSTEM

Stephen W. Taylor
Department of Mathematical Sciences
Montana State University
Bozeman, Montana 59717

1   Introduction

In recent years much attention has been focused on the boundary controllability of beams. In the mathematical treatment of such problems, it is often assumed that the beam is free of loads along its length. This ideal situation is not always attainable. A beam might, for example, have small objects attached to it at various points, perhaps as part of the control mechanism. This paper investigates the exact boundary controllability of such a system. More specifically, we consider an Euler-Bernoulli Beam which is clamped at its left end and either pinned (i.e. supported) or free and attached to a point mass at its right end. We also allow the possibly of the beam being pinned at various points along its length, and of having a finite number of point masses mounted at various points along its length. We show that the system may be controlled (i.e. steered to its equilibrium state) in an arbitrarily short time interval by applying a torque to the right end of the beam.

Such a system is also applicable to the control of a ladder-like structure which consists of two identical, parallel beams connected by a series of stiff rungs. For motions of the ladder that are perpendicular to the plane containing the beams, the equations of motion of the structure are the same as those for a single beam with point masses mounted along its length.

The method used here is one that W. Littman and S. W. Taylor use [2] to investigate the controllability of a beam that is pinned at several points along its length. We first investigate a smoothing property of the corresponding semi-infinite beam-mass system, and then show that this implies the controllability of the finite beam-mass system. This "smoothing implies controllability" method has its origins in an earlier paper [3] by Littman and Taylor.

We begin in Section 2 by investigating the local smoothing and local energy decay properties of the "semi-infinite system" and use these in Section 3 to obtain the controllability results.

2   The Dynamics of the Semi-Infinite Beam and Mass System.

The Physical Problem. Consider a semi-infinite Euler-Bernoulli Beam which coincides with the non-negative x-axis in its equilibrium state. We assume that the beam is clamped at $x = 0$ and that at a finite number of ordinates $0 < b_1 < b_2 < \ldots < b_n < \infty$ the beam is either pinned or it is attached to a point mass. Let b_{α_j}, $1 \leq j \leq m$ be the locations of the masses and let b_{β_j}, $1 \leq j \leq n - m$ be the locations of the pins. We allow the possibilities $m = 0$ and $n = m$. Let $\mathcal{M}$ be the set of indices α_j corresponding to the locations of the masses and let $\mathcal{P}$ be the set of indices β_j corresponding to the points where the beam is pinned. For $j \in \mathcal{M}$, we let m_j be the mass of the particle at point b_j.

If $w(x,t)$ denotes the deflection of the beam from the x-axis at position x and time t, then w should satisfy for $x \neq 0, b_1, b_2, \ldots, b_n$ the Euler-Bernoulli Beam Equation

$$\rho \frac{\partial^2 w}{\partial t^2} + EI \frac{\partial^4 w}{\partial x^4} = 0.$$

Because the beam is clamped at $x = 0$, w must satisfy $w(0,t) = w_x(0,t) = 0$. Further, w, w_x and w_{xx} should all be continuous at the points b_j (the continuity of w_{xx} follows from the assumption that the pins and point masses do not impose any torques on the beam). The shearing force on a cross-section of the beam is EIw_{xxx} and because both the pins and the point masses can impose forces on the beam, w_{xxx} is expected to jump at each $x = b_j$. If $j \in \mathcal{P}$ then $w(b_j,t) = 0$, while if $j \in \mathcal{M}$ then Newton's Second Law of Motion gives

$$m_j \frac{d^2 y_j}{dt^2} = EI(w_{xxx}(b_j^+,t) - w_{xxx}(b_j^-,t)),$$

where $y_j(t) = w(b_j,t)$.

The total mechanical energy of the beam-mass system is given by

$$\mathcal{E}(t) = \frac{1}{2} \int_0^\infty \rho \left(\frac{\partial w}{\partial t} \right)^2 + EI \left(\frac{\partial^2 w}{\partial x^2} \right)^2 dx + \frac{1}{2} \sum_{j \in \mathcal{M}} m_j \left(\frac{dy_j}{dt} \right)^2.$$

A formal calculation shows that the energy is constant if w satisfies the conditions discussed above. This corresponds to the fact that the solution operator of the system is a strongly continuous one parameter family of unitary groups on a Hilbert space, the *finite energy space*.

The Abstract ODE. We assume now that the space and time variables have been scaled so that $\rho = EI = 1$.

Definition 2.1 *We set*

1. $\mathcal{H}_a = \{f \in H^2_{loc} : f(0) = f'(0) = 0, f(b_j) = 0 \text{ for } j \in \mathcal{P}, f'' \in L^2(0,\infty)\}$. *This is a Hilbert space with inner product given by*

$$(u, w)_a = \int_0^\infty u''(x) \bar{w}''(x) \, dx.$$

2. $\mathcal{H}_b = L^2(0, \infty) \times R^m$. *This is a Hilbert space with inner product given by*

$$\left(\begin{pmatrix} u \\ v \end{pmatrix}, \begin{pmatrix} p \\ q \end{pmatrix} \right) = \int_0^\infty u(x)\bar{p}(x)\, dx + \sum_{i=1}^m m_{\alpha_i} v_i \bar{q}_i$$

3. $\mathcal{H} = \mathcal{H}_a \times \mathcal{H}_b$. *This is a Hilbert space with inner product given by the sum of the inner products on $\mathcal{H}_a$ and $\mathcal{H}_b$. We call it the finite energy space.*

The space $\mathcal{H}_a$ may be thought of as the space of displacements for the problem, and the space $\mathcal{H}_b$ may be thought of as the space of velocities. With this in mind, it is easy to see how the energy is related to the inner products defined above.

For convenience, we set $\mathcal{I}_0 = (0, b_1)$, $\mathcal{I}_j = (b_j, b_{j+1})$, $1 \leq j < n$, $\mathcal{I}_n = (b_n, \infty)$, and let $\mathcal{I}$ denote the union of these intervals. We first formally write the equations of motion of the system as

$$\dot{w}(x,t) = v(x,t),$$
$$\dot{v}(x,t) = -w_{xxxx}(x,t), \quad x \in \mathcal{I},$$
$$\dot{v}_j(t) = \frac{1}{m_j}(w_{xxx}(b_j^+, t) - w_{xxx}(b_j^-, t)), \quad j \in \mathcal{M}.$$

Thus, we define an unbounded linear operator $\mathcal{A}$ on $\mathcal{H}$ by

$$\mathcal{A} = \begin{bmatrix} 0 & 1 & 0 & \cdots & 0 \\ -\frac{\partial^4}{\partial x^4} & 0 & 0 & \cdots & 0 \\ \mathcal{J}_{\alpha_1} & 0 & 0 & \cdots & 0 \\ \mathcal{J}_{\alpha_2} & 0 & 0 & \cdots & 0 \\ \vdots & \vdots & \vdots & \cdots & \vdots \\ \mathcal{J}_{\alpha_m} & 0 & 0 & \cdots & 0 \end{bmatrix},$$

where $\mathcal{J}_{\alpha_m} f = \frac{1}{m_j}(f_{xxx}(b_j^+) - f_{xxx}(b_j^-))$ and it is understood that the $\frac{\partial^4}{\partial x^4}$ term above acts only in $\mathcal{I}$. The domain of $\mathcal{A}$ is given by

$$\mathcal{D}(\mathcal{A}) = \{(w, v, v_{\alpha_1}, v_{\alpha_2}, \ldots, v_{\alpha_m})^T \in \mathcal{H} : w^{iv}|_{\mathcal{I}} \in L^2,$$
$$v \in \mathcal{H}_a, v_j = v(b_j), j \in \mathcal{M}\}.$$

In terms of $\mathcal{A}$, the equations of motion may be written as the ode in $\mathcal{H}$

$$\frac{dX}{dt} = \mathcal{A}X. \tag{2.1}$$

A routine verification shows that $\mathcal{A}$ is closed and densely defined. In fact, one can show that $i\mathcal{A}$ is self-adjoint. Thus, by Stone's Theorem, $\mathcal{A}$ is the infinitesimal generator of a strongly continuous 1 parameter family of unitary operators $\mathcal{U}(t)$ on $\mathcal{H}$. $\mathcal{U}(t)$ is the solution operator of the ode (2.1).

Some Useful Properties of $\mathcal{A}$. We now investigate the properties of $\mathcal{A}$ that yield the smoothing and local energy decay properties of the semi-infinite beam.

Lemma 2.2 $\mathcal{A}$ *has no eigenvalues.*

Proof: Suppose that $\mathcal{A}X = \lambda X$, where $X = (w, v, v_{\alpha_1}, v_{\alpha_2}, \ldots, v_{\alpha_m})^T \in \mathcal{D}(\mathcal{A})$. This gives

$$v = \lambda w, \tag{2.2}$$

$$-w^{(iv)} = \lambda v, \tag{2.3}$$

$$\mathcal{J}_j(w) = \lambda v(b_j), \quad j \in \mathcal{M}. \tag{2.4}$$

Suppose first that $\lambda = 0$. Then by (2.4), w''' is continuous at the locations of the point masses and thus $w^{(iv)} = 0$ everywhere except possibly at the pins. Also, since $w \in \mathcal{H}_a$, we have $w'' \in L^2$ so w must be linear for $x \geq b_n$. Integrating by parts, using the clamped end conditions and the conditions at the pins, gives

$$\sum_{j=0}^{n} \int_{\mathcal{I}_j} \bar{w} w^{(iv)} \, dx = \int_0^{\infty} |w''| \, dx.$$

Hence $w'' = 0$. Thus the clamped end conditions and the conditions at the pins imply that $w = 0$. Thus $X = 0$ and 0 cannot be an eigenvalue.

Suppose now that λ is a non-zero eigenvalue. Since $i\mathcal{A}$ is self-adjoint, the spectrum of $\mathcal{A}$ is purely imaginary so we can write $\lambda = \pm i\mu^2$, where $\mu > 0$. We obtain from (2.2), (2.3) and (2.4) that

$$w^{(iv)} - \mu^4 w = 0, \quad x \in \mathcal{I}, \tag{2.5}$$

$$\mathcal{J}_j w = -\mu^4 w, \quad j \in \mathcal{M}. \tag{2.6}$$

But $\lambda w = v \in L^2$ so $w(x) = Ce^{-\mu x}$ for $x \geq b_n$. We might as well assume that $C \geq 0$. Let f_j denote the restriction of w to $\mathcal{I}_j$. We claim that each f_j has the property that its value and all of its even order derivatives are non-negative, while all of its odd-order derivatives are non-positive, or that this can be arranged by multiplying w by -1. This is certainly true for $f_n(x) = Ce^{-\mu x}$. Suppose that the claim is true for f_j, f_{j+1}, $\ldots$, f_n. If the beam is pinned at b_j then $w(b_j) = 0$ and thus $w(x) = 0$ for $x \geq b_j$ because the claim implies that w is a non-negative, non-increasing function in that region. In this case, applying the conditions at the pin, we can write

$$f_{j-1}(x) = A(\sinh(b_j - x) - \sin(b_j - x)),$$

where we might as well assume that $A \geq 0$. Thus f_{j-1} satisfies the claim if the beam is pinned at b_j. Suppose now that b_j is the location of a point mass. We write

$$f_j(x) = c_1(\sinh(b_j - x) - \sin(b_j - x)) + c_2(\cosh(b_j - x) - \cos(b_j - x)) +$$
$$c_3(\sinh(b_j - x) + \sin(b_j - x)) + c_4(\cosh(b_j - x) + \cos(b_j - x)).$$

We remark that this is a linear combination of functions, each of which satisfies the claim on the interval $\mathcal{I}_{j-1}$. Applying the claim to f_j at $x = b_j$ shows that the constants c_1, c_2, c_3, and c_4 are all non-negative. Applying the conditions (2.5), (2.6) and the continuity of w and its first and second derivatives shows that f_{j-1} is given by the same expression defining f_j except that c_1 should be replaced by $c_1 + \mu c_4/m_j$. Thus f_{j-1} also satisfies the claim in this case. Hence, by induction, the claim is true.

Finally, we note that the claim shows that w is a non-negative, non-increasing function. But we also know that $w(0) = 0$, which implies that w must vanish. Thus, λ is not an eigenvalue.

Definition 2.3 *Let $\mu \geq 0$. We say that w satisfies the μ radiation condition if there exists $R \geq b_n$ such that in the interval (R, ∞) w is a linear combination of $e^{-\mu x}$ and $e^{-i\mu x}$ if $\mu > 0$, or w is linear if $\mu = 0$.*

The following result shows that $\mathcal{A}$ does not have a certain kind of generalized eigenvalue.

Lemma 2.4 *Let w satisfy the μ radiation condition and suppose that*

1. $w^{(iv)} - \mu^4 w = 0$ in $\mathcal{I}$,

2. $w(0) = w'(0) = 0$,

3. $w(b_j) = 0$ for $j \in \mathcal{P}$,

4. $\mathcal{J}_j w = -\mu^4 w(b_j)$ for $j \in \mathcal{M}$,

5. w, w' and w'' are all continuous at the points b_j.

Then $w = 0$.

Proof: If $\mu = 0$ then the μ radiation condition implies that w is linear for $x \geq R$. Hence, if w does not vanish then $(w, 0)^T$ is an eigenvector of $\mathcal{A}$, which is impossible by Lemma 2.2.

If $\mu > 0$ then on each interval $\mathcal{I}_j$ we have

$$0 = w^{(iv)}\bar{w} - \bar{w}^{(iv)}w = \frac{d}{dx}(w'''\bar{w} - \bar{w}'''w - w''\bar{w}' + \bar{w}''w')$$

$$= -2i \operatorname{Im}\frac{d}{dx}(w'''\bar{w} - w''\bar{w}').$$

Integrating this from 0 to $x > b_n$, we find that

$$0 = \text{Im}(w'''(x)\bar{w}(x) - w''(x)\bar{w}'(x)) + \text{Im}\sum_{j=1}^{n}(w'''(b_j^+) - w'''(b_j^-))\bar{w}(b_j).$$

But by conditions (3) and (4) of the statement of the Lemma, the summation term in the expression above vanishes. Since w satisfies the μ radiation condition, we can write $w(x) = c_1 e^{-\mu x} + c_2 e^{-i\mu x}$ for $x \geq R > b_n$. Thus

$$0 = \text{Im}(w'''(x)\bar{w}(x) - w''(x)\bar{w}'(x)) = 2\mu^3|c_2|^2.$$

Hence $c_2 = 0$. But this would imply that $(w, i\mu^2 w, i\mu^2 w(b_{\alpha_1}), \ldots, i\mu^2 w(b_{\alpha_m}))^T$ is an eigenvector of $\mathcal{A}$, unless $w = 0$. But be Lemma 2.2, w must vanish.

We now consider the resolvent operator of $\mathcal{A}$, $R(\lambda) = (\lambda - \mathcal{A})^{-1}$. It is clear that

$$R(\lambda)\begin{bmatrix} p \\ q \\ q_{\alpha_1} \\ \vdots \\ q_{\alpha_m} \end{bmatrix} = \begin{bmatrix} w \\ v \\ v_{\alpha_1} \\ \vdots \\ v_{\alpha_m} \end{bmatrix} \Leftrightarrow \begin{bmatrix} w \\ v \\ v_{\alpha_1} \\ \vdots \\ v_{\alpha_m} \end{bmatrix} \in \mathcal{D}(\mathcal{A})$$

and

$$p = \lambda w - v,$$
$$q(x) = \lambda v(x) + w^{(iv)}(x), \quad x \in \mathcal{I},$$
$$q_j = \lambda v_j - \mathcal{J}_j w, \quad j \in \mathcal{M}.$$

Since $v_j = v(b_j)$ for $j \in \mathcal{M}$, we obtain

$$w^{(iv)}(x) + \lambda^2 w(x) = \lambda p(x) + q(x), \quad x \in \mathcal{I}, \tag{2.7}$$
$$\mathcal{J}_j w - \lambda^2 w(b_j) = -\lambda p(b_j) - q_j, \quad j \in \mathcal{M}, \tag{2.8}$$
$$v = \lambda w - p. \tag{2.9}$$

This system may be solved by adding the solutions of two separate problems.

Problem 1

$$w^{(iv)}(x) + \lambda^2 w(x) = \tilde{q}(x), \quad x \in \mathcal{I}, \tag{2.10}$$
$$\mathcal{J}_j w - \lambda^2 w(b_j) = 0, \quad j \in \mathcal{M}, \tag{2.11}$$
$$w(b_j) = 0, \quad j \in \mathcal{P}, \tag{2.12}$$

where w, w' and w'' are continuous at each point b_j, and $\tilde{q} = \lambda p + q$.

Problem 2

$$w^{(iv)}(x) + \lambda^2 w(x) = 0, \quad x \in \mathcal{I}, \tag{2.13}$$
$$\mathcal{J}_j w - \lambda^2 w(b_j) = -\tilde{q}_j, \quad j \in \mathcal{M}, \tag{2.14}$$
$$w(b_j) = 0, \quad j \in \mathcal{P}, \tag{2.15}$$

where w, w' and w'' are continuous at each point b_j, and $\tilde{q}_j = \lambda p(b_j) + q_j$.

As in [2], we attempt to find a solution of each of these problems in terms of the Green's function g for the semi-infinite beam with no pins and no masses. This function g is given by the expression

$$g(x, a, k) = \frac{1}{4k^3}\left(\exp\left(\frac{\pi i}{4} - \frac{1+i}{\sqrt{2}}k|x-a|\right) + \exp\left(\frac{-\pi i}{4} - \frac{1-i}{\sqrt{2}}k|x-a|\right)\right.$$
$$+ \exp\left(\frac{-\pi i}{4} - \frac{1+i}{\sqrt{2}}k(x+a)\right) + \exp\left(\frac{\pi i}{4} - \frac{1-i}{\sqrt{2}}k(x+a)\right)$$
$$\left. - \sqrt{2}\exp\left(-\frac{1+i}{\sqrt{2}}kx - \frac{1-i}{\sqrt{2}}ka\right) - \sqrt{2}\exp\left(-\frac{1-i}{\sqrt{2}}kx - \frac{1+i}{\sqrt{2}}ka\right)\right).$$

This function g satisfies

$$\frac{\partial^4 g}{\partial x^4} + k^4 g = \delta(x - a),$$

$$g(0, a, k) = \frac{\partial g}{\partial x}(0, a, k) = 0,$$

$$g(x, a, k) \to 0 \text{ as } x \to \infty \text{ if } |\arg k| < \pi/4.$$

Further, g has a removable singularity at $k = 0$, which essentially makes it an entire function of k. As usual, g is constructed so that g, g_x and g_{xx} are continuous at all points on the line $x = a$, but

$$\frac{\partial^3 g}{\partial x^3}(a^+, a, k) - \frac{\partial^3 g}{\partial x^3}(a^-, a, k) = 1. \tag{2.16}$$

Recall that the displacement of the beam is also expected to have jumps in its third derivatives at the points b_j. This suggests that we might find a Green's function $h(x, a, k)$ for the solution of Problem 1 (with $\lambda = k^2$) in the form

$$h(x, a, k) = g(x, a, k) + \sum_{j=1}^{n} g(x, b_j, k)c_j(a, k). \tag{2.17}$$

Imposing the conditions (2.11), (2.12) yields the following system of equations for the coefficients c_j.

$$\sum_{j=1}^{n} g(b_l, b_j, k)c_j(a, k) - \frac{1}{m_l k^4} c_l(a, k) = -g(b_l, a, k), \quad l \in \mathcal{M}, \quad (2.18)$$

$$\sum_{j=1}^{n} g(b_l, b_j, k)c_j(a, k) = -g(b_l, a, k), \quad l \in \mathcal{P}. \quad (2.19)$$

In terms of h, the solution of Problem 1 is

$$w(x, k) = \int_0^\infty h(x, a, k)\tilde{q}(a)\, da. \quad (2.20)$$

Similarly, it is natural to seek a solution of Problem 2 in the form

$$w = \sum_{j=1}^{n} g(x, b_j, k)d_j(k). \quad (2.21)$$

Imposing the conditions (2.14), (2.15) yields the following system of equations for the coefficients d_j.

$$\sum_{j=1}^{n} g(b_l, b_j, k)d_j(k) - \frac{1}{m_l k^4} d_l(k) = \frac{\tilde{q}}{k^4}, \quad l \in \mathcal{M}, \quad (2.22)$$

$$\sum_{j=1}^{n} g(b_l, b_j, k)d_j(k) = 0, \quad l \in \mathcal{P}. \quad (2.23)$$

Both systems of equations have the same coefficient matrix

$$H(k) = G(k) - \frac{1}{k^4} J,$$

where $G(k)$ is the matrix with ij entry $g(b_i, b_j, k)$ and J is the diagonal matrix with jth diagonal entry

$$\begin{cases} \frac{1}{m_j}, & j \in \mathcal{M}, \\ 0, & j \in \mathcal{P}. \end{cases}$$

Lemma 2.5 $\det H(k) \neq 0$ for all k satisfying $|\arg k| \leq \pi/4$.

Proof: Let $P(x, k)$ be the matrix function obtained by replacing the first row of $H(k)$ by the vector with jth component $g(x, b_j, k)$ and let $p(x, k) = \det P(x, k)$.

If $i \in \mathcal{P}$ and $i > 1$ then $p(b_i) = 0$ because $p(b_i)$ is a determinant which has identical first and ith rows. Similarly, if $i \in \mathcal{M}$ and $i > 1$ then $\lambda^2 p(b_i) - \mathcal{J}_i p = 0$ because this may be written as a determinant, the ith row of which is a multiple of the first. Thus, p satisfies the conditions

1. $p^{(iv)} + k^4 p = 0$ in $\mathcal{I}$,

2. $p(0) = p'(0) = 0$,

3. $p(b_j) = 0$ for $j \in \mathcal{P}$ satisfying $j > 1$,

4. $\mathcal{J}_j p = k^4 p(b_j)$ for $j \in \mathcal{M}$ satisfying $j > 1$,

5. p, p' and p'' are all continuous at the points b_j.

Suppose that $\det H(k_0) = 0$. If $1 \in \mathcal{P}$ then this is equivalent to $p(b_1) = 0$, while if $1 \in \mathcal{M}$ then it is equivalent to $\mathcal{J}_1 p = k_0^4 p(b_1)$. Thus the restriction $j > 1$ in conditions (3) and (4) may be removed when $k = k_0$.

Now we proceed as in [2]. If $|\arg k_0| < \pi/4$ then

$$(p, k_0^2 p, k_0^2 p(b_{\alpha_1}), k_0^2 p(b_{\alpha_2}), \dots, k_0^2 p(b_{\alpha_m}))^T$$

is either zero or an eigenvector of $\mathcal{A}$. But by Lemma 2.2, it must be zero. Similarly, if $\arg k_0 = \pi/4$ then p satisfies the μ radiation condition, where $k_0 = (1 + i)\mu/\sqrt{2}$. But by Lemma 2.4, p must be zero. The same conclusion holds if $\arg k_0 = -\pi/4$ because then $\overline{p(x, k_0)}$ satisfies the μ radiation condition, where $k_0 = (1 - i)\mu/\sqrt{2}$.

But by (2.16), p must satisfy

$$p'''(b_1^+, k_0) - p'''(b_1^-, k_0) = \det H^{(2)}(k_0),$$

where $H^{(2)}(k)$ is the matrix obtained by deleting the first row and first column of $H(k)$. Since $p(x, k_0) = 0$, we see that $\det H^{(2)}(k_0) = 0$. The matrix $H^{(2)}(k)$ corresponds to the same physical problem without a pin or mass at b_1. We can continue in this way and reduce the problem to the simple case where there are no pins or masses, obtaining a contradiction.

Definition 2.6 *If $\eta > 0$, we let S_η denote the quadrant $|\arg k| \leq \pi/4$ shifted to the left by η units, i.e. $S_\eta = \{k : |\arg(k + \eta)| \leq \pi/4\}$.*

Lemma 2.7 *There are constants η_0, R_1, R_2 and R_3 such that for*

$$k \in S_{\eta_0} \cap \{z : |z| > R_3\}$$

we have

$$R_2 \geq |\det k^3 H(k)| \geq R_1.$$

Proof: By Theorem 1.6 of [2], there is a constant η such that the matrix G satisfies such an estimate in the whole set S_η. But $H(k) = G(k) - J/k^4$, and it is easy to check that given any $\epsilon > 0$ we can choose R_3 sufficiently large so that if $k \in S_\eta \cap \{z : |z| > R_3\}$ then the entries of the matrices $H(k)$ and $G(k)$ satisfy

$$|h_{ij} - g_{ij}| \leq \epsilon |g_{ij}|.$$

Thus we can make $\det k^3 H(k)$ uniformly as close as we please to $\det k^3 G(k)$ in $S_\eta \cap \{z : |z| > R_3\}$ by making R_3 sufficiently large.

Theorem 2.8 *There exists $\eta > 0$ such that systems (2.18), (2.19) and (2.22), (2.23) have solutions which are analytic functions of k in the set S_η (see Definition 2.6). Further, there exist constants C_1 and C_2 such that for all χ satisfying $0 \leq \chi \leq \eta$, k satisfying $|\arg k + \chi| = \pi/4$, x and a in $\mathcal{I}$, $i \geq 0$ and $j \geq 0$*

$$\left| \frac{\partial^{i+j}}{\partial x^i \partial a^j} h(x, a, k) \right| \leq C_1 e^{\chi C_2 (x+a)} (1 + |k|)^{i+j-3}. \tag{2.24}$$

Proof: For convenience, we relabel the points b so that $b_1, b_2, \ldots, b_m$ correspond to the locations of the masses and $b_{m+1}, b_{m+2}, \ldots, b_n$ correspond to the pins. Then the system (2.22), (2.23) may be rewritten

$$\frac{1}{m_l} d_l - \sum_{j=1}^{n} k^4 g(b_l, b_j, k) d_j = -\tilde{q}_l, \qquad 1 \leq l \leq m, \tag{2.25}$$

$$\sum_{j=1}^{n} g(b_l, b_j, k) d_j = 0, \qquad m + 1 \leq l \leq n. \tag{2.26}$$

We write

$$G = \left[\begin{array}{cc} G_{11} & G_{12} \\ G_{21} & G_{22} \end{array} \right],$$

where G_{11} is m by m and G_{22} is $n - m$ by $n - m$, and let

$$R = \text{diag}\{1/m_1, 1/m_2, \ldots, 1/m_m\}.$$

If we set $\tilde{q} = (\tilde{q}_1, \tilde{q}_2, \ldots, \tilde{q}_m)^T$, $x = (d_1, d_2, \ldots, d_m)^T$ and $y = (d_{m+1}, d_{m+2}, \ldots, d_n)^T$, then (2.25), (2.26) may be written

$$(R - k^4 G_{11})x - k^4 G_{12} y = -\tilde{q},$$
$$G_{21} x + G_{22} y = 0.$$

But, by Theorem 1.6 of [2], G_{22}, which corresponds to the same physical problem with $n - m$ pins and no masses, is analytic with an analytic inverse in a neighborhood of the origin. Thus, our system is equivalent to

$$(R - k^4 G_{11} + k^4 G_{12} G_{22}^{-1} G_{21})x = -\tilde{q},$$
$$y = -G_{22}^{-1} G_{21} x.$$

The first of these equations has a solution x which is an analytic function of k in a neighborhood of the origin (because R is invertible). Thus, the solution is analytic in a disk $|k| < \epsilon$. We can treat the system (2.18), (2.19) similarly.

Let η_0 and R_3 be the constants of Lemma 2.7. Then it is clear that the system (2.18), (2.19) has an analytic solution for k in the region $S_{\eta_0} \cap \{z : |z| > R_3\}$.

By Lemma 2.5, the systems have analytic solutions for k in the set $\{z : \epsilon \leq |z| \leq R_3, |\arg z| \leq \pi/4\}$ and thus there must be analytic solutions in a neighborhood $\mathcal{U}$ of this set.

Now we just observe that we can choose η so that S_η is contained in

$$\{z : |z| < \epsilon\} \cup \mathcal{U} \cup (S_{\eta_0} \cap \{z : |z| > R_3\}.$$

To obtain the estimate (2.24), one uses (2.17), (2.18), (2.19), the explicit expression of g, the estimate of Lemma 2.7 and Cramer's rule. We omit the estimate, which is not difficult. This completes the proof.

We now define, as in [2]

$$g_2(x, a, k) = \frac{1}{4k^5}\left(\exp\left(\frac{-\pi i}{4} - \frac{1+i}{\sqrt{2}}k(a-x)\right) + \exp\left(\frac{\pi i}{4} - \frac{1-i}{\sqrt{2}}k(a-x)\right)\right.$$

$$\exp\left(\frac{\pi i}{4} - \frac{1+i}{\sqrt{2}}k(x+a)\right) \quad \exp\left(\frac{-\pi i}{4} - \frac{1-i}{\sqrt{2}}k(x+a)\right)$$

$$\left. -\sqrt{2}i\exp\left(-\frac{1+i}{\sqrt{2}}kx - \frac{1-i}{\sqrt{2}}ka\right) + \sqrt{2}i\exp\left(-\frac{1-i}{\sqrt{2}}kx - \frac{1+i}{\sqrt{2}}ka\right)\right)$$

and $g_1(x, a, k) = \frac{\partial g_2}{\partial a}(x, a, k)$. For $x < a$, we have

$$\frac{\partial^2 g_2}{\partial a^2} = \frac{\partial g_1}{\partial a} = g.$$

We also define

$$h_1(x, a, k) = g_1(x, a, k) + \sum_{j=1}^{n} g(x, b_j, k)c_j^1(a, k),$$

$$h_2(x, a, k) = g_2(x, a, k) + \sum_{j=1}^{n} g(x, b_j, k)c_j^2(a, k),$$

where the coefficients c_j^1 (respectively c_j^2) are obtained by solving system (2.18), (2.19) with the function g_1 (respectively g_2) appearing on the right-hand-side instead of g. Then $\frac{\partial h_2}{\partial a} = h_1$ and for $x < a$ and $b_n \leq a$, we have

$$\frac{\partial^2 h_2}{\partial a^2} = \frac{\partial h_1}{\partial a} = h.$$

While g has a removable singularity at $k = 0$, the function g_1 has a second order pole and g_2 has a third order pole. Consequently, $k^3 h_2$ and $k^2 h_1$ both have removable singularities at $k = 0$. It is easy to check that these functions satisfy an estimate similar to (2.24) in the region S_η of Theorem 2.8. In fact, in the notation of Theorem 2.8,

$$\left| \frac{\partial^{i+j}}{\partial x^i \partial a^j} k^3 h_2(x, a, k) \right| \leq C_1 e^{\chi C_2(x+a)} (1 + |k|)^{i+j-2}, \qquad (2.27)$$

$$\left| \frac{\partial^{i+j}}{\partial x^i \partial a^j} k^2 h_1(x, a, k) \right| \leq C_1 e^{\chi C_2(x+a)} (1 + |k|)^{i+j-2}. \qquad (2.28)$$

Local Smoothing and Energy Decay Properties of the Beam and Mass System. The following theorem asserts that if the initial data decays sufficiently rapidly then the solution of the initial value problem for the beam-mass system is pointwise infinitely differentiable for $t > 0$. This is enough to establish a local energy decay for arbitrarily small $t > 0$ (see the proof of the controllability of the finite beam, Theorem 3.1), but we also establish estimates for the local decay of energy as $t \to \infty$ (see Theorem 2.11). We give only a sketch of the proof of Theorems 2.9 and 2.11. For full details of the proof, the reader is invited to see Theorems 1.8 and 1.9 of [2], which have very similar proofs.

Theorem 2.9 (Local Smoothing Property). *Let* $(p, q, q_{\alpha_1}, q_{\alpha_2}, \ldots, q_{\alpha_m})^T$ *belong to the Finite Energy Space* $\mathcal{H}$ *and suppose that there is a constant* α *such that* $e^{\alpha x} p''(x)$ *and* $e^{\alpha x} q(x)$ *are in* $L^2(0, \infty)$. *Let*

$$\begin{bmatrix} w(., t) \\ v(., t) \\ v_{\alpha_1} \\ v_{\alpha_2} \\ \vdots \\ v_{\alpha_m} \end{bmatrix} = \mathcal{U}(t) \begin{bmatrix} p \\ q \\ q_{\alpha_1} \\ q_{\alpha_2} \\ \vdots \\ q_{\alpha_m} \end{bmatrix}$$

Then for $t > 0$ *and* $i \in \mathcal{M}$ *we have* $v_i(t) = v(b_i, t)$, $v = w_t$, *and* $w(x, t)$ *and* $v(x, t)$ *are smooth functions of* x *and* t *in the following sense:*

1. w *and* v *belong to* $C^\infty(\mathcal{I} \times (0, \infty))$, *and for each bounded set* K *of* $\mathcal{I}$ *and each interval* $[t_0, t_1]$ *with* $0 < t_0 < t_1$ *there exist constants* c *and* θ *such that for* $(x, t) \in K \times [t_0, t_1]$, *and all* r, s

$$\left| \frac{\partial^{r+s} w}{\partial t^r \partial x^s} \right| \leq c\theta^{2r+s} (2r + s)!$$

2. *For all* r, s *such that* $s \neq 3 \bmod 4$, *we have*

$$\frac{\partial^{r+s} w}{\partial t^r \partial x^s}(b_j^-, t) = \frac{\partial^{r+s} w}{\partial t^r \partial x^s}(b_j^+, t), \quad j \in \mathcal{M}, \quad t > 0.$$

Definition 2.10 *Let $R > 0$. For finite energy solutions*

$$
\begin{bmatrix}
w(\cdot, t) \\
v(\cdot, t) \\
v_{\alpha_1} \\
v_{\alpha_2} \\
\vdots \\
v_{\alpha_m}
\end{bmatrix}
= \mathcal{U}(t)
\begin{bmatrix}
p \\
q \\
q_{\alpha_1} \\
q_{\alpha_2} \\
\vdots \\
q_{\alpha_m}
\end{bmatrix}
$$

of the Beam-Mass Problem, we define the local energy

$$
\mathcal{E}_R(t) = \frac{1}{2} \int_0^R |v(x,t)|^2 + |w_{xx}(x,t)|^2 \, dx + \frac{1}{2} \sum_{j \in \mathcal{M}} m_j |v_j(t)|^2.
$$

Theorem 2.11 (Local Energy Decay). *Suppose that the conditions of Theorem 2.9 are satisfied. Then for each $t_0 > 0$ there exists a constant B depending only on R, t_0 and α such that for $t \geq t_0$*

$$
\mathcal{E}_R(t) \leq Bt^{-3}(\|e^{\alpha x}p''\|_{L^1}^2 + \|e^{\alpha x}q\|_{L^1}^2 + \sum_{j \in \mathcal{M}} |q_j|^2).
$$

Proof of Theorems 2.9 and 2.11: (Sketch). We start with the standard Inverse Laplace Transform formula which holds for all strongly continuous semigroups (see [4], for example).

$$
\int_0^t (t-s)\mathcal{U}(s)X \, ds = \frac{1}{2\pi i} \int_{\gamma - i\infty}^{\gamma + i\infty} e^{\lambda t} \lambda^{-2} R(\lambda) X \, d\lambda.
$$

This is valid for all $t > 0$, $X \in \mathcal{H}$, and (since the spectrum of $\mathcal{A}$ lies on the imaginary axis) $\gamma > 0$. The first component of this equation is

$$
\int_0^t (t-s)w(x,s) \, ds = \frac{1}{2\pi i} \int_{\gamma - i\infty}^{\gamma + i\infty} e^{\lambda t} \lambda^{-2} w_\lambda(x) \, d\lambda, \tag{2.29}
$$

where w_λ is obtained by adding together the solutions of Problems 1 and 2 at the beginning of this section. Thus,

$$
w_\lambda(x) = \int_0^\infty h(x, a, \lambda^{1/2})(q(a) + \lambda p(a)) \, da
$$

$$
+ \sum_{j=1}^n g(x, b_j, \lambda^{1/2}) d_j(\lambda^{1/2}), \tag{2.30}
$$

where the coefficient functions d_j are solutions of (2.22), (2.23).

The idea of the proof is to use our estimates for h and the functions d_j to shift the path of contour integration in (2.29) to a path in the left half-plane. This can be done because the image of the region S_η (see Definition 2.6) under the mapping $k \to k^2 = \lambda$ is contained in the left half-plane. Once this change of contour is done, (2.29) can be differentiated as often as we please with respect to t (because the exponential term in the integral decays). Note that the shift in path of the integral cannot be done for arbitrary initial data in the finite energy space, because the spectrum of $\mathcal{A}$ lies on the imaginary axis.

The new contour integrals are used directly to obtain the estimates for the derivatives and the local energy decay. For more specific details, see the proofs of Theorems 1.8 and 1.9 in [2].

3 Boundary Control of the Finite Beam

Consider a finite beam and mass system identical to the semi-infinite system except for the fact that the beam ends at $x = L = b_n$, where there is either a mass attached to the beam, or the beam is pinned. Here, we consider the possibility of controlling the beam-mass system by applying a torque to the end $x = L$. We prove that given any initial data with finite energy, and given $T > 0$, one can find a control torque which, when applied to the free end of the beam, brings the beam to rest by time T.

We continue with the notation of the last section except that we are now working on the bounded set

$$\mathcal{K} = \bigcup_{j=0}^{n-1} \mathcal{I}_j = (0, b_1) \cup \bigcup_{j=1}^{n-1} (b_j, b_{j+1}).$$

The beam-mass system satisfies the following system of equations.

$$\frac{\partial^2 w}{\partial t^2} + \frac{\partial^4 w}{\partial x^4} = 0, \quad x \in \mathcal{K}, \tag{3.31}$$

$$m_j \frac{dv_j}{dt} = \frac{\partial^3 w}{\partial x^3}(b_j^+, t) - \frac{\partial^3 w}{\partial x^3}(b_j^-, t), \quad j \in \mathcal{M}, \quad j \neq n, \tag{3.32}$$

$$w(b_j, t) = 0, \quad j \in \mathcal{P}, \tag{3.33}$$

$$w(0, t) = \frac{\partial w}{\partial x}(0, t) = 0, \tag{3.34}$$

$$w(x, 0) = p(x), \; w_t(x, 0) = q(x), \tag{3.35}$$

$$v_j(0) = q_j, \quad j \in \mathcal{M}, \tag{3.36}$$

$$m_j \frac{dv_j}{dt} = -\frac{\partial^3 w}{\partial x^3}(b_n^-, t), \quad n \in \mathcal{M}, \tag{3.37}$$

$$w(b_n, t) = 0, \quad n \in \mathcal{P}, \tag{3.38}$$

$$\frac{\partial^2 w}{\partial x^2}(b_n, t) = \tau(t). \tag{3.39}$$

Here, only one of equations (3.37) and (3.38) holds, depending on whether there is a mass attached to the end of the beam, or the beam is pinned at its end. Equation (3.39) represents the applied torque at the end of the beam. As before, we have the conditions that w, w_x, w_{xx} are continuous at the points b_j. The energy of the system is given by

$$\mathcal{E}_f(t) = \frac{1}{2} \int_0^\infty \left| \frac{\partial w}{\partial t} \right|^2 + \left| \frac{\partial^2 w}{\partial x^2} \right|^2 \, dx + \frac{1}{2} \sum_{j \in \mathcal{M}} m_j |v_j|^2.$$

Our controllability result concerns weak solutions of the system for which the energy remains bounded. We call these solutions finite energy solutions. We do not give any further details of weak solutions here. [2] has more details of this.

Theorem 3.1 *Given any initial data* $(p, q, q_{\alpha_1}, q_{\alpha_2}, \ldots, q_{\alpha_m})$ *of finite energy, and* $T > 0$, *there exists a torque control function* τ *which is an infinitely differentiable function of* t *for* $t \in (0, T)$, *such that the solution* w, v_j, $j \in \mathcal{M}$, *satisfies* $w(x, T) = w_t(x, T) = 0$, $v_j(T) = 0$, $j \in \mathcal{M}$. *Further, because of the way the control function is chosen, there is a constant* C *such that, for each set of initial data, the energy of such controlled solutions satisfies* $\mathcal{E}_f(t) \leq C\mathcal{E}(0)$ *for* $0 \leq t \leq T$. *The constant* C, *which depends on* T, *tends to 1 as* $T \to \infty$.

Proof: The proof is similar to the corresponding proof in [2], but we sketch it here to show the reader how our properties of the semi-infinite beam and mass system can be used.

First, we consider a subspace $\mathcal{S}$ of the Hilbert space $\mathcal{H}$ of the semi-infinite beam and mass system.

$$\mathcal{S} = \{(p, q, q_{\alpha_1}, q_{\alpha_2}, \ldots, q_{\alpha_m}) \in \mathcal{H} : p''(x) = q(x) = 0 \text{ for } x \geq b_n\}$$

The idea of the proof is to show that we can extend the initial data of the "finite problem" into the set $x \geq b_n$ in such a way that the projection onto $\mathcal{S}$ of the solution of the "semi-infinite problem" vanishes at time T. Since this projection corresponds to the values of the solution for $0 \leq x \leq b_n$, we obtain the desired solution of the control problem by using (3.39) to define the control function. We now show that such an extension of the initial data exists.

First, let g denote the initial data, extended to be in $\mathcal{S}$. Let P denote the projection onto $\mathcal{S}$ and let $U = \mathcal{U}(T)$, where $\mathcal{U}(t)$ is the unitary group which solves the "semi-infinite problem." Consider the equation

$$\tilde{h} - PU^{-1}PUP\tilde{h} = g$$

Suppose that this can be solved and set $h = P\tilde{h} - U^{-1}PUP\tilde{h}$. Then $PUh = 0$ and $Ph = g$. Thus h agrees with the initial data g for $x \leq b_n$ and the solution with initial data h vanishes for $x \leq b_n$ at time T. Thus, h is the desired extension of g. If we solve for h in terms of g, we obtain $h = Rg$, where

$$R = (P - U^{-1}PUP)(I - PU^{-1}PUP)^{-1}.$$

For this to make sense, it is clearly enough to show that PUP is a contraction. Clearly $\|PUP\| \leq 1$. By the smoothing property, Theorem 2.9, PUP is compact. If we assume that $\|PUP\| = 1$, then we can use the compactness to show that the set

$$V = \{z \in \mathcal{S} : Uz \in \mathcal{S}\}$$

is non-empty. V is finite dimensional because it is contained in the kernel of $I - PU^{-1}PUP$ and PUP is compact. Also, if $z \in V$ then $Uz \in \mathcal{D}(\mathcal{A})$ because, by the smoothing property, it is smooth enough to be in $\mathcal{D}(\mathcal{A})$, but since it is in $\mathcal{S}$, it must be in $\mathcal{D}(\mathcal{A})$. $Uz \in \mathcal{D}(\mathcal{A})$ implies that $z \in \mathcal{D}(\mathcal{A})$, so V is a subset of $\mathcal{D}(\mathcal{A})$. Thus, $\mathcal{A}$ is a bounded operator on the finite dimensional space V, and as such, must possess an eigenvalue. But by Lemma 2.2, this is not true. Thus, PUP must be a contraction.

Finally, the fact that $C \to 1$ as $T \to \infty$ follows from the local energy decay (Theorem 2.11), which implies that $\|PUP\| \to 0$ as $T \to \infty$. This completes the proof.

Remark. A small modification of the procedure above gives another proof of the controllability of the "Hybrid System of Elasticity" investigated by Littman and Markus [1], in which a rigid body attached to a flexible mast is studied. Indeed, here we get controllability with just one controller (torque), while the proof of the result in [1], requires two controllers (both force and torque).

Acknowledgement

The author would like to take this opportunity to thank Ken Bowers and John Lund for inviting him to participate in "Computation and Control IV," and to congratulate them and all others involved for making it a very successful event.

References

[1] W. Littman and L. Markus, "Exact Boundary Controllability of a Hybrid System of Elasticity," *Archive for Rational Mechanics and Analysis*, Vol 103, 1988, pp 193–236.

[2] W. Littman and S. W. Taylor, "Local Smoothing and Energy Decay for a Semi-Infinite Beam Pinned at Several Points and Applications to Boundary Control," in "Differential Equations, Dynamical Systems and Control Science", *Lecture Notes in Pure and Applied Mathematics*, Vol. 152, Dekker, NY, 1994, pp. 683–704.

[3] W. Littman, S. W. Taylor. "Smoothing Evolution Equations and Boundary Control Theory", *J. d'Analyse Math..*, 59, 1992, pp 117-131.

[4] A. Pazy, *Semigroups of Linear Operators and Applications to Partial Differential Equations*, Springer-Verlag, 1983.

A MULTIGRID METHOD FOR TOTAL VARIATION-BASED IMAGE DENOISING

C. R. Vogel *

Department of Mathematical Sciences
Montana State University
Bozeman, Montana 59717
e-mail: **vogel@math.montana.edu**

1 Introduction

By *image reconstruction* we mean obtaining the solution u of an operator equation of the form

$$Au = z, \tag{1.1}$$

where A is typically (but not necessarily) linear, and the data z is assumed to be contaminated by error. Two important special cases are: (i) *Denoising*, or noise removal, where $Au = u$; and (ii) *Deblurring*, where A is a Fredholm first kind integral operator

$$(Au)(x) = \int_{\Omega} k(x, y)u(y)\,dy, \quad x \in \Omega, \tag{1.2}$$

which models "blurring" processes involved in image formation. See [9] for details.

Standard linear image reconstruction methods tend to do poorly when u is not smooth, e.g., when it has jump discontinuities. To overcome this difficulty Rudin, Osher, and Fatemi [14] considered the least squares-constrained Total Variation (TV) minimization problem

$$\min_{u} TV(u) \stackrel{\text{def}}{=} \int_{\Omega} |\nabla u| \tag{1.3}$$

subject to

$$\|Au - z\|^2 = \sigma^2, \tag{1.4}$$

where σ is the variance of the error in the data, which is assumed to be known. Computational experiments and mathematical analysis [5] have shown that this approach works well when the solution u is "blocky", e.g., it is almost piecewise constant with jump discontinuities or sharp gradients separating regions where it is nearly constant. This is the case in many important biomedical image processing applications.

Several approaches to solving (1.3)-(1.4) have been proposed. In [14] artificial time evolution was used to solve the PDE obtained from the associated Euler-Lagrange equations. This yields a gradient descent algorithm

---

*Supported in part by the NSF under Grant DMS-9303222.

which is very slow to converge. Li and Santosa [10] have recently proposed
an affine scaling algorithm, which is based on interior point methods in linear programming. Ito and Kunisch [8] have proposed an active set strategy
based on an augmented Lagrangian formulation.

In [13] the authors proposed the following approach: First apply a
penalty method to problem (1.3)-(1.4), minimizing the TV-penalized least
squares functional

$$\frac{1}{2}\|Au - z\|^2 + \alpha \int_\Omega \sqrt{|\nabla u|^2 + \beta}. \tag{1.5}$$

Here α is a positive penalty, or trade-off, parameter, and the integral term
is a smooth approximation to the TV functional, defined in (1.3). β is also a
positive parameter. The well-posedness of this unconstrained minimization
problem is demonstrated in [1].

To minimize (1.5), the authors introduced in [13] the fixed point iteration

$$\left(A^*A + L(u^{(\nu)})\right) u^{(\nu+1)} = A^*z, \quad \nu = 0, 1, \ldots, \tag{1.6}$$

where L is the differential operator

$$L(u)v = -\nabla \cdot (\kappa \nabla v) \tag{1.7}$$

with the diffusion coefficient

$$\kappa = \kappa(u; \alpha, \beta) = \frac{\alpha}{\sqrt{|\nabla u|^2 + \beta^2}} \tag{1.8}$$

and homogeneous Neumann (no flux) boundary conditions. For obvious
reasons, this was dubbed the "lagged diffusivity" fixed point iteration. Although no rigorous analysis has carried out, this iteration seems to be
rapidly convergent provided the parameter β is not too small.

An important subproblem in this fixed point iteration is the solution
to the linear operator equation (1.6). When A is an integral operator, i.e.,
in the case of deblurring, this is an integro-differential equation. For this
case, an efficient preconditioned conjugate gradient (CG) scheme [2] to solve
(1.6) is introduced in [12]. Application of the preconditioner requires the
solution of a linear differential equation of the form

$$(\gamma + L)u = z, \tag{1.9}$$

where γ is an appropriately chosen positive constant. When $\gamma = 1$ and
$Au = u$, i.e., in the case of denoising, equations (1.9) and (1.6) are of
identical form.

In one space dimension, standard Finite Element and Finite Difference
discretizations of (1.7)-(1.9) yield tridiagonal linear systems, whose inversion can be performed with $\mathcal{O}(n)$ computational complexity, where n is

the number of unknowns in the system. In higher space dimensions the choices of both the discretization and iterative methods to solve the resulting linear system become very important issues. In the 2-dimensional biomedical image processing applications that the author has encountered, the number of pixels n in the images is on the order of 10^5 to 10^6. Standard 2-dimensional Finite Element and Finite Difference discretizations of (1.9) yield penta-diagonal systems with bandwidth $\approx \sqrt{n}$. Direct band solvers for these systems have $\mathcal{O}(n^{3/2})$ complexity.

Multigrid methods can be used to solve the discrete systems arising from certain positive elliptic PDE's, like Poisson's equation, with $\mathcal{O}(n)$ complexity (see [11]). The author attempted to use conventional multigrid methods to solve systems (1.7)-(1.9), but found convergence to be extremely slow when u was not smooth. This poor performance is due to the nonsmoothness of the diffusion coefficient. To overcome the difficulties associated with poorly behaved diffusion coeffients, the author applied techniques originally developed for numerical simulation of flow through porous media (see [6] and the references therein)—in particular, a cell-centered finite difference (CCFD) discretization and a variant of the multigrid algorithm introduced by Ewing and Shen [7] to solve the resulting discrete linear system. A unique feature of this multigrid algorithm is the use of nonstandard itergrid transfer operators.

The CCFD discretization method and its implementation are discussed in the following section. This is followed in section 3 by a discussion of Ewing and Shen's multigrid method. With a variant of this multigrid method used as a preconditioner, the CG method [2] is applied to solve the large sparse systems arising from CCFD discretization of (1.7)-(1.9). For numerical results, see [13, 12].

2   Cell-Centered Finite Difference Discretization

Assume the region Ω is the unit square in R^2. The second order elliptic PDE (1.7)-(1.9) can be written in first order system form

$$\frac{1}{\kappa}\vec{q} - \nabla u \;=\; \vec{0}, \tag{2.1}$$

$$-\nabla \cdot \vec{q} + u \;=\; z. \tag{2.2}$$

If the diffusivity κ is discontinuous across a curve Γ, then so is the solution u. On the other hand, the component of the flux $\vec{q} = -\kappa\nabla u$ which is normal to Γ must be smooth. This motivates the use of the CCFD discretization. We adopt some of the notation in [6] (WARNING: our u corresponds to their pressure variable p, while our flux $\vec{q}$ corresponds to their u).

Construct a regular grid in the following manner (See Figure 1): Let n_x. n_y denote the number of cells (i.e., pixels) in the $x-$ and $y-$directions,

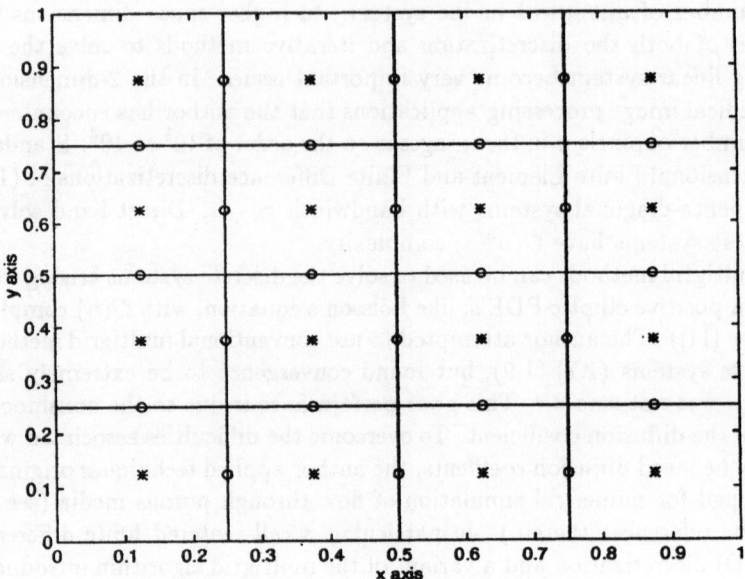

Figure 2.1: CCFD grid in 2 space dimensions. Stars (*) indicate cell centers (x_i, y_j). Circles (o) indicate interior x−edge midpoints $(x_{i\pm1/2}, y_j)$ and y−edge midpoints $(x_i, y_{j\pm1/2})$.

respectively. The total number of cells (pixels) is $n = n_x n_y$. With $h_x = 1/n_x$ and $h_y = 1/n_y$, the cell centers are given by

$$x_i = (i - 1/2)h_x, \quad i = 1, 2, \ldots, n_x, \tag{2.3}$$

$$y_j = (j - 1/2)h_y, \quad j = 1, 2, \ldots, n_y. \tag{2.4}$$

Let

$$x_{i\pm1/2} = x_i \pm h_x/2, \quad y_{j\pm1/2} = x_j \pm h_y/2. \tag{2.5}$$

The ij^{th} cell is given by

$$e_{ij} = \{(x, y) : x_{i-1/2} < x < x_{i+1/2}, \ y_{j-1/2} < y < y_{j+1/2}\} \tag{2.6}$$

and is centered at (x_i, y_j) and has area

$$|e_{ij}| = h_x h_y. \tag{2.7}$$

Take an approximation to u of the form

$$u(x, y) \approx \sum_{i=1}^{n_x} \sum_{j=1}^{n_y} u_{i,j} \chi_i(x) \, \chi_j(y), \tag{2.8}$$

where $\chi_i(x)$, $\chi_j(y)$ denote the indicator functions for the intervals $x_{i-1/2} < x < x_{i+1/2}$ and $y_{j-1/2} < y < y_{j+1/2}$, respectively. Take approximations to the components q^x and q^y of the flux $\vec{q}$ of the form

$$q^x(x, y) \approx \sum_{i=1}^{n_x-1} \sum_{j=1}^{n_y} u_{i,j} \phi_{i+1/2}(x) \chi_j(y), \tag{2.9}$$

$$q^x(x, y) \approx \sum_{i=1}^{n_x} \sum_{j=1}^{n_y-1} u_{i,j} \chi_i(x) \phi_{j+1/2}(y), \tag{2.10}$$

where the $\phi_{i+1/2}(x)$ are continuous, piecewise linear, and satisfy

$$\phi_{i+1/2}(x_{k+1/2}) = \delta_{ik}, \tag{2.11}$$

and similarly for the $\phi_{j+1/2}(y)$. Note that the no flux boundary conditions are built into the components of $\vec{q}$, since $\phi_{k+1/2}(0) = \phi_{k+1/2}(1) = 0$ for all relevant indices k.

Now substitute the approximations (2.8)-(2.10) into (2.1)-(2.2), multiply the first and second components of (2.1) by $\phi_{\mu+1/2}(x)\chi_\nu(y)$ and $\chi_\mu(x)\phi_{\nu+1/2}(y)$, respectively, and multiply (2.2) by $\chi_\mu(x)\chi_\nu(y)$. Then integrate over Ω and apply the midpoint quadrature approximation

$$\int\int_{e_{ij}} f(x, y) \, dx, dy \approx f(x_i, y_j) |e_{ij}|. \tag{2.12}$$

The resulting system can be expressed in matrix form

$$\begin{bmatrix} D^{-1}(\kappa^x) & 0 \\ 0 & D^{-1}(\kappa^y) \end{bmatrix} \begin{bmatrix} Q^x \\ Q^y \end{bmatrix} - \begin{bmatrix} B^x \\ B^y \end{bmatrix} U = \begin{bmatrix} 0 \\ 0 \end{bmatrix} \tag{2.13}$$

$$\begin{bmatrix} (B^x)^T & (B^y)^T \end{bmatrix} \begin{bmatrix} Q^x \\ Q^y \end{bmatrix} + EU = F, \tag{2.14}$$

where the entry of F corresponding to the ij^{th} cell is $z_{ij}|e_{ij}|$, E is the $n \times n$ diagonal matrix with diagonal entries $|e_{ij}|$, and $D(\kappa^x)$ is an $(n_x - 1) \times n_y$ diagonal matrix whose diagonal entries $\kappa_{i\pm1/2,j}$ correspond to the diffusivity κ evaluated at the interior x-edge midpoints $(x_{i\pm1/2}, y_j)$ (see Figure 1), and similarly for $D(\kappa^y)$. Eliminating Q^x, Q^y yields

$$AU = F, \tag{2.15}$$

where

$$A = (B^x)^T D(\kappa^x) B^x + (B^y)^T D(\kappa^y) B^y + E. \tag{2.16}$$

Although it is sparse (in fact, penta-diagonal), the matrix A need not be explicitly computed. All that is necessary in the implementation of the iterative methods discussed in the following section is the application of A. Given a grid function V with V_{ij} representing its value at the center of the ij^{th} cell,

$$
\begin{aligned}
[AV]_{ij} &= (\kappa^x_{i-1/2,j} + \kappa^x_{i+1/2,j} + \kappa^y_{i,j-1/2} + \kappa^y_{i,j+1/2} + |\epsilon_{ij}|)V_{ij} \qquad (2.17) \\
&\quad - \kappa^x_{i-1/2,j}V_{i-1,j} - \kappa^x_{i+1/2,j}V_{i+1,j} - \kappa^y_{i,j-1/2}V_{i,j-1} - \kappa^y_{i,j+1/2}V_{i,j+1}.
\end{aligned}
$$

3   The CCFD Multigrid Algorithm

We first discuss a simple two-grid version of the algorithm. The multigrid algorithm can then be defined recursively (see [11], [7]). Let the lower case h, n, i, j denote the fine grid and the upper case H, N, I, J denote the coarse grid. Hence fine grid cells are e^h_{ij}, $i = 1, \ldots, n_x$, $j = 1, \ldots, n_y$ and coarse grid cells are e^H_{IJ}, $I = 1, \ldots, N_x$, $J = 1, \ldots, N_y$.

Let Π^h_H denote the (coarse-to-fine grid) prolongation operator induced by the piecewise constant Finite Element approximation (2.8), i.e.,

$$
[\Pi^h_H V^H]_{ij} = V^H_{IJ} \quad \text{whenever} \quad e^h_{ij} \subset e^H_{IJ}. \qquad (3.1)
$$

The adjoint of this operator, which we denote by Π^H_h, maps fine grid functions to coarse grid functions. In terms of matrix representations,

$$
\Pi^H_h = c\left(\Pi^h_H\right)^T, \quad c = c(h, H) = \frac{|e^h_{ij}|}{|e^H_{IJ}|} = \frac{h_x h_y}{H_x H_y}. \qquad (3.2)
$$

Now let

$$
A^h V^h = F^h \qquad (3.3)
$$

denote the fine grid version of (2.16)-(2.17). The corresponding coarse grid equation is

$$
A^H V^H = F^H, \qquad (3.4)
$$

where

$$
A^H = \Pi^H_h A^h \Pi^h_H \qquad (3.5)
$$

and

$$
F^H = \Pi^H_h F^H. \qquad (3.6)
$$

The components of F^H are simply arithmetic averages of components of F^h. For example, when $h_x = h_y = 1/4$ (as in Figure 1) and $H_x = H_y = 1/2$, then

$$
F^H_{1,1} = \frac{1}{4}\left(F^h_{1,1} + F^h_{1,2} + F^h_{2,1} + F^h_{2,2}\right). \qquad (3.7)
$$

The application of the coarse-grid operator A^H is virtually identical to the application of the fine grid operator, c.f., equation (2.17). One must replace

$|e^h_{ij}| = h_x h_y$ in (2.17) by $|e^H_{IJ}| = H_x H_y$, and one must transfer values of κ^x and κ^y from the fine to the coarse grid. This is done for κ^x by injection in the x-indices and simple arithmetic averaging in the y-indices, and vice versa for κ^y.

We can now formulate a two-grid algorithm:

1. Perform ν_1 pre-smoothing sweeps for the system $A^h U^h = F^h$.

2. Compute the residual $R^h = F^h - AU^h$.

3. Restrict the residual to the coarse grid, $R^H = \Pi^H_h R^h$.

4. Solve the coarse-grid error equation $A^H E^H = R^H$.

5. Prolongate the error to the fine grid and update the solution, $U^h := U^h + \Pi^h_H E^H$.

6. Perform ν_2 post-smoothing sweeps for the system $A^h U^h = F^h$.

A multigrid V-cycle is obtained by replacing the coarse-grid solution step 4 with a recursive call to the the algorithm.

In [7] numerical results are presented for a number of combinations of pre- and post-smoothers and smoothing steps ν_1 and ν_2. We tried most of these combinations for our systems (1.7)-(1.9) but had only mixed success. We then decided to use preconditioned CG with a CCFD multigrid algorithm as a preconditioner. This required that the combination of pre- and post-smoothing steps preserve symmetry (Note that A is symmetric positive definite, c.f., (2.16)). We settled on red-black Gauss-Seidel (see [11]) as the smoother, with red-then-black sweeps for the pre-smoother and the same number of black-then-red sweeps (so $\nu_1 = \nu_2$) for the post-smoother. This approach produced an algorithm which, while sometimes not as efficient as the "pure multigrid" approach in [7], was still quite efficient and very robust.

Computational Complexity. Due to the sparsity of A (c.f., equation (2.17)), its application is an $\mathcal{O}(n)$ operation. For the same reason, the pre- and post-smoothing steps have $\mathcal{O}(N)$ complexity. Hence, the application of a single V-cycle and a single (preconditioned) CG iteration requires $\mathcal{O}(n)$ arithmetic operations. If one could verify that the number of CG iterations required to solve system (1.7)-(1.9) to within a specified tolerance is independent of mesh size, then the solution of this system would also require $\mathcal{O}(n)$ operations (see [11]). The analysis in [3, 4] suggests that this could be done provided one could obtain bounds

$$0 < \kappa_{min} \le \kappa(x, y) \le \kappa_{max} < \infty, \quad \text{for all } (x, y) \in \Omega. \qquad (3.8)$$

From (1.8), $\kappa(x, y) \leq \alpha/\beta$, but no uniform lower bound can exist without an upper bound on $|\nabla u|$. Such a bound cannot exist when u has jump discontinuities. On the other hand, the minimizer of (1.5) may be smooth, or perhaps a mesh-independent rate of convergence can be obtained without a uniform lower bound. In any event, our numerical results have been very positive (see [12]).

Acknowledgment

The author wishes to thank Jian Shen of the Institute for Scientific Computing at Texas A & M University. Without his ground-breaking work and helpful comments, this paper would not have been possible.

References

[1] R. Acar and C. R. Vogel, *Analysis of Total Variation penalty methods for ill-posed problems*, Inverse Problems, to appear.

[2] O. Axelsson and V.A. Barker, *Finite Element Solution of Boundary Value Problems*, Academic Press, 1984.

[3] J. H. Bramble, J. E. Pasciak, and J. Xu, *The analysis of multigrid algorithms with non-nested spaces or non-inherited quadratic forms*, Math. Comp., vol. 56 (1991), pp. 1-34.

[4] J. H. Bramble, R. E. Ewing, J. E. Pasciak, and J. Shen, *The analysis of multigrid algorithms for cell centered finite difference methods*, preprint, Institute for Scientific Computation, Texas A & M University (1994).

[5] D. Dobson and S. Santosa, *Recovery of blocky images from noisy and blurred data*, Tech. Report No. 94-7, Center for the Mathematics of Waves, University of Delaware (1994).

[6] R. E. Ewing and J. Shen, *A discretization scheme and error estimate for second-order elliptic problems with discontinuous coefficients*, preprint, Institute for Scientific Computation, Texas A & M University.

[7] R. E. Ewing and J. Shen, *A multigrid algorithm for the cell-centered finite difference scheme*, in the Proceeding of the 6th Copper Mountain Conference on Multigrid Methods, April 1993.

[8] K. Ito and K. Kunisch, *An active set strategy for image restoration based on the augmented Lagrangian formulation*, preprint, Center for Research in Scientific Computing, North Carolina State University (1994).

[9] A. K. Jain, *Fundamentals of Digital Image Processing*, Prentice-Hall, 1988.

[10] Y. Li and F. Santosa, *An affine scaling algorithm for minimizing Total Variation in image enhancement*, preprint (1994).

[11] S. F. McCormick, ed., *Multigrid Methods*, SIAM, 1987.

[12] M. E. Oman, *Fast multigrid techniques in Total Variation-based image reconstruction*, preprint, Montana State University (1994).

[13] M. E. Oman and C. R. Vogel, *Iterative methods for Total Variation denoising* Preprint, Montana State University (1994).

[14] L. I. Rudin, S. Osher, and E. Fatemi, *Nonlinear Total Variation Based Noise Removal Algorithms*, Physica D, vol 60 (1992), pp. 259-268.

PARAMETER ESTIMATION IN SURFACE CATALYSIS WITH REACTION-DIFFUSION FRONTS

J. G. Wade*

Department of Mathematics and Statistics
Bowling Green State University
Bowling Green, OH 43403-0221

1 Introduction

Mathematical modeling of chemical kinetics in systems in which all chemical species are assumed to be "well mixed" involves systems of nonlinear ordinary differential equations. If this assumption is dropped, then the modeling leads to partial differential equations which resemble the ODE's of the well mixed case but have additional "spatial mixing" terms representing diffusion and possibly advection. Here we focus on non-well mixed chemical phenomena which do not include advection, so that the relevant mathematical models are systems of reaction-diffusion equations.

One such class of phenomena is the catalytic reaction of gases at gas-metal interfaces. These problems are currently of interest materials science [1, 3, 5, 10], due largely to recent advances in electron microscopy techniques which facilitate real-time imaging of the reactions on a spatial scale of 10^{-4} meters or less [3, 7, 8, 9, 10]. Typically, certain spatial and temporal pattern formation is observed as these reactions progress, such as travelling waves, "target patterns", and spirals. A goal of researchers is a detailed analytical description of these phenomena, in hopes of gaining better understanding of the basic physical processes involved.

The simplest instance of such patterns is the "travelling wave" such as is observed in the oxidization of carbon monoxide on platinum surfaces. An heuristic description this is as follows. (See [1, 5, 10] for a more rigorous and detailed description.) The experimental arrangement is a chamber containing a metal wafer and two species of gas, A and B, the partial pressures of which are controlled by the experimentalist. The molecular structure of the surface and the molecules of A and B are such that the gas molecules can "adsorb" onto the surface at "lattice sites", and desorb from it, and such that when a molecule of A is at an adjacent site to a B, then these two molecules react and form a third molecule of type C which then desorbs from the surface leaving two empty sites. A new A or B molecule can adsorb at these sites, with chances being greater for, say, an A being adsorbed if the partial pressure of A is greater then that of B. In this way, eventually the entire surface becomes covered with a layer of A.

---

*Research was supported in part by the BGSU Faculty Research Council, under FRC Project #MA9507.

Now suppose that at time $t = 0$ the surface is covered with A, and that the partial pressures are changed so that that of B is greater than that of A. At points of impurity on the surface, if a molecule of B is adsorbed, then it will react with a neighboring A molecule. Since the pressure of B is now greater, chances are that the two new empty sites will become occupied by B molecules. This will induce further reaction with yet other A molecules, producing more empty sites which in turn become occupied with B molecules, and so on. In this way a propagating front will be established, and eventually the whole surface will be covered with B.

Systems of reaction-diffusion models of the form

$$\mathbf{u}_t = D\mathbf{u} + \mathbf{f}(\mathbf{u})$$

are used to describe these kind of phenomena, where the dependent variable $\mathbf{u}$ has one or more components, each corresponding to a chemical species. The operator D here is elliptic and represents the diffusion, and the chemical kinetics are embodied in $\mathbf{f}$. The components of $\mathbf{u}$ represent the "coverages" of the various gas species on the surface. The coverage of a species at a given spatial position is essentially the local fraction of the total lattice sites occupied by that species.

Using the electron microscopy techniques mentioned above, experimentalists can obtain snapshots in time of the coverages of the various species over entire regions, and are thus able to produce images of the pattern formation. An interesting class of inverse problems, then, is: with the images of the reaction fronts as experimental data, estimate various quantities pertaining to D and $\mathbf{f}$. The focus of this paper is such an inverse problem.

2 A model problem

Consider the following system of reaction-diffusion equations defined for $x \in (0,1)$:

$$u_t(x,t) \quad = \quad \epsilon u_{xx}(x,t) + f(u(x,t),v(x,t)), \qquad (2.1)$$

$$v_t(x,t) \quad = \quad \epsilon v_{xx}(x,t) + g(u(x,t),v(x,t)), \qquad (2.2)$$

with Neumann boundary conditions at $x = 0$ and $x = 1$ for both u and v, and some initial conditions $u(x,0) = u_0(x), v(x,0) = v_0(x)$. The functions f and g have the form

$$f(u,v) \quad = \quad k_1(1 - u - v) - k_2 uv, \qquad (2.3)$$

$$g(u,v) \quad = \quad k_3(1 - u - v) - k_2 uv, \qquad (2.4)$$

with $0 < k_1 < k_3$, and $k_2 > 0$.

In terms of the discussion in §1, u and v are, respectively, the coverages of A and B (and hence only solutions which are everywhere nonnegative have physical meaning). This model is a simplification of one which has been proposed for the simulation of oxidization of carbon monoxide on platinum surfaces — see [5] for example. The coefficients k_1 and k_3 are proportional to the partial pressures of the two gases. However, they still may be subject to estimation, because they are products of factors other than the pressures, notably the "sticking coefficients" describing how readily a molecule will to adsorb if it approaches an empty site. The coefficient k_2 is the kinetic rate constant for the reaction of A with B.

3 The well mixed case

Note that if u and v are both constant in x, then the diffusion terms are identically zero and the solution remains constant in x for all t. This is the "well mixed" case in which (2.1-2.2) reduces to a system of ODE's. Moreover, $(1,0)$ and $(0,1)$ (denoted by CP1 and CP2, respectively) are clearly critical points for this system. A standard stability analysis can be performed on these critical points, using the Jacobian $J(u,v)$, which is given by

$$J(u,v) = - \left[\begin{array}{cc} k_1 + k_2 v & k_1 + k_2 u \\ k_3 + k_2 v & k_3 + k_2 u \end{array} \right]$$

The eigenvalues $\lambda_{i,\pm}$ of J at CPi, $i = 1, 2$ are

$$\lambda_{1,\pm} = \left(-\sigma \pm \sqrt{\sigma^2 + 4k_2(k_3 - k_1)} \right)/2, \tag{3.1}$$

$$\lambda_{2,\pm} = \left(-\sigma \pm \sqrt{\sigma^2 - 4k_2(k_3 - k_1)} \right)/2, \tag{3.2}$$

with $\sigma \overset{\text{def}}{=} k_1 + k_2 + k_3$. Clearly, both $\lambda_{1,\pm}$ are real and of opposite signs, since $k_3 > k_1$. Hence CP1 is a saddle point. Also, since

$$\sigma^2 > \sigma^2 - 4k_2(k_3 - k_1) = \tilde{\sigma}^2 + 4k_1(k_2 + k_3) > 0 \tag{3.3}$$

with $\tilde{\sigma} \overset{\text{def}}{=} (k_1 + k_2 - k_3)$, we see that $\lambda_{2,\pm}$ are real, negative and distinct. Hence CP2 is an asymptotically stable improper node.

On the basis of these observations, we expect that, in the well mixed case, (u,v) trajectories of (2.1-2.2) originating near CP1 depart from that point and approach CP2 as $t \to \infty$.

For use below, we also note that the eigenvectors $\mathbf{a}_{i,\pm}$ of J are given by

$$\mathbf{a}_{1,\pm} = \left(-\left(\frac{k_1 + k_2}{k_1 + \lambda_{1,\pm}} \right), 1 \right)^T, \tag{3.4}$$

$$\mathbf{a}_{2,\pm} = \left(-\left(\frac{k_3 + \lambda_{2,\pm}}{k_2 + k_3} \right), 1 \right)^T. \tag{3.5}$$

4   Travelling wave solutions

Now suppose the well mixed assumption is dropped. Suppose further that the initial condition for (u, v) is a perturbation from the unstable equilibrium at CP1. Specifically, for some $0 < \beta \le 1$ and small $\delta > 0$, define $I_0 = [1 - \delta, 1]$, and suppose that

$$u_0(x) \;=\; 1 - \beta\chi(I_0), \tag{4.1}$$
$$v_0(x) \;=\; \beta\chi(I_0) \tag{4.2}$$

where χ is the characteristic function. Then intuitively we can believe that for $x \in I_0$, the solution (u, v) will tend toward $(0, 1)$. Then, the diffusion term in (2.1-2.2) will cause the solution (u, v) to move away from $(1, 0)$ at nearby x's, at which (u, v) will then tend toward $(0, 1)$ also. This will induce the same effect at x's which are yet further from I_0, and so on. Thus it is plausible that $\big(u(x, t), v(x, t)\big)$ will assume the form of a *propagating front*, moving, in this case, from right to left.

This behavior models the reaction fronts described in §1. It is also observed numerically, as seen below in Figure 1. In this example, $(k_1, k_2, k_3) = (0.1, 10, 1)$, and $\epsilon = 1 \times 10^{-4}$. The initial condition is given by (4.1-4.2) with $\beta = 1$ and $\delta = 1/10$. The profiles of $u(x, t)$ shown in Figure 1 are at $t \approx 29.06$ and $t \approx 29.36$. (The computational details are described in §5.)

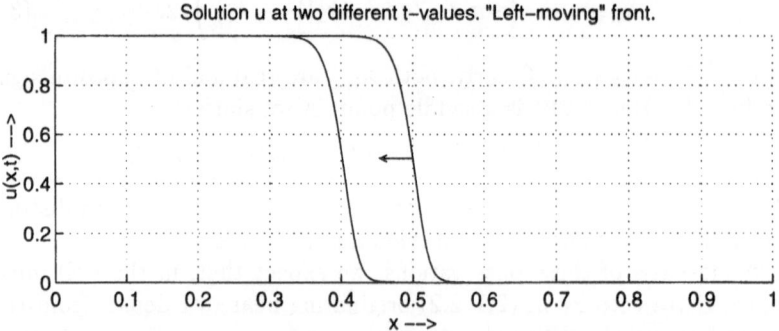

Figure 1. Profiles of $u(x, t)$ for two different t's.

On the basis of this heuristic and numerical evidence, it is reasonable to seek a *travelling wave* solution, for which, with fixed t, u increases monotonically (in x) ¿from 0 to 1, and v decreases monotonically from 1 to 0. To this end, consider (2.1-2.2), but posed on $x \in (-\infty, \infty)$, and seek a (u, v)

of the form

$$u(x,t) = w_1(x + ct), \quad v(x,t) = w_2(x + ct) \tag{4.3}$$

for some "wave speed" c. In light of (2.1–2.2), and denoting $s = x + ct$, we have

$$
\begin{aligned}
c\dot{w}_1(s) &= \epsilon\ddot{w}_1(s) + f(w_1(s), w_2(s)), \\
c\dot{w}_2(s) &= \epsilon\ddot{w}_2(s) + g(w_1(s), w_2(s)),
\end{aligned}
$$

with the "dot" denoting d/ds. Rewriting this as a first order system in the standard way yields

$$
\begin{aligned}
\dot{w}_1(s) &= w_3(s), & (4.4) \\
\dot{w}_2(s) &= w_4(s), & (4.5) \\
\dot{w}_3(s) &= \big(-f(w_1(s), w_2(s)) + cw_3(s)\big)/\epsilon, & (4.6) \\
\dot{w}_4(s) &= \big(-g(w_1(s), w_2(s)) + cw_4(s)\big)/\epsilon, & (4.7)
\end{aligned}
$$

with $\mathbf{w}(s) \to (1,0,0,0)^T \overset{\text{def}}{=} \widetilde{CP}1$ as $s \to -\infty$, and $\mathbf{w}(s) \to (0,1,0,0)^T \overset{\text{def}}{=} \widetilde{CP}2$ as $s \to \infty$ as side conditions. The points $\widetilde{CP}1$ and $\widetilde{CP}2$ are critical points for the system (4.4–4.7).

Thus the search for a travelling wave solution for the (u,v)-system amounts to seeking a value c such that there is a trajectory for the $\mathbf{w}$-system connecting one critical point to another. Let $\bar{\mathbf{w}}(s)$ denote this trajectory. Also, denote by $\mathbf{t}_i$ the unit tangent vectors of $\bar{\mathbf{w}}$ at $\widetilde{CP}i$, $i = 1, 2$.

A stability analysis of the two critical points $\widetilde{CP}i, i = 1, 2$ is useful. In terms of the eigenvalues $\lambda_{i,\pm}$ of J given by (3.1–3.2), the eigenvalues $\gamma_{i,\pm}^{\pm}$ of the Jacobian of system (4.4–4.7) are

$$\gamma_{i,\pm}^{\pm} = \frac{1}{2\epsilon}\left(c \pm \sqrt{c^2 - 4\epsilon\lambda_{i,\pm}}\right). \tag{4.8}$$

The eigenvectors $\mathbf{b}_{i,\pm}^{\pm}$ are given in terms of the eigenvectors $\mathbf{a}_{i,\pm}$ of J in (3.4–3.5) by

$$\mathbf{b}_{i,\pm}^{\pm} = \begin{bmatrix} \mathbf{a}_{i,\pm} \\ \gamma_{i,\pm}^{\pm}\mathbf{a}_{i,\pm} \end{bmatrix}. \tag{4.9}$$

Now, the unit tangent $\mathbf{t}_1$ must be parallel to one of the four eigenvectors $\mathbf{b}_{1,\pm}^{\pm}$, and since the trajectory $\bar{\mathbf{w}}$ moves away from $\widetilde{CP}1$ with increasing s, the tangent eigenvector must correspond to a $\gamma_{1,\pm}^{\pm} > 0$. Inspection reveals that $\gamma_{1,-}^{-} < 0$ but that the other three are positive. The choice among the remaining three $\mathbf{b}_1$'s can be further restricted by imposition of the monotonicity requirements stated just above (4.3). The monotonicity requirements imply that the "upper half" of $\mathbf{t}_1$ must point into the second quadrant of the (w_1, w_2) plane. By (4.9), the upper half of $\mathbf{t}_1$ must also be parallel to the one of $\mathbf{a}_{1,\pm}$, since the two $\lambda_{1,\pm}$ are distinct. Again, inspection reveals that $\mathbf{a}_{1,+}$ points into the second quadrant, while $\mathbf{a}_{1,-}$ points into the first. Hence the upper half of $\mathbf{t}_1$ must be parallel to $\mathbf{a}_{1,+}$, and $\mathbf{t}_1$ itself must be parallel to one of $\mathbf{b}_{1,+}^{\pm}$. Since $\gamma_{1,+}^1 \leq \gamma_{1,+}^+$, it must be that $\mathbf{t}_1$ is parallel to $\mathbf{b}_{1,+}^+$. (Note that the case in which $\gamma_{1,+}^1 = \gamma_{1,+}^+$ poses no difficulty, since that case the two eigenvectors also coincide and hence the direction of $\mathbf{t}_1$ is still well-defined.)

Since $\bar{\mathbf{w}}(s)$ (if it exists) approaches $\widetilde{CP}2$ as $s \to \infty$, it must be that $\mathbf{t}_2$ is parallel to a $\mathbf{b}_{2,\pm}^{\pm}$ for which $\gamma_{2,\pm}^{\pm} < 0$. Reasoning similar to that above leads to the conclusion that $\mathbf{t}_2$ must in fact be parallel to $\mathbf{b}_{2,+}^-$.

Returning to the fact that $\mathbf{t}_1$ must be parallel to $\mathbf{b}_{1,+}^+$, we see that, in light of (3.1) and (4.8), we must have

$$c \geq c_{MIN} \overset{\text{def}}{=} 2\sqrt{\epsilon \lambda_{1,+}}, \tag{4.10}$$

so that the discriminant in (4.8) is nonnegative. For if it were negative, then $\widetilde{CP}1$ would be a spiral point and the monotonicity would be violated. This gives a lower bound on the possible values of c.

The results of this discussion are summarized as follows:

Theorem 4.1 *A necessary condition for the existence of a travelling wave solution $\bar{\mathbf{w}}(s)$ is that $c \geq c_{MIN}$. If $\bar{\mathbf{w}}(s)$ exists, then $\mathbf{t}_1$ is parallel to $\mathbf{b}_{1,+}^+$ and $\mathbf{t}_2$ is parallel to $\mathbf{b}_{2,+}^-$, both of which are well-defined.*

It is interesting to note that, in simpler cases where the wave speed c can be computed analytically, it turns out that $c = c_{MIN}$; see [2] and [4, §11.2.4], for example. Numerical evidence suggests that this is also the case here. In the example represented in Figure 1, $c_{MIN} \approx 1.74 \times 10^{-2}$, whereas the numerical value of c obtained from the approximate solution of the system of PDE's (2.1–2.2) was about 1.70×10^{-2} — somewhat less than the theoretical minimum. An analytical proof of the plausible conjecture that the true wave speed always equals c_{MIN} would be of significant value; however, it is not pursued here.

For the example represented in Figure 1, a numerical approximation of $(w_1(s), w_2(s))$ was formed from the computed solution $(u(x,t), v(x,t))$ at $t \approx 29$ (when the wave is about at $x = 1/2$). Then, $(w_3(s), w_4(s))$ were computed via second-order numerical differentiation. In this way, a numerical $\bar{\mathbf{w}}$ was obtained. Its phase portraits are shown in Figure 2, along with (scaled versions of) the tangent vectors $\mathbf{t}_i, i = 1, 2$.

5   The inverse problem

As noted in §1, experimentalists are able to obtain "snapshots" of travelling waves in physical systems, and an interesting inverse problem is to use this

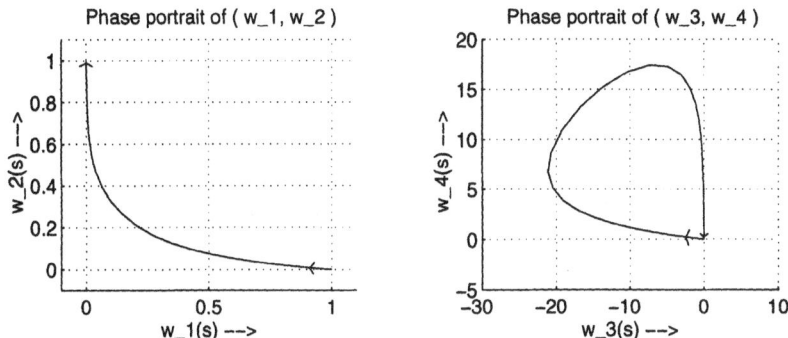

Figure 2. Phase portraits for the **w** system, showing the analytically computed tangent vectors.

data to estimate the diffusion and kinetic terms. Here, we assume that the solution of (2.1–2.2) is a travelling wave, and that for some for some t_* and δt we are given "profiles" $\big(u(x_j, t_*), v(x_j, t_*) \big)$ and $\big(u(x_j, t_* + \delta t), v(x_j, t_* + \delta t) \big)$, for $x_j \overset{\text{def}}{=} j/N, j = 0, \ldots, N$, for some N. From this we wish to estimate $\mathbf{q} \overset{\text{def}}{=} (k_1, k_2, k_3, \epsilon)$.

The strategy is as follows. With the given data, compute an approximation of the wave speed c, and use one of the given profiles to compute a $\{\bar{\mathbf{w}}(s_j)\}_j$ in the manner described just above Figure 2. Since $d/ds = \partial/\partial x$, we can numerically differentiate $\bar{\mathbf{w}}(s_j)$ to obtain an approximate

$$\{\boldsymbol{\tau}(s_j)\}_{j=0}^{N} \overset{\text{def}}{=} \{\dot{\bar{\mathbf{w}}}(s_j)\}_{j=0}^{N}.$$

With $G(\mathbf{w}; \mathbf{q})$ denoting the right-hand side of (4.4–4.7), we then seek $\mathbf{q}$ by minimizing

$$\Phi(\mathbf{q}) \overset{\text{def}}{=} \frac{1}{2} \|G(\bar{\mathbf{w}}; \mathbf{q}) - \boldsymbol{\tau}\|^2,$$

where $\|\cdot\|$ is the Euclidean norm in $\mathbf{R}^{N+1}$. This strategy was carried out; the results are reported below.

The data for the inverse problem were computed for various $\mathbf{q}$ via the numerical solution of (2.1–2.2) with initial conditions given by (4.1–4.2) (with $\beta = 1$ and $\delta = 1/10$), a linear spline Galerkin scheme with $N = 256$ for the spatial variable, and the trapezoidal or Crank-Nicholson method in t. The time step Δt was computed by $\Delta t = 1/(4Nc_{MIN})$, with c_{MIN} given by (4.10), thus accounting for the wave speed and keeping the spatial and temporal discretization of the same order of accuracy. At each time level, the Crank-Nicholson method requires the solution of a nonlinear resolvent equation. This was performed via fixed point iteration with relative error tolerance of 10^{-5}. Figure 3 shows the inverse problem data computed for the case represented in Figures 1 and 2.

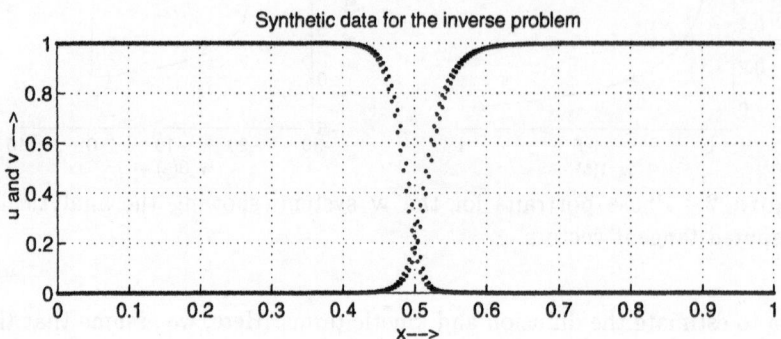

Figure 3. Computed data for the inverse problem, two (u, v) "profiles". This is the data for "Case 4" below.

The minimization of Φ was performed numerically using the subroutine "lmdif" from the public-domain MINPACK package. This subroutine uses a Levenberg-Marquardt method with finite difference Jacobians.

For various choices of $\mathbf{q}$, data were computed in the manner described above, and $\Phi(q)$ was minimized in attempts to recover it. In all of the results reported below, the initial guess for $\mathbf{q}$ was taken as $(1, 1, 1, 1/10)$. The least-squares minimization proved quite robust with respect to initial guess.

Here we present the numerical results. The first row of each table shows the "true" $\mathbf{q}$ which was used to compute the data, and the second row shows the result of the inverse problem. In the column labelled "Wave speed", the second entry is the numerically observed wave speed, $\bar{c}$.

<u>Case 1</u>. Maximum relative error in computed $\mathbf{q}$ is $8.9e - 3$.

k_1	k_2	k_3	ϵ	Wave speed
1	2	3	$1e - 4$	$c_{MIN} = 1.56e - 2$
0.991	1.989	2.987	$1.009e - 4$	$\bar{c} = 1.50e - 2$

Case 2. Maximum relative error in computed $\mathbf{q}$ is $1.8e-2$.

k_1	k_2	k_3	ϵ	Wave speed
1	1	10	$1e-4$	$c_{MIN} = 1.68e-2$
0.998	0.997	9.98	$1.01e-4$	$\bar{c} = 1.64e-2$

Case 3. Maximum relative error in computed $\mathbf{q}$ is 0.19.

k_1	k_2	k_3	ϵ	Wave speed
0.1	1	10	$1e-4$	$c_{MIN} = 1.82e-2$
$8.1e-2$	0.998	10.005	$1.02e-4$	$\bar{c} = 1.78e-2$

Case 4. Maximum relative error in computed $\mathbf{q}$ is $4.9e-2$.

k_1	k_2	k_3	ϵ	Wave speed
0.1	10	1	$1e-4$	$c_{MIN} = 1.74-2$
$9.51e-2$	9.91	0.994	$1.048e-4$	$\bar{c} = 1.70e-2$

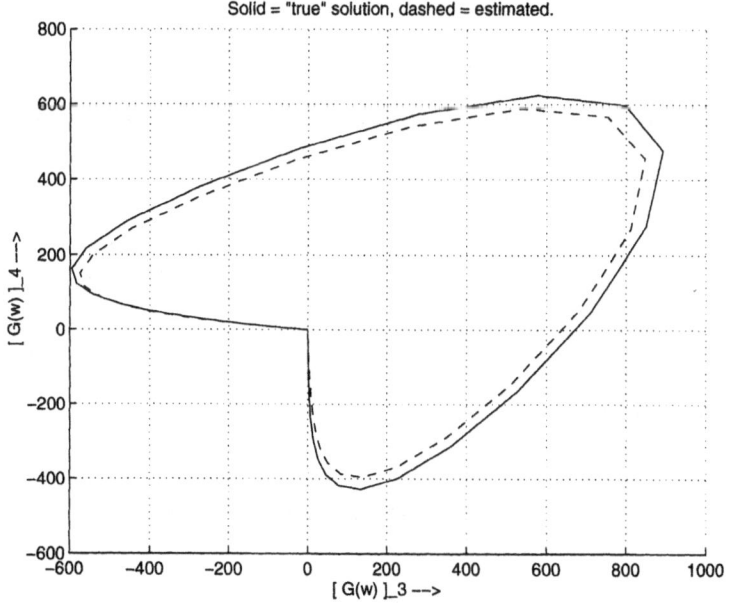

Figure 4. Results of the "Case 4". Shown are the third and fourth components of $G(\bar{\mathbf{w}}; \mathbf{q})$, with the "true" $\mathbf{q}$ and the estimated $\mathbf{q}$, computed as the minimizer of Φ.

6   Conclusions and future directions

On the basis of these numerical results, the strategy described here for estimating the diffusion and kinetic terms in a reaction-diffusion system seems viable, if each of the chemical species has the same diffusion coefficient (ϵ in (2.1–2.2)), and if profiles of each are available.

However, some improvements are in order. First, the strategy explored here requires differentiation of data. Usually, this would require some sort of smoothing, which is likely to reduce the accuracy of the final results. Also, it is highly restrictive from a physical point of view to assume that each of the chemical species has the same diffusion coefficient, as we did.

More substantially, it may be that profiles of each individual species are *not* available, but some only linear combination of them [6]. In this case, one strategy would be to develop a robust numerical scheme for solving the w system (4.4–4.7), and to use this in a least-squares fit-to-data approach. This would of course be more numerically intensive, yet still significantly cheaper than working directly with (2.1–2.2). However, developing a robust numerical scheme for solving (4.4–4.7) is nontrivial, since the solution is a trajectory connecting two saddle points in a nonlinear system. These ideas will be the subject of future research.

References

[1] ERTL, G., "Oscillatory kinetics and spatio-temporal self-organization in reaction at solid surfaces", *Science*, Vol. 254, Dec. 1991, p. 1750–1755.

[2] GRINDROD, P., *Patterns and Waves,* Oxford Applied Mathematics and Computing Series, Oxford University Press, New York, 1991.

[3] GARCIA, A. and KORDESCH, M.E., "Surface reaction-diffusion fronts observed with photoelectron emission microscopy during deposition on Mo(310)," submitted Sept., 1994.

[4] GRAY, P. and SCOTT, S.K., *Chemical Oscillations and Instabilities, Non-linear Chemical Kinetics,* Oxford University Press, New York, 1990.

[5] KRISCHER, K., EISWIRTH, M. ERTL, G., "Oscillatory CO oxidation on Pt(100): Modeling of temporal self-organization" *Journal of Chemical Physics* **96**(12) June 1992, p. 9161–9172.

[6] KORDESCH, M.E., private communication.

[7] MUNDSCHAU, M., KORDESCH, M.E., RAUSENBERGER, B., EN-GEL, W., BRANDSHAW, A.M., and ZEITLER, E., "Real-time observation of the nucleation and propagation of reaction fronts on surfaces using photoemission electron microscopy", *Surface Science* **227**, 1990, p. 246–260.

[8] MUNDSCHAU, M., and RAUSENBERGER, B., "Chemical reaction fronts on platinum surfaces", *Platinum Metals Review* **35**(4), 1991, pp 188–195.

[9] RAUSENBERGER, B., ŚWIĘCH, W., RASTOMJEE, C.S., MUND-SCHAU, M., ENGEL, W., ZEITLER, E., and BRANDSHAW, A.M., "Imaging reaction-diffusion fronts with low-energy electron microscopy", *Chemical Physics Letters* **215**(1,2,3), 26 November, 1993, pp 109–113.

[10] ŚWIĘCH, W., RASTOMJEE, C.S., IMBIHL, R., EVANS, J.W., RAUSENBERGER, B., ENGEL, W., SCHMID, A.K., BRAND-SHAW, A.M., ZEITLER, E., "On the complex structure of reaction diffusion fronts observed during CO oxidation on PT{100}" Fritz-Haber-Institut der Max-Planck-Gesellschaft, submitted Sept., 1993 to *Surface Science*.

SIGNAL PROCESSING AND THE JACOBI GROUP

Dorothy I. Wallace

Department of Mathematics and Computer Science
Dartmouth College
Hanover, New Hampshire 03755

1 Introduction

The recent concern with advancing our understanding of both analog and digital signal processing has lead to two approaches which have been in competition for several years. One approach, the "window Fourier transform", has the advantage of being amenable to theoretical analysis to the extent that we now understand its workings quite well. In fact we have a full description of its shortcomings too, largely evident in the ill conditioning of the inverse transform. The newer approach, "wavelets", seems, at least experimentally, to avoid the conditioning difficulties of its predecessor. It suffers, however, from a less well developed theoretical basis, and furthermore has no easily describable inverse transform, so that one relies heavily on numerical techniques if one chooses this path. This paper introduces a new method which, combining the sampling techniques of both of its predecessors, promises someday to offer the advantages of both, as well as an added advantage of generalizability discussed in the conclusion of this paper.

The wavelet Fourier transform stems from the action of the Heisenberg group on functions of a single real variable. The wavelet transform similarly comes from the affine group. A semidirect product of these two groups lies inside the classical Jacobi group, which is itself the semidirect product of the Heisenberg group and $SL(2, \mathbb{R})$. This group has been studied in the context of automorphic forms, such as Jacobi's theta function, for a century. It is the heart of this paper.

2 The Heisenberg group and the metaplectic representation.

The smallest version of the Heisenberg group $\mathbb{H}$, is as vectors in $\mathbb{R}^3$,

$$(\lambda, \mu, k) \circ (\lambda_1, \mu_1, k_1) = (\lambda + \lambda_1, \mu + \mu_1, k + k_1 + \lambda \mu_1 - \lambda_1 \mu).$$

This group acts on $L^2(\mathbb{R})$ by sending

$$f(x) \overset{p(\lambda, \mu, k)}{\longmapsto} e^{2\pi i k h} e^{2\pi i q x + \pi i h p q} f(x + hp)$$

which gives a unitary representation of $\mathbb{H}$ for each choice of h. Different choices of h yield different representations of $\mathbb{H}$. The standard practice in the literature seems to be to set $h = 1$. The alert reader will notice that the map

$$g \in \mathcal{L}^2(\mathbb{R}) \mapsto \langle g, \rho(\lambda, \mu, k)f \rangle$$

gives a transform $g \to \tilde{g}(\lambda, \mu, k)$ where

$$\tilde{g}(\lambda, \mu, k) = e^{2\pi i k h} e^{\pi i h \lambda \mu} \int\limits_{\mathbb{R}} e^{2\pi i \mu x} f(x + h\lambda) g(x) dx$$

and thus with appropriate choices of q and p represents a coherent state system with basic building block f.

Assume $f \not\equiv 0$ in $\mathcal{L}^2(\mathbb{R})$.

Proposition 2.1. If (λ, μ, k) vary over all possible choices then the system (x) is *complete*, that is $\tilde{g} \equiv 0 \Rightarrow g \equiv 0$ in the $\mathcal{L}^2$ sense.

Proof. Let $p = k = 0$. Then

$$0 = \tilde{g}(0, \mu, 0) = \int\limits_{\mathbb{R}} e^{2\pi i \mu x} f(x) g(x) dx = \Im(fg)(\mu)$$

Therefore $fg \equiv 0$. This can only happen if supp $f \cap$ supp g is a set of measure 0. Now as λ varies, the support of $\rho(\lambda, \mu, k)$ moves along the real line, so for some $c_1 \leq \lambda \leq c_2$ we have

$$\text{supp } p(\lambda, \mu, 0)f \cap \text{supp } g \neq 0.$$

Now for all

$$
\begin{aligned}
\tilde{g}(\lambda, \mu, 0) &= 0 \\
&= e^{\pi i h \lambda \mu} \int\limits_{\mathbb{R}} e^{2\pi i \mu x} f(x + h\lambda) g(x) dx \\
&= e^{\pi i h \lambda \mu} \Im(\rho(0, \lambda, 0)f \cdot g)(\mu),
\end{aligned}
$$

where $\Im$ equals the Fourier transform. Therefore $\rho(0, \lambda, 0)f \cdot g \equiv 0$ in $\mathcal{L}^2$ which contradicts our choice of λ.

Corollary 2.2. Any set of functions containing the $\rho(\lambda, \mu, k)f$ must also be complete.

Of course one would wish to find a discrete set of (λ, μ, k) for which the above system is complete and for certain choices of f this has been done.

Gabor, [3], showed that if $f = e^{-\pi c x^2}$ then the system is complete for $(\lambda, \mu) \in \mathbb{Z}^2$ and remains complete if one lattice point is deleted. Unfortunately the transform is ill-conditioned, that is, g does not in general possess an $\mathcal{L}^2$-convergent expansion of the form

$$g = \sum_{\lambda, \mu \in \mathbb{Z}^2 - (0,0)} C_{\lambda, \mu}(\rho(\lambda, \mu, 0)f).$$

There are two explanations for this phenomenon. One is that this irreducible representation of the Heisenberg group is equivalent to one on $\mathcal{L}^2(\mathbb{H}/\mathbb{H}_{\mathbb{Z}})$ which decomposes naturally into invariant subspaces on which the center of $\mathbb{H}$ acts according to some character. Since the representation we have been discussing has a central character, it is equivalent to some subrepresentation of $\mathbb{H}$ or $\mathcal{L}^2(\mathbb{H}/\mathbb{H}_{\mathbb{Z}})$ and therefore its invariant subspace can't form a basis (in $\mathcal{L}^2$) for the whole space. In particular, since the central character is non-trivial, the function 1 cannot be written as an $\mathcal{L}^2$-convergent sequence of coherent states coming from the Heisenberg group via this representation.

There is a technical reason that the above arguments hold which is that the functions on $(\mathbb{H}/\mathbb{H}_{\mathbb{Z}})$ which transform according to the central character $k \to e^{\pi i k}$ must all have zeros. The presence of the zero gives rise to the ill-conditioning of our series.

The physical explanation for the ill-conditioning is that the coherent states do not vary in localization. That is, supp $(\rho(\lambda, \mu, k)f)$ is of the same size as supp f. Therefore this family of functions cannot be expected to pick out high frequencies well. One possible way around this would be to use more than one basic function and let them vary in how localized they are. Various *ad hoc* attempts have been made at such a scheme with promising results. In this paper we will provide a theoretical framework for such an approach. Our choice of building blocks will come from the action of yet another group on $\mathcal{L}^2(\mathbb{R})$.

The natural candidate for such a group is the group of automorphisms of $\mathbb{H}$. One would expect to be able to coerce this group into an action on $\mathcal{L}^2(\mathbb{R})$ which intertwines that of $\mathbb{H}$. This group contains $SL(2, \mathbb{R})$ or if $\mathbb{H} = \mathbb{H}^n$ it contains $S_p(n)$, and one can form a semi-direct product with multiplication given by

$$(M_1, (\lambda_1, \mu_1), k_1) \circ (M_2, (\lambda_2, \mu_2), k_2)$$

$$= (M_1 M_2, (C_1, \mu_1)M_1 + (\lambda_2, \mu_2), k k_1 + det \begin{pmatrix} (\lambda_1, \mu_1)M_2 \\ (\lambda_2, \mu_2) \end{pmatrix}).$$

Thus we can write

$$(M_1, (\lambda_1, \mu_1,), k_1) = (M_1, (0,0), 0) \circ (I, (\lambda_1, \mu_1), k_1)$$

and can search for a representation, $w(M, (\lambda, \mu), k)$ on $\mathcal{L}^2(\mathbb{R})$, which is equal to $\mu(M)\, \rho(\lambda, \mu k)$ where ρ is the representation of $\mathbb{H}$ discussed thus far. It turns out that there is such a representation which is unitary and unique up to isomorphism but double valued. That is, in order to make a genuine representation out of it one must pass to a double cover of $SL(2, \mathbb{R})$ (again, unique up to isomorphism) called the metaplectic group. The representation μ thus constructed is characterized by what it does to certain subgroups of $SL(2, \mathbb{R})$. It is called the metaplectic representation, while w is the extended metaplectic representation. We have

$$\mu \begin{pmatrix} a & 0 \\ 0 & a^{-1} \end{pmatrix} f = a^{-\frac{1}{2}} f(a^{-1}x)$$

$$\mu \begin{pmatrix} 1 & 0 \\ c & 1 \end{pmatrix} f = \pm e^{-\pi i x C x} f(x)$$

$$\mu \begin{pmatrix} 0 & 1 \\ -1 & 0 \end{pmatrix} f = i^{\frac{1}{2}} \mathfrak{F}^{-1}.$$

where $\mathfrak{F}$ is the Fourier transform. See, for example, [2] for more details.

Now one readily sees that, due to the dilations arising from the diagonal group, the localization of f changes. Therefore if one were to use a family of functions such as

$$\{w(\gamma)f(x), \gamma \in \Gamma\}.$$

where Γ is a discrete subgroup of G, one might expect not only completeness but perhaps decent conditioning as well.

There are two problems with this approach. One is the representation of $\begin{pmatrix} 0 & 1 \\ -1 & 0 \end{pmatrix}$ as the inverse Fourier transform of f. If f is well localized in space then the $\mathfrak{F}^{-1}f$ will not be localized in the frequency domain due to the Heisenberg uncertainty principle and therefore computing integrals against $\mathfrak{F}^{-1}f$ is going to be unfeasable. One could attempt to find a Γ not containing this element but, short of discrete subgroups of the affine group, this is not so easy.

The second problem is that G, which for number theorists is classically called the Jacobi group, has many actions on $\mathcal{L}^2(\mathbb{R})$, of which this is but one. If we change the central character of the Heisenberg group it is possible to give other, more convenient actions of G where the role of discrete subgroups is far better understood.

3   The action of the Jacobi group

A good basic reference for what follows is Eichler & Zagier, [1]. The action we will consider in this section is defined on functions of two variables (z, τ)

where $Z \in \mathbb{C}$ and $\tau \in \mathcal{H}$, the Poincaré upper half plane. It is characterized by two numbers. An integer k called the *weight* and another integer m called the *index*. The standard notation for the action uses a slash, so that for $\phi(\tau, z)$ and $\exp(z) = e^{2\pi i z}$,

$$\phi\left|\left[\begin{pmatrix} a & b \\ c & d \end{pmatrix}, (\lambda_1 \mu), k\right]\right|(\tau, z) = e^{2\pi i m k}(c\tau + d)^{-k} \cdot$$

$$\exp(m(\frac{-c(z + \lambda\tau + \mu)^2}{c\tau + d} + \lambda^2\tau + 2\lambda z + \lambda\mu)) \circ \phi\left(\frac{a\tau + b}{c\tau + d}, \frac{z + \lambda\tau + \mu}{c\tau + d}\right).$$

Theorem 1.4 of [1] says this is indeed a group action. If one were to choose ϕ to be independent of τ and $\text{Im}z$, one would then generate a family of functions on $\mathbb{R}$. It is readily apparent that this family will be modulated in ways similar to those discussed in the last section, that is, multiplication by an exponential, dilation, translation are all included in this new group action as well. Only the Fourier transform is omitted. Viewing $\mathbb{R}$ as the real part of z and choosing a generator ϕ we then have a transform

$$g \mapsto \hat{g}(\gamma_1 \lambda_1 \mu, k)$$

where $\gamma = \begin{pmatrix} a & b \\ c & d \end{pmatrix}$, given by $\hat{g}(\gamma, \lambda, \mu, k) = \int_{\mathbb{R}} \phi|(\gamma, (\lambda, \mu), k)(\tau_1 x_1 y)g(x)dx$.

Lemma 3.1. Let $\phi(\tau, z) = \alpha(\tau)\beta(y)(x)$ with $\alpha(i) \neq 0$, $\beta \neq 0$ anywhere and supp ξ nontrivial. Then the system given by

$$\{\phi|_\gamma, \gamma \in G\}$$

is complete for any choice of k (index) and in (weight).
Proof. Let $\gamma = (I, \lambda, \mu, 0)$. Then

$$\langle \phi|_\gamma(x), g \rangle = \int_\chi e^{2\pi i m(\lambda^2\tau + 2\lambda z + \lambda\mu)} \cdot \phi(\tau, z + \lambda\tau + \mu)g(x)dx$$

$$= e^{2\pi i m(\lambda^2\tau + 2i\lambda y + \lambda\mu)}\alpha(\tau)\beta(y + \lambda\Im\tau)$$

$$\cdot \int_\chi e^{2\pi i m(2\lambda x)}\xi(x + \lambda\text{Re } \tau + \mu)g(x)dx.$$

Let $\tau = i$ and we have

$$\langle \phi|_\gamma(x), g \rangle = C(\tau, \lambda, \mu, y) \int_x e^{2\pi i(2\lambda x)}\xi(x + \mu)g(x)dx.$$

Since $\alpha(i) \neq 0$ and $\beta \neq 0$ anywhere, we know that

$$c(\tau, \lambda, \mu, y) = e^{2\pi i m(\lambda^2 i + 2i\lambda y + \lambda \mu)} \alpha(i) \beta(y + \lambda) \neq 0.$$

Thus $\langle \phi|_\gamma(x), g \rangle \equiv 0$ forces $\int_x e^{2\pi i m(2\lambda x)} \xi(x + \mu) g(x) dx = 0$. By Corollary 2.2 the system is complete.

The reader can easily see that not only is the system described in lemma 3.1 complete, it is extremely over-complete because we have restricted the group action to a small subgroup of that at hand. This will prove to be an advantage in the end, because when we pass to discrete sampling we will wish to avoid zeros of a certain Poincaré series constructed from ϕ. This will be possible only if we can sample over the full Jacobi group.

4   Poincaré series and the Brezin-Weil-Zak Transform

The proof, due to Bargmann et al, of the completeness of the system discussed in §2, makes use of the Brezin-Weil-Zak transform which maps a function on the real line to a transform depending on two variables in the Heisenberg group. this transform is given by

$$T_0 f(\lambda, \mu) = \sum_{j \in \mathbb{Z}} f(\mu + j) e^{\pi i j \lambda}$$

We refer the reader to Folland, [2], for the proof of the following fact.
Lemma 4.1. T_0 is an isometry from $\mathcal{L}^2(\mathbb{R})$ to $\mathcal{L}^2(\mathbb{R}/\mathbb{Z}^2)$.

A glance at the proof of 4.1 shows immediately that it remains true if we replace the exponential function by

$$e^{2\pi i m j \lambda}$$

which corresponds to our Jacobi form of index m. This generalization of the transform, now given as

$$T_m f(\lambda, \mu) = \sum_{j \in \mathbb{Z}} f(\mu + j) e^{2\pi i m j \lambda}$$

is natural, in that $T_0 = T_{\frac{1}{2}}$, thus placing the metaplectic group in the role of acting via index 1. So we have an easy lemma.
Lemma 4.2. T_m is an isometry from $\mathcal{L}^2(\mathbb{R})$ to $\mathcal{L}^2(\mathbb{R}/\mathbb{Z}^2)$.

Now, T_m fits into the picture we are about to construct via Poincaré series. We wish to start with some well behaved function, f, and map f in a reasonable way to a function on $\mathcal{H} \times \mathbb{C}/\Gamma$ where Γ is the semidirect product of $\mathcal{SL}(2,\mathbb{Z})$ and $\mathbb{Z}^2$. The standard method is via a Poincaré series. Usually

$$P_f = \sum_{\gamma \in \mathcal{SL}(2,\mathbb{Z}\alpha\mathbb{Z}^2)} f|\gamma(\tau,z),$$

however in this case we are forced to modify the series because we are starting with a function depending on x alone. ($x = Re z$)

Let $f(\tau,z) = f(x)$ and let $\gamma = \begin{pmatrix} a & b \\ c & d \end{pmatrix} \alpha(\lambda,\mu)$. Then

$$f|_\gamma(\tau,z) = (c\tau+d)^{-k} e^{2\pi i m(\frac{-c(z+\lambda\tau+\mu)}{c\tau+d}+\lambda\tau+2\lambda z+\lambda\mu)} f\left(Re\left(\frac{z+\lambda\tau+\mu}{c\tau+d}\right)\right).$$

Taking $\gamma = \begin{pmatrix} 1 & b \\ 0 & 1 \end{pmatrix} \alpha(0,r)$ we have $f|_\gamma(\tau,z) = f(x+r)$ and

$$T_m f = \sum_{j\in\mathbb{Z}} f(\mu+j)e^{2\pi i\lambda jm} = g(\mu).$$

Setting $z = \mu + i\lambda$, we have

$$\begin{aligned} T_m f|_\gamma &= \left(\sum_{j\in\mathbb{Z}} f(\mu+j+r)e^{2\pi im\lambda j}\right) \\ &= \left(\sum_{\hat{j}=j+r\in\mathbb{Z}} f(\mu+\hat{j})e^{2\pi i\lambda\hat{j}}\right)e^{-2\pi im\lambda r} \\ &= e^{-2\pi im\lambda r}T_m f. \end{aligned}$$

Furthermore $T_m(f)(\mu,\lambda) = \sum_r f|_{(I,\lambda,\mu+r)}(i,0)$. Thus, viewing $T_m(f)(\mu,\lambda)$ as a function of the complex variable $\mu + i\lambda$, we have shown the following facts.

Lemma 4.3. $T_m f(\mu,\lambda) = \sum_{r\in\mathbb{Z}} f\Big|_{(I,\lambda,\mu+r)}(i,0)$.

Lemma 4.4. Let $\Gamma_\varphi = \{\begin{pmatrix} 1 & b \\ 0 & 1 \end{pmatrix} \alpha(0,r) \in \mathcal{SL}(2,\mathbb{Z})\alpha\mathbb{Z}^2\}$. Then, under Γ_φ, $e^{2\pi im\lambda\mu}T_m f(z)$ is invariant.

Proof. Let $\gamma \in \Gamma_\varphi$. Then

$$\begin{aligned} e^{2\pi im\lambda\mu}T_m f\Big|_\gamma(z) &= e^{2\pi im\lambda(\mu+r)}\sum_{j\in\mathbb{Z}} f(\mu+r+j)e^{2\pi im\lambda j} \\ &= e^{2\pi im\lambda\mu}\sum_{j\in\mathbb{Z}} f(\mu+r+j)e^{2\pi im\lambda(j+r)} \end{aligned}$$

$$= e^{2\pi i m \lambda \mu} T_m f(\mu, \lambda).$$

Now, since the function $e^{2\pi i m \lambda \mu} T_m f$ is invariant under T_φ, we can form the relative Poincaré series for $\tau = u + iv$

$$P_{T_m f, g} = \sum_{\gamma \in \Gamma/\Gamma_\varphi} g(v) e^{2\pi i m \lambda \mu} T_m f \Big|_\gamma (\tau, z).$$

Note that $g(v)$ is invariant under $\Gamma_{\cdot}\varphi$.

We have the following easy fact.

Theorem 4.5. *Suppose g and $T_m f$ are bounded. Then $P_{T_m f, g}$ converges for k sufficiently large.*

Proof. If $T_m f$ is bounded so is $e^{2\pi i m \lambda \mu} T_m f$. For k sufficiently large the Eisenstein series

$$\sum_{\gamma \in \Gamma/\Gamma_\varphi} 1 \Big|_\gamma^{k,m} (\tau, z)$$

converges. Therefore, by comparison so does $P_{T_m f}$.

As an aside, we also have the following useful lemma

Lemma 4.6. $e^{2\pi i m \lambda \mu} T_m f = \sum_j e^{2\pi i \lambda \mu m} f \Big|_{(I,0,j)}$.

Lemma 4.7. $f \to e^{2\pi i m \lambda \mu} T_m f$ *is an isometry of $\mathcal{L}^2(\mathbb{R})$ to a bundle over $\mathbb{C}/\mathbb{Z}^2$ in the sense that*

$$\int_{\mathbb{C}/\mathbb{Z}^2} e^{2\pi i m \lambda \mu} T_m f \overline{e^{2\pi i m \lambda \mu} T_m g} \, d\lambda d\mu = \int_{c/\mathbb{Z}^2} T_m f \overline{T_m g} \, d\lambda d\mu$$

$$= \int f \overline{g} \, dx dy$$

and is independent of fundamental domain.

Now, the function given by

$$P_{T_m f, g}(\tau, z) = \sum_{\Gamma/\Gamma_\varphi} g(v) (e^{2\pi i m x y} T_m f) \Big|_\gamma (\tau, z)$$

is invariant under all of Γ. This is the lift we will be using for test functions of $x \in \mathbb{R}$ to functions of (τ, z).

5   Rankin-Selberg method for $P_{T_m f, g}$

Let ϕ and ψ be Jacobi forms of weight k and index m. Then the Petersson scalar product

$$\langle \phi, \psi \rangle = \int_{\Gamma \backslash \mathcal{H}_x \mathbb{C}} J(v) \phi(v) \overline{\psi(v)} dv$$

is invariant under Γ and thus independent of choice of fundamental region for Γ. Here,

$$J(v) = v^k e^{-4\pi m y^2 / v}$$

and

$$dv = v^{-3} dx dy du dv$$

where

$$x + iy = z, u + iv = \tau.$$

Lemma 5.1. $J(\gamma V) = \overline{j(\gamma, v) j(\gamma, v)}^{-1} J(v).$
Proof. This follows easily from invariance

$$
\begin{aligned}
J(v) \phi(v) \overline{\psi(v)} &= J(\gamma v) \phi(\gamma v) \overline{\psi(\gamma v)} \\
&= J(\gamma v) j(\gamma, v) \overline{j(\gamma, v)} \phi(v) \overline{\psi(v)}.
\end{aligned}
$$

Now, let $h(v)$ be a function on $\mathcal{H} \times \mathbb{C}$ which is invariant under the operators $|_\gamma$ for all $\gamma \in \Gamma_\varphi$. For example, ϕ could be $P_{T_m f, g}$ for some f and g as in the preceding section. Let

$$\phi(v) = \sum_{\gamma \in \Gamma / \Gamma_\varphi} h|_\gamma(v).$$

Let $\theta(v)$ be some other Jacobi form. We have this theorem.

Theorem 5.2. For $\langle, \rangle$ the Petersson scalar product, and ϕ, θ as described above,

$$\langle \phi, \theta \rangle = \int_{z \in \mathbb{C}/\mathbb{Z}_2} \int_{-\frac{1}{2} < u \leq \frac{1}{2}} \int_{v > 0} h(v) \overline{\theta(v)} J(v) dv$$

providing $\langle \phi, \theta \rangle$ exists.

Proof. This is just Rankin-Selberg unwinding for this situation.

$$\langle \phi, \theta \rangle = \int_{\Gamma \backslash \mathcal{H} \times \mathbb{C}} \phi(v) \overline{\theta(v)} J(v) dv$$

$$= \int_{\mathfrak{S}} \left(\sum_{\gamma \in \Gamma_\varphi \backslash \Gamma} h|_\gamma(v) \right) \overline{\theta(v)} J(v) dv$$

$$= \int_{\mathfrak{S}} \left(\sum_{\gamma \in \Gamma_\varphi \backslash \Gamma} j(\gamma, v)^{-1} h(\gamma v) \right) \overline{\theta(v)} J(v) dv$$

$$= \sum_{\gamma \in \Gamma_\varphi \backslash \Gamma} \int_{\mathfrak{S}} j(\gamma, v)^{-1} h(\gamma v) \overline{\theta(v)} J(v) dv \quad \text{letting } W = \gamma v$$

$$= \sum_{\gamma \in \Gamma_\varphi \backslash \Gamma} \int_{\gamma^{-1} \mathfrak{S}} j(\gamma, \gamma^{-1} w)^{-1} h(w) \overline{\theta(\gamma^{-1} w)} J(\gamma^{-1} w) dw$$

$$= \sum_{\gamma \in \Gamma_\varphi \backslash \Gamma} \int_{\gamma^{-1} \mathfrak{S}} j(\gamma, \gamma^{-1} w)^{-1} h(w) \overline{j(\gamma^{-1} w)} \ \overline{\theta(w)} J(\gamma^{-1} w) dw.$$

Where $\mathfrak{S}$ stands for any choice of fundamental region. Now, $\overline{j(\gamma^{-1}, w)} = \overline{j(\gamma^{-1}, w)} j(\gamma^{-1}, w) j(\gamma^{-1} w)^{-1}$ and $j(\gamma, \gamma^{-1} w)^{-1} j(\gamma^{-1}, w)^{-1} = 1$, so the above integral equals

$$\sum_{\gamma \in \Gamma_\varphi \backslash \Gamma} \int_{\gamma^{-1} \mathfrak{S}} h(w) \overline{\theta(w)} \| j(\gamma^{-1}, w) \|^2 J(\gamma^{-1} w)$$

and by lemma 5.1,

$$= \sum_{\gamma \in \Gamma_\varphi \backslash \Gamma} \int_{\gamma^{-1} \mathfrak{S}} h(w) \overline{\theta(w)} J(w) dw.$$

Gluing the $\gamma^{-1} \mathfrak{S}$ together for all $\gamma \in \Gamma_\varphi \backslash \Gamma$ gives the region $\mathcal{H} / \begin{pmatrix} 1 & 1 \\ 0 & 1 \end{pmatrix} \times \mathbb{C}^2 / \mathbb{Z}^2$ so the above integral equals

$$\int_{z \in \mathbb{C}^2 / \mathbb{Z}^2} \int_{v > 0} \int_{-\frac{1}{2} < u \le \frac{1}{2}} h(w) \overline{\theta(w)} J(w) dw.$$

In the event that $\phi(w) = P_{T_m f, g}$ and θ is a traditional cusp form or Eisenstein series, we can say more.

Corollary 5.3. Let $\phi(w) = P_{T_m f, g}(w)$ and let θ be a cusp form. Then

$$\langle \phi, \theta \rangle = 0.$$

Proof. $\langle \phi, \theta \rangle = \int_{z \in \mathbb{C}^2 / \mathbb{Z}^2} \int_{v > 0} \int_{-\frac{1}{2} < u \le \frac{1}{2}} g(v) e^{2\pi i m X y} T_m f(z) \overline{\theta(\tau, z)} J(v) dv.$

Noticing that only θ depends on u, we have the inside integral

$$\int_{-\frac{1}{2}<u\leq\frac{1}{2}} \theta(u+iv,z)dx = 0$$

because θ is cuspidal. Thus $\langle\phi,\theta\rangle = 0$.

6 A Numerical Experiment

Now that we have built this machine, gentle reader, let us see if anything sensible comes of it. Our ϕ will be the indicator function of the unit interval, lifted so that $\tau = i$. We apply ten elements of the discrete integer subgroup of G to this function to obtain a list of basis functions. Then we took the real part. Then we normalized so everything has $\mathcal{L}^2$-norm equal to one. Here is the list:

$e_1(x) = e^{-2\pi(x^2-2x)}\cos(2\pi(-x^2+2x))/\sqrt{63495.3}I_{(1/2,1)}$

$e_2(x) = e^{-(.8)\pi(x^2-12x)}\cos((.8)\pi(-2x^2+24x))/10^6\sqrt{8.19483}I_{(1/5,3/5)}$

$e_3(x) = e^{-(.8)\pi(4x^2-4x)}\cos((.8)\pi(-2x^2+2x))/\sqrt{3.00391}I_{(2/5,3/5)}$

$e_4(x) = e^{-(2)\pi(x^2-4x)}\cos((2)\pi(-x^2+4x))/10^9\sqrt{4.60361}I_{(1,3/2)}$

$e_5(x) = e^{-(.4)\pi(x^2-6x)}\cos((.4)\pi(-3x^2+18x))/\sqrt{6.91462}I_{(1/10,4/10)}$

$e_6(x) = e^{-(.4)\pi(9x^2-6x)}\cos((.4)\pi(-3x^2+2x))/\sqrt{1.0087}I_{(3/10,4/10)}$

$e_7(x) = e^{-(.4)\pi(x^2-4x)}\cos((.4)\pi(-3x^2+36x))/\sqrt{4.43423}I_{(2/10,5/10)}$

$e_8(x) = e^{-(.4)\pi(9x^2-12x)}\cos((.4)\pi(-3x^2+4x))/\sqrt{35.2531}I_{(6/10,8/10)}$

$e_9(x) = e^{-(.8)\pi(x^2-12x)}\cos((.8)\pi(-2x^2+24x))/10^13\sqrt{1.17715}I_{(3/5,6/5)}$

$e_{10}(x) = e^{-(\frac{4}{13})\pi(4x^2-12x)}\cos((\frac{4}{13})\pi(-6x^2+18x))/\sqrt{54.9201}I_{(2/13,5/13)}$

We computed the relevant matrix of inner products that one would customarily use to numerically invert this transform. We had Mathematica compute the eigenvalues, which were: 2.04956, 0.21806, 1.4135, 0.5851, 0.5881, 1.23238, 0.84257, 1.06839, 0.99893, 1.00107. Thus the matrix in question was quite well conditioned. Finally, we used this basis to reconstruct the function $y = x^2$ on the unit interval. Although the algorithm seemed to be working, the sample size was too small to infer any conclusions as to completeness, so the picture is omitted.

7   Conclusions

A few points are worth mentioning in conclusion. First of all, we conclude from this exercise that a sampling system of this sort might indeed have good conditioning properties. Second, we can see from the various preliminary lemmas and theorems that it does appear to be amenable to the sorts of analysis that led to understanding the window Fourier transform. These alone would make further research interesting. But beyond these two issues is a third, namely the way such a scheme generalizes to higher dimensional sampling.

The Jacobi group is the lowest rank of a family of similar structures, all of which are semidirect products of a symplectic group and a Heisenberg group. The rank can be chosen to derive a similar sampling scheme for two, three or higher dimensional signals. As in this paper, a discrete subgroup can generate a countable family of window functions from a single mother function in such a way that translation, dilation and modulation are all used. Any theorems that can be proved in the case described in this paper can likely be generalized to the higher dimensional case. Therefore we believe that our sampling scheme provides an important prototype for more difficult types of sampling.

Acknowledgements

Many thanks go to Ken Bowers and John Lund for organizing another wonderful event.

References

[1] M. EICHLER and D. ZAGIER, *The Theory of Jacobi Forms*, Progress in Mathematics **55**, Birkhauser, 1985
[2] G. B. FOLLAND, *Harmonic Analysis in Phase Space*, Ann. of Math. Stud. 122, Princeton Univ. Press, 1989
[3] D. GABOR, "Theory of communication", *J. Inst. Elec. Eng* **93**(III) (1946), 429-457.